Proceedings

D1785285

14th International Conference on

Scientific and Statistical Database Management

Proceedings

14th International Conference on
Scientific and Statistical Database Management

24-26th July 2002 • Edinburgh, Scotland

Sponsored by

Napier University
Edinburgh University
Heriot-Watt University

In cooperation with

VLDB Endowment
International Association for Statistical Computing

Edited by

J Kennedy

Los Alamitos, California

Washington • Brussels • Tokyo

IEEE Computer Society Order Number PR01632
ISBN 0-7695-1632-7
ISSN 1099-3371

Additional copies may be ordered from:

IEEE Computer Society
Customer Service Center
10662 Los Vaqueros Circle
P.O. Box 3014
Los Alamitos, CA 90720-1314
Tel: + 1 714 821 8380
Fax: + 1 714 821 4641
http://computer.org/
csbooks@computer.org

IEEE Service Center
445 Hoes Lane
P.O. Box 1331
Piscataway, NJ 08855-1331
Tel: + 1 732 981 0060
Fax: + 1 732 981 9667
http://shop.ieee.org/store/
customer-service@ieee.org

IEEE Computer Society
Asia/Pacific Office
Watanabe Bldg., 1-4-2
Minami-Aoyama
Minato-ku, Tokyo 107-0062
JAPAN
Tel: + 81 3 3408 3118
Fax: + 81 3 3408 3553
tokyo.ofc@computer.org

Editorial production by A. Denise Williams
Cover art production by Joseph Daigle/Studio Productions
Printed in the United States of America by The Printing House

Table of Contents

14th International Conference on Scientific and Statistical Database Management

Invited Speaker

    Prof. Susan B. Davidson, University of Pennsylvania

Demonstrations

    H. Shan, K.G. Herbert, W.H. Piel, D. Shasha, and J.T.L. Wang

    A.C. Jones, J.S. Robinson, W.A. Gray, J.P. Giddy, and N.J. Fiddian

    R. Bose

    E. Kapetanios, D. Baer, P. Groenewoud, and P. Mueller

Distributed Scientific Databases

    M. Peim, E. Franconi, N.W. Paton, and C.A. Goble

    I. Foster, J. Vöckler, M. Wilde, and Y. Zhao

    M.C. Cavalcanti, M. Mattoso, M.L. Campos, E. Simon, and F. Llirbat

Scientific Metadata Management

    M. Gertz, K-U. Sattler, F. Gorin, M. Hogarth, and J. Stone

    D. Higgins, C. Berkley, and M.B. Jones

    C. Raguenaud and J. Kennedy

Scientific Query Optimisation

    D. Shasha, J.T.L. Wang, H. Shan, and K. Zhang

    K. Wu, E.J. Otoo, and A. Shoshani

    C.A. Lang and A.K. Singh

Foreword

The fourteenth International Conference on Scientific and Statistical Database Management was held in Edinburgh, Scotland from the 24-26th July 2002. This book contains all of the material prepared for SSDBM 2002 and as such represents a valuable compendium of current database research and challenges in scientific and statistical database applications.

We would like to thank all of those who submitted a paper to this conference and commend them for the quality of the submissions regardless of their acceptance. Forty-six papers were submitted, representing 18 different countries. We would also like to express our appreciation to the programme committee, consisting of leading international researchers in fields related to scientific and statistical database management, who did an outstanding job of reviewing the papers. These proceedings contain the nineteen full papers, four demonstration papers, one short paper and six three poster papers which were accepted. The conference has two priority themes, Bioinformatics and Statistical Metadata, which is reflected in the proceedings along with other papers addressing research issues in scientific databases.

The conference itself however includes more than is reflected in the proceedings and the work of several people deserve particular note. Cedric Raguenaud of Napier University developed the software to support the web-based submission and review site and managed the web-based submission process which greatly aided the programme committee review process. Albert Burger did a splendid job of organising the demos and posters and panel sessions. Thanks also go to Marcia Wright for organising the registrations and social events and to Carolyn Newton of Edinburgh University who managed the main web site and helped with the conference administration. Helen McCue, Napier University helped realise the design ideas for the SSDBM logo and Gisela Schuster, Buenos Aires and David McClusky, Napier University helped in transforming the logo into our proceedings artwork.

Many other people helped in the preparation of the conference behind the scenes. In particular thanks go to the Steering Committee members (in particular Arie Shoshani (USA, Chair) and Hans-Joachim Lenz), for their encouragement and advice and to Larry Kerschberg for passing on the wisdom gained from running SSDBM 2001. Finally we would like to acknowledge the institutional support of Napier University, Edinburgh University and Heriot-Watt University.

If you are reading these proceedings as an SSDBM 2002 delegate, we would like to thank you for coming, welcome you and hope that you will enjoy the many attractions of Edinburgh and of course the conference presentations for which your participation in discussions will be most valued. If you are reading them after the conference then we hope they will prove beneficial to your research and encourage you to participate in a future SSDBM.

Programme Chair

Jessie Kennedy
Napier University

Organising Chair

Joanne Lamb
Edinburgh University

Committees

Programme Chair

Jessie Kennedy, *Napier University, UK*

Programme Committee

Peter Buneman, *University of Edinburgh, UK*
Susan Davidson, *University of Pennsylvania, USA*
Jim Frew, *University of California, Santa Barbara, USA*
Johann Christoph Freytag, *Humboldt-Universität zu Berlin, Germany*
Karl Froeschl, *Universität Wien, Austria*
Dan Gillman, *Bureau of Labor Statistics, USA*
Alex Gray, *Cardiff University, UK*
Ralf Hartmut Güting, *University of Hagen, Germany*
Jiawei Han, *Simon Fraser University, USA*
Mikalis Hatzopolous, *University of Athens, Greece*
Yannis Ioannidis, *University of Athens, Greece*
Christian Jensen, *Aalborg University, Denmark*
Toni Kazic, *University of Missouri, USA*
Graham Kemp, *Chalmers University of Technology, Sweden*
Sally McClean, *University of Ulster, UK*
Moira Norrie, *ETH Zurich, Switzerland*
Frank Olken, *LBNL, USA*
Haralampos Papageorgiou, *National & Capodistrian University Of Athens, Greece*
Norman Paton, *University of Manchester, UK*
Sean Wang, *George Mason University, USA*
Andrew Westlake, *Survey & Statistical Computing, UK*

Organising Chair

Joanne Lamb, *Edinburgh University, UK*

Organising Committee:

Local Organising:	Karen Brannen, *Edinburgh University, UK*
Panel/Demos/Posters:	Albert Burger, *Heriot Watt University, UK*
Submissions Web site:	Cedric Raguenaud, *Napier University, UK*
Administration:	Marcia Wright, *Edinburgh University, UK*

Steering Committee:

Menas Kafatos, *USA*
Hans-Joachim Lenz, *Germany*
Tekin Ozsoyoglu, *USA*
Maurizio Rafanelli, *Italy*
Arie Shoshani, *USA—Chair*

Additional Reviewers

Y. Manolopoulos	T. Bach Pedersen	N. Lorentzos
S. Saltenis	M. Vardaki	D. Katsaros
C. Wang	F. Pentaris	A. Papadopoulos

Invited Speaker

Tale of Two Cultures: Are there database research issues in bioinformatics?

Susan B. Davidson
Center for Bioinformatics
Dept. of Computer and Information Science
University of Pennsylvania
susan@cis.upenn.edu

Abstract

It was the best of times, it was the worst of times, it was the age of biology, it was the age of computer science, it was the epoch of whole-genome sequencing, it was the epoch of high performance computing, it was the season of ab-initio discoveries, it was the season of in-silico predictions, it was the spring of mass data production, it was the winter of interoperablity, we had everything before us, we had little real knowledge, we were all going to discover the origins of Life, we were all going to die for lack of a cure–in short, in this period bioinformatics has, for good or for evil, been born from the union of biology and computer science.[1]

1 Overview of Talk

Over the past decade, bioinformatics has become crucial to any biomedical research enterprise. Most of the top schools already have, or are developing, research as well as educational programs in bioinformatics. Funding agencies have mounted special opportunities in bioinformatics. Groups of people who formerly would not – or could not – have spoken about common research interests have begun to do so. However, in all the excitement over the interplay between computer science – the "London" subject of provability and precision – and biology – the "Paris" subject of hypotheses and experimentation that is being revolutionized by computation – what is the role of databases? Is it merely the hand-servant of biology, the producer of schemas and keeper of data? Or is it a subject which is being challenged to expand its research horizons and make new discoveries?

In some ways, the answer to this question is negative: Within molecular biology/bioinformatics journals and conferences, papers dealing with database issues are few and far between. Those that are published tend to describe data resources of interest to biologists. Within database journals and conferences, despite repeated calls for papers, remarkably few papers are submitted and even fewer accepted. (Note: VLDB2001 is a happy departure from this statistic, as is the SSDBM conference series.)

Given these observations, one is tempted to believe that although other areas of computer science (such as algorithms and data mining) are expanding their research horizons, databases as a subject has not found similar challenges in the union of biology with computer science.

Against the backdrop of this negative evidence, I will argue that there are in fact exciting database research problems to be addressed within bioinformatics. Many of these research problems arise from the fact that vast quantities of data are being produced in different formats and modalities, but that are connected through some process. For example, one wave of data has been genomic sequence – DNA and RNA, mainly textual information – while the next wave has been proteomic data characterized by 2D-gel image, protein spot and peptide characterizations. These data are obviously related through the central dogma of biology: DNA $\longrightarrow$ RNA $\longrightarrow$ protein. Other waves of data deal with differential expression, both tissue and in developmental stages, phenotypic data (often captured in images), phylogenetic characterizations, and patient and disease-oriented data. Clearly, all this information is related, and the process by which it is related is in large part the real knowledge being sought.

Some of the database challenges arising from this landscape that will be discussed are: data and program interoperability; workflow and tracking; data provenance; annotation systems; indexing the genome; superimposing biological data and knowledge.

---

[1] Adapted from *Tale of Two Cities* by Charles Dickens.

3

Demonstrations

A Structure-Based Search Engine for Phylogenetic Databases[*]

Huiyuan Shan[†]   Katherine G. Herbert[‡]   William H. Piel[§]   Dennis Shasha[¶]

Jason T. L. Wang[‖]

Abstract

Phylogenetic trees are essential for understanding the relationships among organisms or taxa. Many of the current techniques for searching phylogenetic repositories allow the user to perform a keyword-type search or an aligned sequence data search, or to browse a hierarchical list of taxa. Here we describe a new search engine that allows the user to present an example phylogeny, or a query tree, and then searches a phylogenetic database for trees that contain the query structure. The presented search engine is fully operational and is available on the World Wide Web.

1. Introduction

Phylogenetic trees, usually represented by a dendrogram, model the evolutionary history of a set of taxa that have a common ancestor. The internal nodes within a particular tree represent older organisms from which their child nodes descend. The children represent divergences in the genetic composition in the parent organism. Since these divergences cause new organisms to evolve, these organisms are shown as children of the previous organism.

Currently, in studying the phylogenetic data, most systems use search methods that do not exploit the structure of the data. These systems usually adopt a keyword-based search tool, a sequence alignment, or a manually operated browser [1, 4, 5, 7]. The keyword-based tool allows the user to enter the name of a taxon or some identifying quality such as an identification number to search the database. It then returns all the data it has stored on that particular taxon. Some systems even allow the user to search more than one taxon at a time, using the traditional "AND" and "OR" operators to decide what set of data to return to the user. In sequence alignment, the taxon uses the genetic code from the sequence to align it with homologous sequences. If the sequences are similar, most likely they share a common ancestor. A phylogenetic tree can then be inferred from the pattern of nested instances of shared, derived mutations. Once the taxa are selected that align with the query taxon, links to other information about the returned taxa can be provided. Finally, browsing a hierarchical list allows users to examine one taxon at a time, learning only about that taxon.

Many of the systems that employ the search methods described above include visualization techniques that allow the user to view an entire section of a phylogenetic tree, or the entire tree, as well as interact with it. These interactions may involve visual inspection of the tree as well as linking to other trees that contain a particular taxon, or viewing isomorphism trees so that relationships between the taxa can be better understood.

However, none of the existing systems provides the user with the ability to search a database for the structure of a phylogenetic tree or structures similar to a query structure (as far as we know). Since the structure of a phylogenetic tree models very important information about the taxa contained within the tree, structure search becomes a very helpful as well as important tool for researchers studying phylogeny. This could, for example, help a researcher who desires to identify alternative evolutionary hypotheses for a set of taxa and how their evolutionary histories differ.

We describe here a new search engine, called TreeSearch, that allows the user to query by the structure of a phylogenetic tree. This search engine is fully operational and has been integrated into the phylogenetic information system, TreeBASE, developed at Harvard, UC Davis, Leiden University, and the University at Buffalo. Section 2 reviews TreeBASE. Section 3 presents TreeSearch. Section 4 concludes the paper and describes some future work.

---

[*] Work supported in part by U.S. NSF grants IIS-9988345 and IIS-9988636.

[†] Department of Computer Science, NJIT.

[‡] Department of Computer Science, NJIT.

[§] Dept. Biol. Sci., 608 Cooke, University at Buffalo, Buffalo, NY 14260, USA.

[¶] Courant Institute of Mathematical Sciences, New York University, New York, NY 10012, USA.

[‖] Contact author: College of Computing Sciences, New Jersey Institute of Technology, Newark, NJ 07102, USA (wangj@oak.njit.edu).

2. TreeBASE

TreeBASE [5, 7], accessible at `http://www.treebase.org`, is a relational database containing phylogenetic information from research papers submitted to the Web site. This site then allows users to search the database freely according to various keywords, and see visual representations of the trees. Moreover, it allows the user to gain access to information concerning a tree as well as use comparison tools to learn more about various taxa contained within the tree and their relationships with other taxa within the database.

The dataset TreeBASE maintains consists essentially of phylogenetic trees submitted to TreeBASE by the authors of the papers that present the trees. The site accepts for review any peer-reviewed and published paper that presents information on any type of phylogenetic trees. For the paper to be contained within the database, it must be submitted to the site. The paper then goes through a review process before the submitted data are officially put within the database.

The schema and relational tables in TreeBASE contain various types of data including the citations of the papers stored within the database, the abstracts from the papers, the information about the authors, the algorithm used to obtain the phylogenetic trees, the titles and types of the trees, the software used to perform the relevant analysis, the association of the trees and matrices with the study through which they were obtained, and the information about the taxa.

TreeBASE allows the user to search its database by various keywords, including taxa, author, citation, study accession and matrix accession. Search results contain the information about the study that an input keyword was found in. This information includes the publishing date, the author, the title of the study in which the keyword was found, and the periodical in which the study appeared. Also, accompanying the study are analyses of the data presented within the study. These analyses can include the matrix from which the phylogenetic trees are generated [6], a link to drawing the tree in a frame within the Web site, a link to download the tree so that the user can view it on his or her own viewer, and a link to "mark" a tree which allows the user to store the tree for quick retrieval later. TreeBASE is also equipped with various visualization tools for drawing and displaying trees, and allows the user to "tree surf" and download the matrix of a particular tree. Detailed descriptions of TreeBASE can be found in [5].

3. TreeSearch

The source code of TreeSearch, accessible at `http://cs.nyu.edu/cs/faculty/shasha/papers/treesearch.html`, can be applied to non-phylogeny applications including XML querying. The underlying algorithms and their applications to querying XML data can be found in [8, 9]. We focus here on applying the code to phylogenetic database search. This structure-based search engine is now fully operational, accessible at `http://aria.njit.edu/~biotool`, and has been integrated into TreeBASE. To access it from TreeBASE, please visit the Web site `http://www.treebase.org/treebase/console.html`, and click "Structure" in the pull down menu (search window) on the Web site.

The existing search mechanisms in phylogeny databases such as TreeBASE are generally keyword based and performed on tuples in relational tables. By contrast, TreeSearch offers the researcher the ability to begin to search for structural relationships within trees. TreeSearch allows the user to specify more than one taxon to search upon in any given search. It also allows the user to specify relationships among taxa through the use of desired relationships in tree format. The user can customize this query through the use of variable length don't cares and fixed length don't cares, together with a distance value so that either an exact match to the query tree or an approximate match to the query tree will be returned [8, 9]. As of today, TreeSearch has been accessed, via TreeBASE, at least 1000 times by scientists around the world. From user's feedback, this search engine appears useful in conducting phylogenetic studies.

3.1. System Architecture

Figure 1 shows the software architecture of TreeSearch. The system is composed of four components: Web-based Interface, Query Processor, Structure Viewer and Performance Log. From the Web-based Interface, the user is able to type in his/her own query (an example tree), upload the query tree from a file, or use and modify a sample query provided by the system. The Query Processor searches TreeBASE for phylogenetic trees containing the query tree. The Structure Viewer displays the trees using either a parenthesized string notation or a dendrogram format, which are presented to the user via Web-based Interface. User queries and their timestamps are maintained in the Performance Log, which helps to analyze user's needs and better tune the system for working more effectively. TreeSearch is connected to TreeBASE on the Web and therefore it uses the visualization tools available in TreeBASE for displaying trees graphically. The system is implemented using Java, HTML, Perl CGI, and K (`http://www.kx.com`). The system can run under both UNIX and Microsoft Windows.

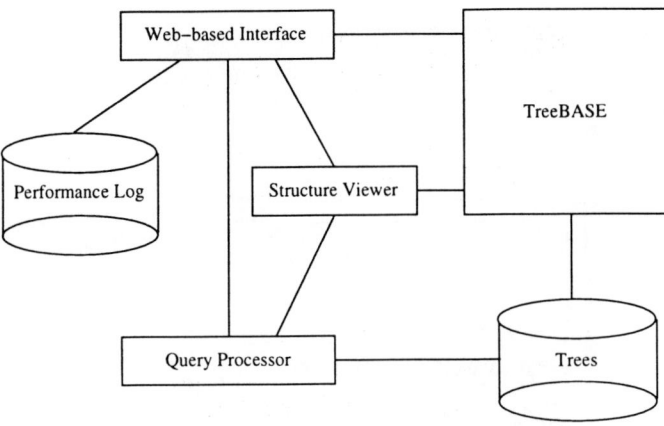

Figure 1. The software architecture of TreeSearch.

3.2. Querying the System

Figure 2, in the leftmost window, shows the main screen from which the user can query the system. The uppermost frame displays the main menu for the system. The leftmost bottom frame allows the user to input a query. The rightmost bottom frame allows the user to gain access to Tree-BASE; when a search is performed, this frame also displays the search results.

In Figure 2, a query tree, expressed in the parenthesized string notation, and search results are shown in the main screen. The two windows to the right of the main screen display the trees graphically; the top window shows the query tree and the bottom one shows a matching tree. To view a tree in a dendrogram format, the user would need to click the icon with the pencil overlaid upon the phylogenetic tree in the main screen. To view the parenthesized string notation, the user would need to click on the "Text" link in the main screen.

Referring to Figure 2, in the matching tree dendrogram, the entire phylogenetic tree is drawn and the user can use scroll bars to view portions of the tree. The specific taxa specified within the query tree are highlighted in the matching tree using underscored red font with a green circle next to the taxon's name. Also, each taxon has a number next to its name. This number represents the number of studies in TreeBASE the taxon is found within. If the user clicks on this number, he or she is linked to TreeBASE so that he or she can search on that taxon about those studies.

The query tree in Figure 2 contains a variable length don't care (VLDC), denoted "*", and a fixed length don't care (FLDC), denoted "?". When matching with a data tree, the VLDC "*" in the query tree may substitute for a path of length zero or more in the data tree. (The VLDC may also match with several paths, connected together with a shape

like an umbrella of nodes in the data tree [10, 11].) The FLDC "?" in the query tree may substitute for a single node in the data tree.

In addition, the system allows the user to specify a distance value for performing approximate searches. Given an integer d (typed in the "set maxdist" window in the main screen), the search engine finds all the phylogenetic trees T in TreeBASE that approximately contain the query tree Q within distance d. That is, T contains a substructure T' and the distance from Q to T' is at most d. We measure the distance from Q to T' by the total number of root-to-leaf paths in Q that do not appear in T'; the nodes in T' that do not appear in Q can be freely removed. This distance measure differs from the editing distance between ordered [10, 11] and unordered [12] trees. Notice that calculating the editing distance between unordered phylogenetic trees in which the order among siblings is unimportant is NP-hard [12].

In contrast to existing tree comparison algorithms in mathematical phylogeny [2, 3], our approach measures the distance between two trees by counting the mismatching paths in the two trees. Furthermore, our approach focuses on efficient searches in a database of trees by utilizing a suffix array index structure [9], as opposed to pairwise comparisons of trees in the existing algorithms. In general, TreeSearch can perform a search on TreeBASE with approximately 1600 trees in about 2 seconds on a SUN Ultra 20 workstation.

4. Conclusion and Future Work

TreeSearch provides the user with a powerful tool to allow the user to query phylogenetic information. By allowing the user to enter a structure as a query, he or she can obtain information about relationships between taxa that would otherwise require visual analysis.

The TreeSearch project is ongoing. Future work includes

- developing better user interface and visualization tools to help the user enter the query tree more easily and better understand the trees returned from the system;

- developing more effective schemes that mix and combine TreeSearch with the different search methods available in TreeBASE as well as other phylogenetic information systems;

- understanding the kinds of analysis that scientists do and extending TreeSearch in that direction;

- studying various techniques for ranking and scoring search results for approximate searches and exploring the possibilities of using XML and metadata concerning phylogenetic information for improving searches.

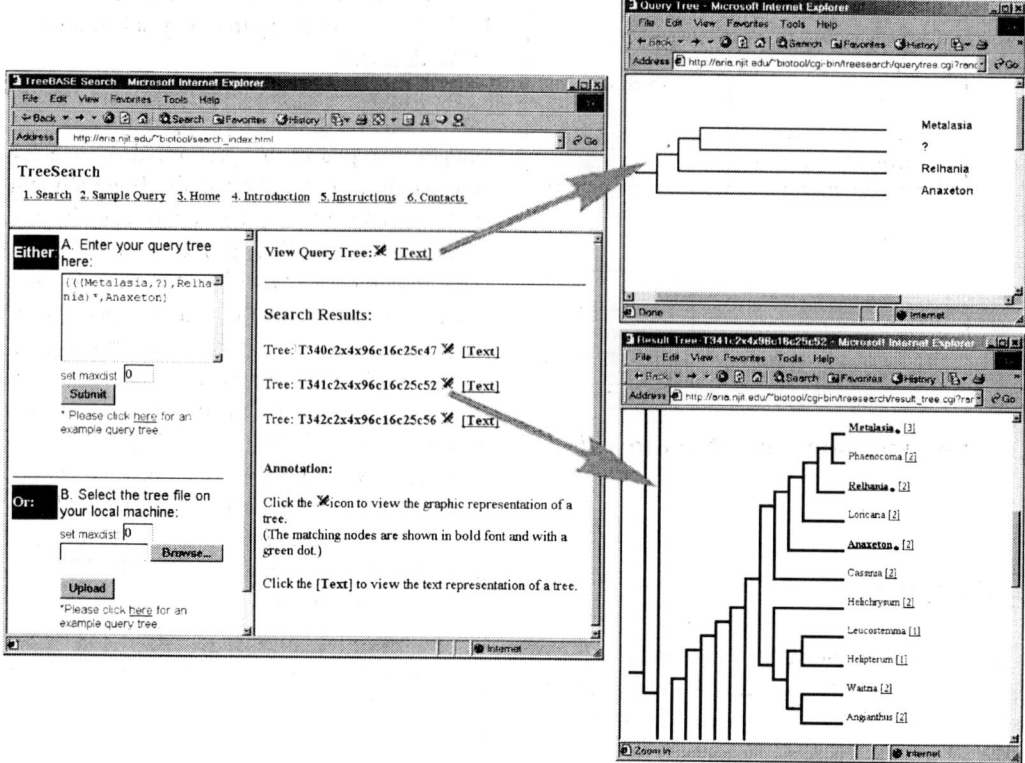

Figure 2. The TreeSearch **interface for showing an example query and search results.**

References

[1] B. L. Cohen, J. A. Sheps, and M. Wilkinson. Archiving molecular phylogenetic alignments as nexus files. *Systematic Biology*, 47:495–496, 1998.

[2] B. DasGupta, X. He, T. Jiang, M. Li, J. Tromp, and L. Zhang. On distances between phylogenetic trees. In *Proceedings of the 8th Annual ACM-SIAM Symposium on Discrete Algorithms*, 1997.

[3] M. Kao, T. Lam, T. Przytycka, W. Sung, and H. Ting. General techniques for comparing unrooted evolutionary trees. In *Proceedings of the 29th Annual ACM Symposium on Theory of Computing*, 1997.

[4] D. R. Maddison. Tree of life. `http://phylogeny.arizona.edu/tree/phylogeny.html`.

[5] W. H. Piel, M. J. Donoghue, and M. J. Sanderson. TreeBASE: A database of phylogenetic information. In *Proceedings of the 2nd International Workshop of Species 2000*, 2000.

[6] M. A. Ragan. Phylogenetic inference based on matrix representation *Molecular Phylogenetics and Evolution*, 1:53–58, 1992.

[7] M. J. Sanderson, M. J. Donoghue, W. H. Piel, and T. Eriksson. TreeBASE: A prototype database of phylogenetic analyses and an interactive tool for browsing the phylogeny of life. *Am. J. Bot.*, 81(6), 1994.

[8] D. Shasha, J. T. L. Wang, and R. Giugno. Algorithmics and applications of tree and graph searching. In *Proceedings of the ACM SIGACT-SIGMOD-SIGART Symposium on Principles of Database Systems*, 2002.

[9] D. Shasha, J. T. L. Wang, H. Shan, and K. Zhang. ATreeGrep: Approximate searching in unordered trees. In *Proceedings of the 14th International Conference on Scientific and Statistical Database Management*, 2002.

[10] J. T. L. Wang, K. Jeong, K. Zhang, and D. Shasha. A system for approximate tree matching. *IEEE Transactions on Knowledge and Data Engineering*, 6(4):559–571, 1994.

[11] K. Zhang, D. Shasha, and J. T. L. Wang. Approximate tree matching in the presence of variable length don't cares. *Journal of Algorithms*, 16(1):33–66, 1994.

[12] K. Zhang, R. Statman, and D. Shasha. On the editing distance between unordered labeled trees. *Information Processing Letters*, 42:133–139, 1992.

Using the GRID for Biodiversity Research: the GRAB Demonstrator

Andrew C. Jones[1], J.S. Robinson[2], W. Alex Gray[1], Jonathan P. Giddy[1] and N.J. Fiddian[1]

[1]Department of Computer Science
Cardiff University
PO BOX 916
Cardiff, CF24 3XF, UK
{andrew|alex|jonathan|nick}@cs.cf.ac.uk

[2]Biodiversity & Ecology Research Division
School of Biological Sciences
University of Southampton
Southampton, SO16 7PX, UK
J.S.Robinson@soton.ac.uk

Abstract

In the GRAB (GRid And Biodiversity) project we are developing a prototype to illustrate some aspects of the GRID's potential for collaborative Biodiversity research. A catalogue of life, two Species Information Systems (SISs) and a climate database are made available in a problem solving environment that demonstrates how bioclimatic modelling can be performed by bringing together such resources. We use Globus to provide access to these resources, in a secure environment. We also discuss more generally the GRID's potential for Biodiversity research, and identify the main areas of development within the GRID that are needed to support such research.

1. Introduction

The GRAB (GRid And Biodiversity) prototype illustrates some aspects of the GRID's [5] potential for Biodiversity research. The background to GRAB is that the UK Department of Trade and Industry has provided funding for a number of short GRID demonstrator projects, including GRAB, in order to explore the relevance of GRID concepts in various application domains, using existing GRID software such as Globus [4] and SRB (Storage Resource Broker[1]).

In this paper we first outline a small selection of previous, relevant Biodiversity informatics and related research, including our own, and discuss the relevance of the GRID to the kinds of tasks that have to be performed. We then describe the architecture of GRAB, and how it is used. In conclusion, we assess the current limitations of GRAB and then we discuss future work and GRID developments that are desirable to support this work.

2. Biodiversity informatics

Biodiversity informatics is concerned with organising and processing knowledge about living things and covers a range of areas including the provision of tools to aid in studying organisms (e.g. collection & analysis of data; classification) and their relationships to each other and to their environment (e.g. organising species of organisms into a taxonomic hierarchy; assessing biodiversity richness).

Central to such activities is the provision of a reliable species catalogue. The fundamental component of a species catalogue is a list of scientific names, synonyms and associated information for the species it covers – a taxonomic checklist. The names in a checklist reflect a scientific classification of organisms, and research has been carried out into how the analysis of specimens and the process of classification can be supported (e.g. ReTAX [1], PROMETHEUS [11]). In the LITCHi project we developed techniques for analysis of taxonomic checklists for consistency, and for postulating taxonomic reasons for conflicts identified [9, 3]. We have also developed techniques, in the SPICE project [10, 12], for co-ordinating a federated catalogue of life, comprising globally distributed, heterogeneous databases covering individual sectors of a taxonomic hierarchy.

A stable system of groupings of organisms (normally referred to as *taxa*, of which *species* are an example) and of names for these groupings is essential if scientific questions of relevance to biodiversity are to be answered accurately. For example, resources such as gene sequence databases (e.g. the EMBL Nucleotide Sequence Database[2]) and bioclimatic modelling tools (e.g. BIOCLIM [2]) can only be used effectively to explore phenomena associated with individual species if the data is organised in a taxonomically consistent manner. The biological significance of the checklist provided by SPICE is that it comprises individual check-

---

[1]http://www.npaci.edu/DICE/SRB/

[2]http://www.ebi.ac.uk/embl/index.html

11

lists that are subject to expert scrutiny, and each species has an *accepted name*, but also has alternative *synonyms* which allow species to be located under other names, be it in a Species Information System (SIS) providing geographical information about the regions inhabited by individual species, or in other resources like those mentioned above.

3. The GRID

Typically, tools such as those mentioned above offer important individual functions, but are difficult or impossible to use in combination. The GRID [5, 7] is being developed to improve exploitation of distributed resources – particularly for applications requiring high performance computing, e.g. high energy physics. Many of the biodiversity informatics application areas do not require high performance computing, but support for distributed systems and collaborative working, in which tools and data can be brought together to solve new problems, is desirable. To achieve this, resources must be described using appropriate metadata and heterogeneity must be accommodated – including, e.g., heterogeneity of data formats and differences in the schemes according to which knowledge has been organised and named. We also need tools to find appropriate resources and synthesise new knowledge. The GRAB prototype demonstrates how some of these facilities could be used.

4. The GRAB prototype

The GRAB prototype is built from the following resources:

- the SPICE catalogue of life;

- two SISs – the International Legume Database & Information Service (ILDIS)[3] and FISHBASE[4] – which act as sources for images and geographical information about individual species;

- a database we have built from public domain climate information obtained from the US National Climate Data Centre (NCDC),[5] and

- a problem solving environment that we have built to co-ordinate access to these resources.

Because of the nature of the previous experience of the programmer employed for GRAB, our first step was to build wrappers around these resources in order to allow them to be accessed using HTTP, the results of requests being returned as XML files. This has now been augmented so that a

Web front-end is provided to an application that uses Globus to communicate with the various resources as follows:

- The *Globus Resource Allocation & Management (GRAM)* service is used to invoke the GRAB services;

- the *Globus Access to Secondary Storage (GASS)* facility is used to retrieve the results of invoking a GRAB service, and

- the *Globus Security Infrastructure (GSI)* allows single logon to the resources used across the environment.

Having thus brought together some biodiversity-related resources, there are various ways they could be used. For the purposes of the present demonstrator we have adopted a fixed sequence of operations:

1. The user enters a search string and all matching scientific names in the SPICE catalogue are retrieved and displayed. Alongside the names retrieved that are synonyms, the accepted name is also displayed.

2. The user selects an accepted name from this list, then full information on this species is retrieved from the appropriate SIS and displayed, including geographical distribution and images (see figure 1).

3. The user selects a region from the geographical distribution, then associated climate data is retrieved from the NCDC database and displayed.

4. At this point the user can make use of the climate envelope obtained, or modify it to find other regions with similar climatic characteristics (see figure 2).

5. The user selects a region from the new list of regions displayed, and species native to that region are retrieved from the SISs.

6. The sequence can be repeated from step 3, if desired, by selecting one of the native species displayed.

Thus we have been able to find species native to some region having similar climate to at least one of the regions the originally-submitted species was native to (step 5), and to retrieve information relating to the originally-submitted species (step 2) and the additional species found (step 6). There is also a real-time monitor (not illustrated) that shows graphically the communication between components.

5. Discussion and future work

The above scenario illustrates that the catalogue of life plays an important role in supporting access to other biodiversity related systems and resources, by giving scientific names

---

[3]http://www.ildis.org/
[4]http://www.fishbase.org/
[5]http://lwf.ncdc.noaa.gov/oa/documentlibrary/pdf/climatesoftheworld.pdf

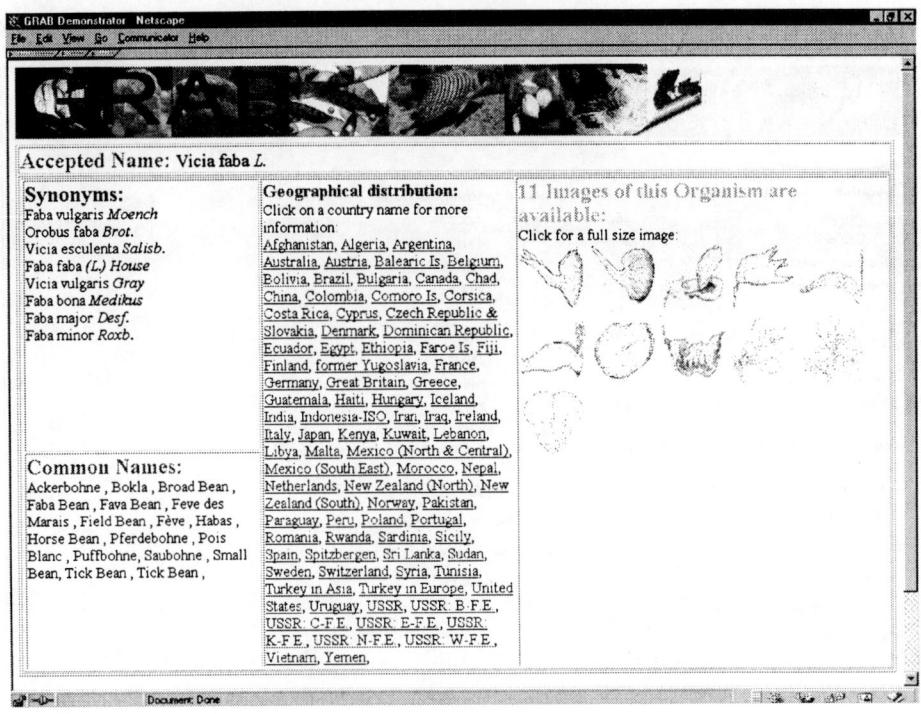

Figure 1. GRAB species display

that should be used to access these systems. The situation is somewhat simplified in our demonstrator because the catalogue uses the same taxonomy as the SISs: in general the synonyms retrieved from the catalogue would also be of use in accessing SISs. Indeed, the catalogue could be supplemented with knowledge enabling *correct* selection of scientific name(s) for individual resources, using information about the taxonomies concerned. Another issue that would be much more important in a full-scale system would be resource discovery using facilities such as the Globus Metacomputing Directory Service (MDS). Although appropriate metadata could be added to the present system to illustrate this, it would be a somewhat artificial exercise as the resources needed at each stage are already known. Nevertheless, we plan to experiment with this in the immediate future.

As mentioned above, our brief was to investigate the use of Globus and SRB in our biodiversity demonstrator. On close examination, SRB did not appear to provide facilities relevant to the software architecture adopted: it is primarily a platform-independent, distributed file access system. For this reason we decided for our prototype to use Globus only. Yet the Globus toolkit proved to be complex to install and use, but much more importantly, Globus is somewhat limited in providing facilities primarily only at the computation/data level. There is an increasing interest in viewing the GRID as having three layers: *Computation/Data, Information* and *Knowledge* [8]. But the Information and Knowl-

edge layers have yet to be fully built: currently, for example, when we add metadata as described above, it will have to be fairly primitive, being LDAP (Lightweight Directory Access Protocol) - based and hence essentially hierarchical. Sophisticated GRID tools and middleware, for such tasks as finding resources, using metadata to interpret data, and scientific visualisation, are needed in biodiversity informatics – as in other disciplines – if the potential of the GRID is to be realised. The Open Grid Services Architecture [6] is an important development, since it provides a framework for not only locating appropriate services but also for making explicit how they should be used.

We shall be participating in a major project investigating a number of biodiversity informatics applications on the GRID. This project (BiodiversityWorld) will give us the opportunity to contribute to GRID developments, working with a much wider range of biodiversity resources than we used for our demonstrator, including resources of the kinds listed in section 2. The essential aim will be to build a realistic system to support collaborative research in this important area, providing flexible access to biodiversity data and relevant analytic tools.

6. Acknowledgements

This work was funded by a grant from the UK Department of Trade and Industry. We are grateful to ILDIS, Fishbase and Species 2000 for access to their systems and data.

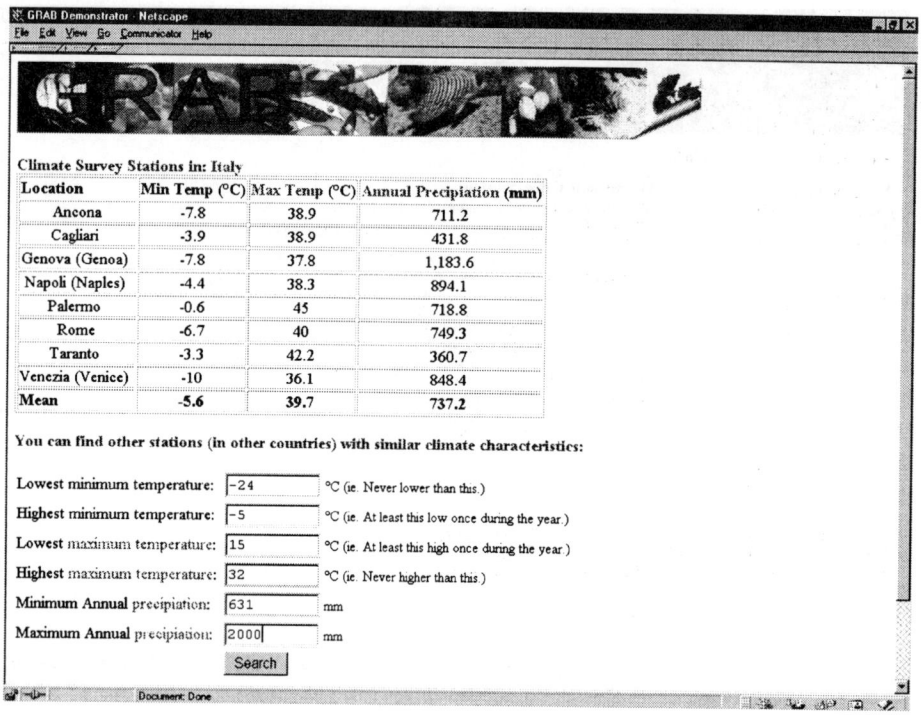

Figure 2. GRAB climate data display

References

[1] E. Alberdi and D. Sleeman. ReTAX: A step in the automation of taxonomic revision. *Artificial Intelligence*, 91(2):257–279, 1997.

[2] J. Busby. BIOCLIM – a bioclimatic analysis and predication system. In C. Margules and M. Austin, editors, *Nature Conservation: Cost Effective Biological Surveys and Data Analysis*, pages 64–68. CSIRO, 1991.

[3] S. M. Embury, S. M. Brandt, J. S. Robinson, I. Sutherland, F. A. Bisby, W. A. Gray, A. C. Jones, and R. J. White. Adapting integrity enforcement techniques for data reconciliation. *Information Systems*, 26(8):657–689, 2001.

[4] I. Foster and C. Kesselman. The Globus toolkit. In I. Foster and C. Kesselman, editors, *The Grid: Blueprint for a New Computing Infrastructure*, pages 259–278. Morgan Kaufmann, San Francisco, CA, 1999. Chap. 11.

[5] I. Foster and C. Kesselman, editors. *The Grid: Blueprint for a New Computing Infrastructure*. Morgan Kaufmann, San Francisco, CA, 1999.

[6] I. Foster, C. Kesselman, J. Nick, and S. Tuecke. The physiology of the Grid: An open Grid services architecture for distributed systems integration. January 2002. http://www.globus.org/research/papers/ogsa.pdf.

[7] I. Foster, C. Kesselman, and S. Tuecke. The anatomy of the Grid: Enabling scalable virtual organization. *The International Journal of High Performance Computing Applications*, 15(3):200–222, Fall 2001.

[8] K. G. Jeffery. GRIDs in ERCIM. *ERCIM News*, April 2001.

[9] A. C. Jones, I. Sutherland, S. M. Embury, W. A. Gray, R. J. White, J. S. Robinson, F. A. Bisby, and S. M. Brandt. Techniques for effective integration, maintenance and evolution of species databases. In O. Günther and H.-J. Lenz, editors, *12th International Conference on Scientific and Statistical Databases*, pages 3–13. IEEE Computer Society Press, 2000.

[10] A. C. Jones, X. Xu, N. Pittas, W. A. Gray, N. J. Fiddian, R. J. White, J. S. Robinson, F. A. Bisby, and S. M. Brandt. SPICE: a flexible architecture for integrating autonomous databases to comprise a distributed catalogue of life. In M. Ibrahim, J. Küng, and N. Revell, editors, *11th International Conference on Database and Expert Systems Applications (LNCS 1873)*, pages 981–992. Springer Verlag, 2000.

[11] M. R. Pullan, M. F. Watson, J. B. Kennedy, C. Raguenaud, and R. Hyam. The Prometheus taxonomic model: a practical approach to representing multiple taxonomies. *Taxon*, 49(1):55–75, February 2000.

[12] X. Xu, A. C. Jones, N. Pittas, W. A. Gray, N. J. Fiddian, R. J. White, J. S. Robinson, F. A. Bisby, and S. M. Brandt. Experiences with a hybrid implementation of a globally distributed federated database system. In X. S. Wang, G. Yu, and H. Lu, editors, *2nd International Conference on Web-Age Information Management (LNCS 2118)*, pages 212–222. Springer Verlag, 2001.

A Conceptual Framework for Composing and Managing Scientific Data Lineage

Rajendra Bose

Donald Bren School of Environmental Science and Management
University of California, Santa Barbara, CA 93106
rbose@bren.ucsb.edu

Abstract

Scientific research relies as much on the dissemination and exchange of data sets as on the publication of conclusions. Accurately tracking the lineage (origin and subsequent processing history) of scientific data sets is thus imperative for the complete documentation of scientific work. However, the lack of a definitive data model for lineage, and the poor fit between current data management tools and scientific software, effectively prevent researchers from determining, preserving, or providing the lineage of the data products they use and create. Based on a comprehensive review of lineage-related research and previous prototype systems, a conceptual framework is presented to help identify and assess basic lineage system components. Within this framework, a direction is outlined for future work on general methods for composing and managing lineage for scientific data.

1. Introduction

Conducting scientific research requires the dissemination and exchange of data sets as well as the publication of conclusions. Through both further processing within the originating organization and propagation to other research groups, a particular assemblage of data may contribute to other, derivative data products over time. The chain or pipeline of processing steps that generate standard "levels" of NASA remote sensing data products [1] provides one common example. As scientists become online data providers, the availability and transmission of various levels of digital data products over computer networks only makes the possible connections between related data sets more complex.

Tracking this processing and propagation of data is imperative. Consider a conceptual view of the flow of data products that consists of possibly concurrent threads of processing among two different research groups (Figure 1). At each step along each processing thread,

those working with the data may need to know about the history of the data set for a variety of reasons. Their potential needs fall within a spectrum that ranges from a simple narrative explaining why a data set exists, to machine-readable details for the complete re-execution of a process step. Knowing the origin of data may help to assess its quality or usefulness, while repeating a process step may be required for independently verifying another's computational research results, or for modifying one's own processing "recipes." The flow in Figure 1 shows the origins and subsequent processing history, or *lineage*, of the data products involved.

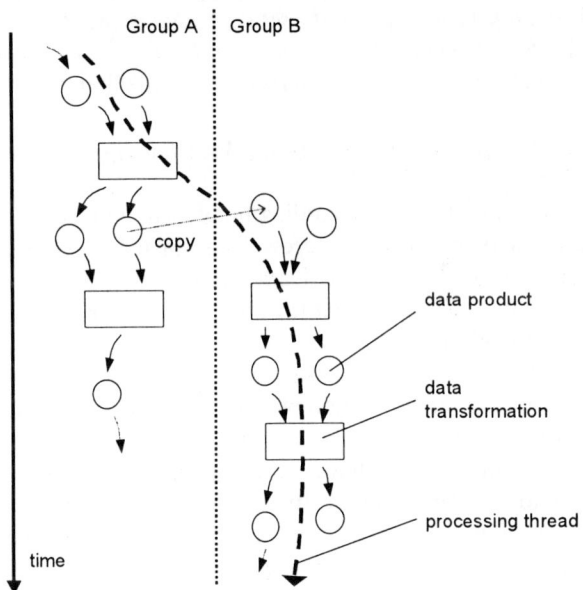

Figure 1. Data product flow

1.1. Data lineage defined

As genealogical charts reveal successive generations of parents for an individual, the lineage of an item describes how it was derived from its source. Data lineage generally refers to the sources and derivation of a data set or product [2, 3], or as summarized by Eagan and Ventura [4]: "all the processes and transformations of data from original measurements to current form." Thus, in addition to source observations or materials, data lineage encompasses data acquisition and compilation methods, conversions, transformations, and analyses, along with the assumptions and criteria applied at any stage of the data set life cycle [3].

Data lineage may also refer to items that have evolved from a data set or product. Two forms of navigating data lineage are thus implied: moving backward to discover ancestor items or events, or forward to discover descendant items or events.

A further refinement of data lineage is possible. I define computational data lineage to be those portions of data lineage that represent, or are directly relevant to, the scripted or coded processing of data on a computer. For example, in the environmental sciences, recording the steps for the acquisition of data may largely be a matter of writing a text narration or summary of field work performed, and thus may not involve computation. This discussion is primarily concerned with computational data lineage.

Data provenance and *data pedigree* [5, 6] are nearly equivalent terms for data lineage. Additionally, *derivation history* [7], *data set dependence* [8], *filiation* [9], *data genealogy* [10], *data archeology*, and *audit trail* [11] are other, related terms used in the literature.

1.2. The problem: composing data lineage

The importance of recording fundamental metadata to help others discover and access disseminated or archived data products is now basically recognized. However, composing metadata about the lineage of data products is usually neglected.

Data lineage is not compiled for two reasons. First, no definitive representation of computational data lineage exists in the literature. Designers of scientific information systems have no archetype or standard to refer to for implementing lineage features. Second, the ability of researchers to generate data in contemporary computing environments can quickly exceed their ability to track how it was created [12]. This is due to an absence of data lineage functionality in the software applications used for scientific work. Recording the processes used to create data products, ironically, has become more difficult and tedious as computational tools have become more sophisticated [3].

Scientists thus face the problem that, although they have growing responsibilities as online data providers [13], they are often unable to provide complete documentation of the data products they create. If data lineage were accessible, scientific researchers would have the opportunity to both improve their own computational work, and communicate more complete information about their data products to others.

2. Research prototypes

Research involving data lineage suggests two general areas of concern. One area recognizes the proliferation of scientific data transfer between disparate groups and systems, and seeks to use data lineage to protect downstream data users from unintended consequences resulting from these transfers. This goal is related to the broader topic of data quality. The second area focuses on how providers of scientific data can themselves benefit from tracking the lineage of their computational work. A detailed review of these research areas is presented in [14].

Several prototype systems stemming from this work directly feature some type of data lineage capabilities. The Earth System Science Workbench (ESSW) [15] is one of the few operational systems to date, and has been successful in capturing lineage metadata for objects involved in scientific processing performed with general scripts. Currently, data lineage can be queried through a Web application, and results are displayed diagrammatically using the Webdot web service interface included with the Graphviz set of graphing tools [16]. Data lineage is not yet involved with other system functionality.

Geolineus [17] was a commercial product at one time, featuring a graphical interface to present the data lineage of Arc/Info coverages, but is no longer available. GOOSE [18], a prototype system associated with the Amazonia project, uses data object attributes to store pointers to original and latest versions of any inputs and outputs. With this information, a cooperation graph of objects and transformations can be constructed to both track object versions and trace varying resolutions of lineage in a graphical interface. The fine grained lineage functionality of Tioga [2] was proposed only.

Numerous other prototype systems do not directly include lineage functionality, but use related concepts. A discussion of these prototypes, resulting from research in areas such as collaborative computing environments and experiment management, workflow, version control, and metadata management, is included in [14].

3. A conceptual framework for lineage systems

Lineage-related prototypes prescribe the elements required to either create special purpose systems, or

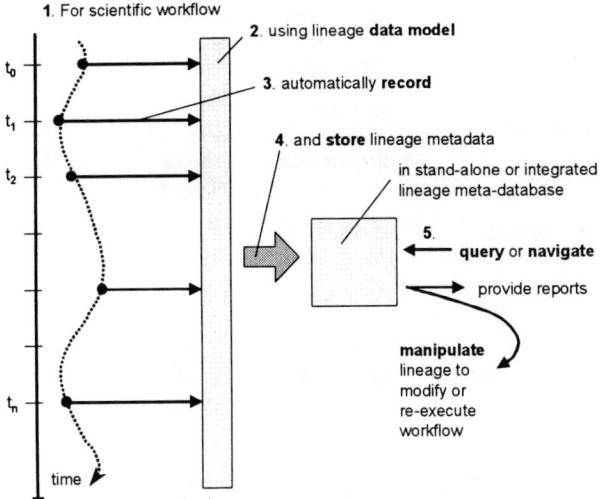

Figure 2. Conceptual framework for lineage system

imbue existing scientific information systems with new functionality, to record, store, query and manipulate data lineage. Figure 2 presents a conceptual framework for lineage systems synthesized from previous research efforts:

1. For scientific workflow,
2. in conjunction with an appropriate **data model** incorporating lineage,
3. the system automatically **records** lineage metadata
   a. through a process definition provided at t_0,
   b. by receiving metadata as tasks occur from t_0 through t_n, or
   c. through a log of completed tasks at t_n, and
4. **stores** the lineage metadata in a stand-alone or integrated meta-database (file system or DBMS).
5. The stored lineage metadata can then be accessed to
   a. answer ad hoc **queries**,
   b. browse or **navigate** (backwards or forwards) through the lineage,
   c. **manipulate** lineage information (for example, query results) to
      i. modify future workflow,
      ii. re-execute past workflow, or
   d. provide lineage reports to communicate data quality, possibly with flexible levels of detail, to other groups or domains.

The bold phrases in the narration above are used as headings in Table 1, providing the basis for a review of the technologies used in previous prototype data lineage systems, and a discussion of the components of future systems that will facilitate data lineage. Some prototype systems that do not track lineage, but feature potentially useful methods for future systems, are included in the table.

Lineage data model: The systems in Table 1 use a variety of data models. Graphs mimic dataflow diagrams of the past, handle varying levels of complexity well, and are supported by many general algorithms. Mention is made of the CSDGM [19] lineage report in some papers, but no prototypes in the table directly support this standard.

Automated lineage recording: Most systems in the table use object concepts to model programs and files. When metadata for an object is contained in its attributes, two issues arise. The first issue is how best to define attributes for an object class or template, and the second is how to use automated methods to populate attributes of each object.

Regarding the definition of attributes for an object class, most prototype systems either design graphical interface tools or rely on "expert" users to accomplish this task. For example, ESSW requires users to create XML DTDs for each object they require for their processing. This procedure was esoteric enough to create difficulties in prototype case studies and required the intercession of staff programmers. Systems with XML components may need to develop tools similar to Morpho, a graphical interface to assist ecologists to create XML metadata for their data sets using standard, domain specific terms, developed as part of the KNB project [20].

Once object classes are created, methods for metadata capture are required. Many prototypes require manual metadata entry prior to or during processing. Since much scientific research computing relies on running programs or scripts, one automated approach is using application programming interfaces (APIs) and software wrappers to send lineage metadata values to a metadata storage system. The ESSW project is evaluating the inclusion of API commands directly in Perl processing scripts, under the assumption that the scripts can invoke functions to deliver useful lineage metadata values.

Another approach involves using existing mechanisms in various applications and programming environments to create log files or simple text files in an appropriate format for periodic import by a lineage metadata system.

Storing lineage metadata: A few basic choices for storing lineage metadata exist, based on either a DBMS or an XML repository. If metadata is stored separately from the original file or program, the issue of canonical identifiers for external objects arises. The storage

System	Data model	Lineage function				
		record	store	query	navigate	manipulate
ESSW	graph with data and experiment objects	Perl API creates XML document	parsed XML in RDBMS	SQL via Web form CGI	Web page with dynamic DAG generation	
Tioga	graph with user-defined functions	DBMS registration, GUI dataflow editor	ORDBMS	PostQUEL	Dataflow diagram GUI	Modify or re-execute dataflow
Geolineus	semantic network	system-prompted manual entry of node attributes	file system	pre-defined custom queries created for file-based meta-database	Dataflow diagram GUI	
GOOSE	graph with objects and transfor-mations	manual template entry, GUI dataflow editor	file system	GUI model view/query tool	Dataflow diagram GUI	Modify or re-execute dataflow
Gaea	Petri Net-based	manual template entry, GUI dataflow editor	ORDBMS	PostQUEL with extensions	Dataflow diagram GUI	Modify or re-execute dataflow
ESP2Net	XML (SEML) document	manual XML (SEML) document creation	XML database	XML queries (also semantic parsing and multicast)	XML (SEML) document browser GUI	
CCDB	ER model with proxy entity	manual template entry, auto text output	OODBMS	vendor-supplied browser/query facility		

Table 1. Conceptual framework components of prototype systems

solution may need to accommodate other data models in addition to the lineage data model chosen.

Querying: Query tools are largely determined by the storage method chosen for lineage metadata. Web interfaces for queries are prevalent in metadata systems.

Navigation: Navigation refers to moving through hierarchic levels, for example when using the "drill down" (and backing out) paradigm for browsing DBMS records with a graphical interface. Most prototypes simply show a processing chain in its entirety in a browser or other interface, although this solution is not scalable.

Manipulation: Manipulation of data lineage refers to re-using or modifying processing sequences or "recipes" that have been saved or reconstructed from lineage information.

4. Future work

Incomplete lineage documentation for scientific data products is a problem that exists due to the absence of a definitive data model incorporating lineage concepts, and the lack of lineage recording and management functionality in the multitude of software applications used for scientific work. Some projects and associated prototype systems have begun to address the various requirements for composing and managing lineage for scientific data, but there is a need for additional prototypes to continue exploring solutions to the problem, and extending relevant concepts from related research areas in scientific information systems.

Work in this vein is planned over the next year, with the objective of developing automated approaches to create and access human- and machine-readable data lineage. These approaches will be tailored to the work of environmental science researchers who create digital data products through computational processing and provide these products to others.

The first phase of research will synthesize ideas and standards from the reviewed literature and use the conceptual framework presented in this paper to arrive at a logical model for computational data lineage. The logical model must be able to describe any items that contribute to the life span of a data set, the relationships between these items, and operations on these items relevant to a complete representation of data lineage. Abiteboul et. al. [21] suggest that semistructured data models emerging from the database research community will converge with XML to facilitate electronic data publishing. In recognition of the trend toward publishing scientific data on the Web, the use of XML with semistructured and traditional DBMS models will be evaluated for implementing a logical data lineage model.

During Phase 2, automated techniques to record lineage metadata will be presented and assessed. Phase 3 will feature strategies for querying and manipulating lineage metadata to support scientific data processing. A last phase of work will examine the ability of machine-readable data lineage to facilitate versatile, flexible lineage reporting to potential data users and decision makers. All phases will refer to the relevant and real example of creating ocean productivity data products from satellite imagery and ocean color algorithms for the Santa Barbara Channel Long Term Ecological Research project.

5. References

[1] National Aeronautics and Space Administration (NASA). 1986. Report of the EOS Data Panel, Volume IIa: Earth Observing System Data and Information System. Technical Memorandum 87777, National Aeronautics and Space Administration (NASA), Washington, DC.

[2] A. G. Woodruff and M. Stonebraker. 1997. Supporting fine-grained data lineage in a database visualization environment. *Proceedings of the 13th International Conference on Data Engineering*, Birmingham, UK, 91-102.

[3] D. G. Clarke and D. M. Clark. 1995. Lineage, in *Elements of Spatial Data Quality*, S. C. Guptill and J. L. Morrison, Eds. Oxford: Elsevier Science, 13-30.

[4] P. D. Eagan and S. J. Ventura. 1993. Enhancing value of environmental data: data lineage reporting. *Journal of Environmental Engineering*, Vol. 119, No. 1, 5-16.

[5] P. Buneman, D. Maier, and J. Widom. 2000. Where was your data yesterday, and where will it go tomorrow? Data Annotation and Provenance for Scientific Applications. White Paper, February 28, 2000.

[6] J. C. French. 1995. What is Metadata? *Proceedings of the SDM-92 Workshop*, Richland, WA, 3-8.

[7] N. I. Hachem, M. Gennert, and M. Ward. 1993. The Gaea System: A Spatio-Temporal Database System for Global Change Studies. *Proceedings of the AAAS Workshop on Advances in Data Management for The Scientist and Engineer*, Boston, MA, 84-89.

[8] G. Alonso, C. Hagen, H.-J. Schek, and M. Tresch. 1997. Towards a Platform for Distributed Application Development, in *Workflow Management Systems and Interoperability*, Vol. 164, *NATO ASI Series*, A. Dogac, L. Kalinichenko, M. T. Ozsu, and A. Sheth, Eds. Berlin: Springer, 195-221.

[9] L. Spery, C. Claramunt, and T. Libourel. 1999. A lineage metadata model for the temporal management of a cadastre application. *Proceedings of the Tenth International Workshop on Database and Expert Systems Applications*, Florence, Italy, 466-474.

[10] B. R. Barkstrom. 1998. Digital Archive issues from the Perspective of an Earth Science Data Producer. *Proceedings of the International Standards Organization (ISO) Archiving Workshop Series: Digital Archive Directions (DADs) Workshop*, College Park, MD.

[11] P. Brown and M. Stonebraker. 1995. Big Sur: A system for the management of Earth science data. *Proceedings of the 21st International Conference of Very Large Data Bases*, Zurich, Switzerland, 720-728.

[12] D. P. Lanter. 1990. Lineage in GIS: The Problem and a Solution. Technical Report 90-6, National Center for Geographic Information and Analysis (NCGIA), UCSB, Santa Barbara, CA.

[13] National Research Council. 1999. *Global Environmental Change: Research Pathways for the Next Decade*. Washington, DC: National Academy Press.

[14] R. Bose. 2002. Composing and Managing Lineage for Scientific Data: A Review. Technical report, Bren School of Environmental Science and Management, University of California, Santa Barbara, CA, 17 Jan 2002.

[15] J. Frew and R. Bose. 2001. Earth System Science Workbench: A Data Management Infrastructure for Earth Science Products. *Proceedings of the 13th International Conference on Scientific and Statistical Database Management*, Fairfax, VA, 180-189.

[16] AT&T. 2001. Graphviz. AT&T Labs - Research, <http://www.research.att.com/sw/tools/graphviz/> (2 Feb 2001).

[17] Geographic Designs. 1993. *Geolineus Version 3.0 User Manual*. Santa Barbara, CA.

[18] G. Alonso and A. El Abbadi. 1993. GOOSE: Geographic Object Oriented Support Environment. *Proceedings of the ACM Workshop on Advances in Geographic Information Systems*, Arlington, VA, 38-49.

[19] Federal Geographic Data Committee. 1998. Content Standard for Digital Geospatial Metadata (CSDGM). FGDC-STD-001-1998, Federal Geographic Data Committee, Washington, DC, June 1998. Available at <http://www.fgdc.gov/metadata/csdgm/>.

[20] C. Berkley, M. Jones, J. Bojilova, and D. Higgins. 2001. Metacat: a Schema-Independent XML Database System. *Proceedings of the 13th International Conference on Scientific and Statistical Database Management (SSDBM '01)*, Fairfax, VA, 171-179.

[21] S. Abiteboul, P. Buneman, and D. Suciu. 2000. *Data on the Web: From Relations to Semistructured Data and XML*. San Francisco, CA: Morgan Kaufmann.

The Design and Implementation of a Meaning Driven Data Query Language

E. Kapetanios, D. Baer, P. Groenewoud, P. Mueller
Dept. of Computer Science
Swiss Federal Institute of Technology
Zurich, Switzerland
kapetanios@inf.ethz.ch

Abstract

We present the design and implementation of a Meaning Driven Data Query Language - MDDQL - which aims at the construction of queries through system made suggestions of natural language based query terms for both scientific application domain terms and operator/operation ones. A query construction blackboard is used where query language terms are suggested to the user in its preferred natural language and in a name centered way, together with their connotation. This helps in understanding the meaning of the terms and/or operators or operations to be included in the query. Furthermore, the construction of the query turns out to be an incremental refinement of the query under construction through semantic constraints, where only those domain language terms and/or operators/operations are suggested which result into meaningful combinations of query terms as related to the scientific application domain semantics. Therefore, semantically meaningless queries can be prevented during the query construction. Such a semantics aware mechanism is not available in conventional database query languages such as SQL, where one is allowed to execute a query calculating, for example, the average of numerical data values whereas they represent the codes of categorical values. Moreover, no familiarity with the semantics of complex database schemes or interpretation of the symbols (names of classes/tables/attributes, value codes) underlying the storage model, as well as familiarity with the syntax of a database specific query language are needed by the end-user. The constructed query can be submitted to the MDDQL query interpretation and transformation engine, where the corresponding SQL-query is generated and delegated to a DBMS (e.g., Oracle, MS-Access, SQL-Server). Generation of SQL-statements addressing NF2 data models such as those provided by the object-relational Oracle DBMS is also enabled. The query result is presented in a table based form where all storage model symbols are interpreted and can be exported for the usage with statistical software packages (e.g., SPSS).

1. Motivation

Query languages as provided by database management systems, e.g., SQL, are mainly designed for programmers. Therefore, when end-users need to pose queries to a database, it is required that they have a substantial knowledge of syntax formalisms and the storage model semantics in order to formulate meaningful queries. The problem becomes more acute when scientific application domains or experiments are considered which address a large number of parameters (large database schemes). Moreover, the needs of formulating queries by using the familiar scientific terminology as expressed in their familiar natural language posed additional requirements for the design and implementation of the query language for accessing well-structured data from data repositories.

In order to meet these requirements, the implemented system aims at assisting the domain scientist to construct a query through system made suggestions of terms from the MDDQL vocabulary, which consists of both *application domain* and *operational terms*. Application domain terms are provided by the *application ontology*. The latter refers to an abstract model of the phenomena characterizing the particular application domain, i.e., *hospitalisation based treatment of myocardial infarction*, as well as the relevant concepts and the constraints on their use [12, 13]. Furthermore, the terms of the domain ontology are mapped to the *storage model symbols* standing for classes/tables, attributes and values.

All MDDQL vocabulary terms can be specified in words from more than one natural language. This is also applicable to their informal definition or connotation. The system made suggestions of terms, as provided by an *inference engine*, takes into consideration not only the semantic constraints as related to the application ontology but also to the operational terms. In other words, since no learning of a query language syntax is requested in order to construct an ad-hoc query, the incremental construction of the query takes place through end-user choices from those system inferred subsets of vocabulary terms which are semantically consistent with the set of terms already considered in the query. Distinction among *homonyms* such as *medication* is, therefore, enabled through the context of the term given either by its relationships with other terms or the underlying informal definition of the term.

This mechanism aims at assisting in constructing *meaningful queries* in terms of a) including only those well–understood terms - natural language words versus acronyms and codes - within the query and, therefore, e.g., only relevant relations, attributes and/or values at the storage model, b) applying those operations or operators which are consistent with the application domain semantics, e.g., only the *equals* operator would be suggested when a categorical value is considered within a restriction clause or the *average* operation will be excluded from suggestion when an attribute is classified as *categorical variable*, c) avoiding combination of mutually exclusive query terms such that they might lead to an empty set as a query result (unnecessary execution of query), d) providing a natural language based interpretation of the values as included in the query result.

The demonstration aims at showing how it is possible to construct queries by having the system driving the end-user to the construction of ad–hoc meaningful queries, i.e., queries which comply with the *application domain semantics*. Thereby, an object-relational database schema with approximately 120 attributes referring to *the treatment of acute myocardial infarctions* in Swiss hospitals will be used. The generation of the query result succeeds in that the constructed query is transformed at the server side into SQL-statements. Notice that *subqueries* addressing *nested tables* are also taken into account.

Given the mapping from the natural language words, in which the MDDQL query has been constructed, to acronyms and codes of the storage model, the same query result will be retrieved and sent to the client, even if the query has been constructed by using words from different natural languages. Presentation of the query result is given in a tuple-oriented form, where data items can be set-oriented values. The query result can be exported as an ASCII file with user specified delimiters such that the query result data can be imported into statistical software packages and/or front-end databases such as SPSS, Excel, MS-Access, etc. In such a case, multi-valued attributes are exported after having been flattened. All elements included in the query result can be viewed interchangeably, either by using the natural language based interpretations or the acronyms and codes as in the database.

Related work: The system does not make use of any diagrammatic presentation of conceptual schemes or any visual formalisms, and, therefore, contrasts with the approaches taken by visual query languages [2, 3, 7, 8, 9, 6] and systems (VQSs). They are mainly classified into *form-based* such as [9], *diagrammatic* such as [7] and *iconic* VQSs [8] according to the representation of the domain of interest for query formulation and/or for the query result. Some hybrid systems such as [6] have also been proposed. In all these high-level querying approaches, the application domain semantics based construction of a query, as described above, does not happen to be of major concern. In addition, the end–user is expected to make him/herself familiar with *visual formalisms* instead of *syntax formalisms*.

Semantics have been extensively used as a user guiding mechanism in interactive query formulation techniques. They are either database schema bound in order to provide incremental or associative query answering [15, 14], or they are ontology driven [11, 10, 1]. The goal of [15, 14] is to provide an end-user with context-sensitive assistance based on database modeling and probabilistic reasoning techniques combined with linguistic-based ones. To this extend, query formulation takes place in terms either of query completion of incomplete queries or suggestions of further predefined queries to reach a complex query goal [5, 4]. To this extent, no advanced semantics in terms of an advanced vocabulary or application ontology is user as a guiding mechanism for the construction of the query.

Instead, an ontology driven mechanism is provided by [11, 10, 1] which enables the clarification of the semantics of terms, such as in biology, to be used for the construction of the query. However, the main focus of the query construction technique is the exploration of taxonomies or classification structures rather than a querying mechanism capable of providing the expressive power as known by conventional database specific query languages. To this extent, usage of logical, comparison or statistical operators is not enabled. Moreover, semantic inconsistencies among terms cannot be prevented.

2. The MDDQL System Components

The system consists of the following components:

1. the *MDDQL query construction blackboard* (a snapshot is given by figure 1) used as the user interface, where the query is being assembled through user/system interaction in terms of requested suggestions of query terms and their signification as related to the specific application domain. The underlying data structure which gets manipulated through the user/system interaction on the blackboard and at the client site is the MDDQL query tree. The nodes of the

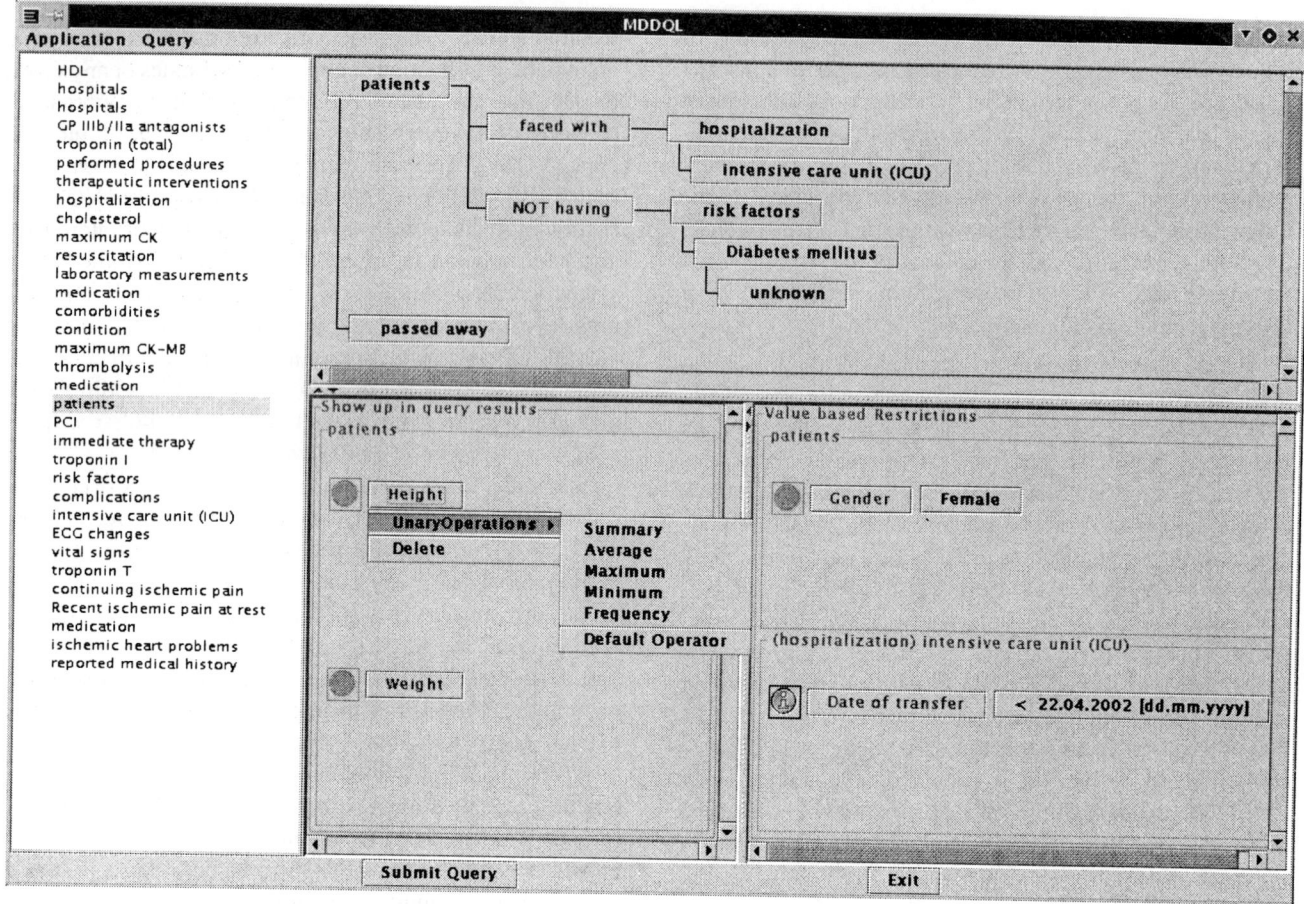

Figure 1. An example of a potential query

query tree are conceived as interconnected objects carrying on additional semantic knowledge needed by the interpretation/transformation component.

2. the *terminology base* which acts as the representation platform for the MDDQL query terms (vocabulary) together with their constraints on use which constitute the application ontology.

3. the *MDDQL inference engine* in order to respond to the user's requests for meaningful suggestions of query terms during query construction. The inference engine is contacted each time suggestions for the consideration of semantically consistent subsets of query terms are requested by the end-user in order to refine the query. The suggestions can be posed from selection menus which can be activated from each term already included within the query under construction. The inferences rely on both the semantics as represented by the *terminology base* and the current set of already considered query terms.

4. the *MDDQL query tree interpretation and transformation component*, where each submitted query in terms of an MDDQL query tree is transformed into an SQL-query.

The system relies on a three-tier system architecture. Assuming that the database resides at the *back–end layer*, components (2) and (4) are assigned to the *middle layer* whereas components (1) and (3) are assigned to the *front-end layer* (client site). However, a query construction session at the client site makes use of component (2) a copy of which is downloaded from the application server at the middle layer. Therefore, there is no network communication overhead between *inference engine* and *terminology server* during a query construction session.

Moreover, given this system architecture, there is a shift of the query construction logic from the middle layer to the clients. The application server deals only with the query transformation logic (component (4)). Given also that all system components are implemented in the *Java* programming language, the components (1) and (3) as needed for the interactive query construction sessions can be installed as a software package on any client running under various operating systems such as Windows 98/NT/2000, Linux, Solaris, HP-AIX, etc.

Furthermore, maintenance of the *terminology base* is done at the application server independently of the interactive query construction software once installed at the client site. Any changes to the terminology space become immediately visible to all clients having installed the query construction components. Hence, the query vocabulary can be adapted dynamically according to changes of the application domain semantics.

2.1 An overview of the SQL generation algorithm

Since the submitted query takes the form of an MD-DQL query tree which carries on all the semantic information needed for the mappings between natural language words and symbols to be used within the SQL statement, the generation algorithm has been implemented on the basis of traversing with a *depth–first* strategy the submitted MDDQL query tree. Thereby, the inference of the tokens and/or clauses to be put into the **select**, **from** and **where** slots relies a) upon the nature of the visited query nodes as well as the nature of the visited path, b) upon the nature of the corresponding storage model symbols, i.e., standing for a table, an attribute or attribute path (nested tables), a value or set of values.

Notice that the MDDQL query nodes are, mainly, classified as *Entity Set*, *Relationhip*, *Property* and *Value* nodes. Following constraints hold for the structure of an MDDQL query tree: an *Entity Set node* might have edges to either one or more *Entity Set* nodes, or one or more *Relationhip* nodes, or one or more *Property* nodes. A *Relationhip* node might have edges to one or more *Entity Set* nodes, or one or more *Property* nodes. A *Property* node might have edges to either one or more *Property* nodes, or to one or more *Value* nodes. Finally, a *Value* node might have edges to one or more *Value* nodes.

Currently, it is possible to infer JOIN operations with or without additional tables implementing a relationship. This also holds for tables used for classification structures. A distinction is made by the algorithm when categorical values are used for classification within the same table (no JOIN operation needed). Moreover, subqueries referring to nested tables such as those used for multi-valued attributes are also generated. This is crucial when *AND-connected* restrictions on values coming from nested tables are addressed.

Acknowledgments: We would like to thank Prof. Dr. M. Norrie and Prof. Dr. H. Hinterberger for their support to make this system reality.

References

[1] S. Bechhofer, R. Stevens, G. Ng, A. Jacoby, and C. Goble. Guiding the User: An Ontology Driven Interface. In N. W. Paton and T. Griffiths, editors, *Proc. User Interfaces to Data Intensive Systems (UIDIS99)*, pages 158–161, Edinburgh, September 1999. IEEE Press.

[2] J. Cardiff, T. Catarci, and G. Santucci. Semantic query processing in the VENUS environment. *International Journal of Cooperative Information Systems*, 6(2):151–192, June 1997.

[3] T. Catarci, M. Costabile, S. Levialdi, and C. Batini. Visual query systems for databases: A survey. *Journal of Visual Languages and Computing*, 8(2):215–260, April 1997.

[4] W. Chu and Q. Chen. A Structured Approach for Cooperative Query Answering. *IEEE Transactions on Knowledge and Data Engineering*, 6(5):738–749, October 1994.

[5] W. W. Chu, H. Yang, K. Chiang, M. Minock, G. Chow, and C. Larson. CoBase: A scalable and extensible cooperative information system. *Journal of Intelligent Information Systems*, 1996.

[6] L. Cinque, S. Levialdi, and F. Ferloni. An expert visual query system. *Journal of Visual Languages and Computing*, 2:101–113, 1991.

[7] Y. Dennebouy, M. Anderson, A. Auddino, Y. Dupont, E. Fontana, M. Gentile, and S. Spaccapietra. SUPER: Visual interfaces for object + relationship data models. *Journal of Visual Languages and Computing*, 6:74–99, 1995.

[8] A. Massari, S. Pavani, L. Saladini, and P. Chrysanthis. QBI: Query by icons. In *Proc. of the ACM SIGMOD Conf. on Management of Data*, page 477, San Jose, USA, 1995. ACM Press.

[9] G. Ozsoyoglu and H. Wang. Example-based graphical database query languages. *COMPUTER*, 26(5):25–38, May 1993.

[10] N. Paton, R. Stevens, P. Baker, C. Goble, S. Bechhofer, and A. Brass. Query Processing in the TAMBIS Bioinformatics Source Integration System. In *Proc. 11th Int. Conf. on Scientific and Statistical Databases (SSDBM)*, pages 138–147 1999. IEEE Press, 1999.

[11] R. Stevens, P. Baker, S. Bechhofer, G. Ng, A. Jacoby, N. Paton, and C. Goble. TAMBIS: Transparent Access to Multiple Bioinformatics Information Sources. *Bioinformatics*, 16(2):184–186, 2000.

[12] R. Studer, R. Benjamins, and D. Fensel. Knowledge Engineering: Principles and Methods. *DKE*, 25(1-2):161–197, 1998.

[13] M. Uschold and M. Grueninger. Ontologies: Principles, Methods and Applications. *Knowledge Engineering Review*, 2, 1996.

[14] G. Zhang. *Interactive Query Formulation Techniques for Databases*. PhD thesis, University of California, Los Angeles, 1998.

[15] G. Zhang, W. W. Chu, F. Meng, and G. Kong. Query Formulation from High-Level Concepts for Relational Databases. In N. Paton and T. Griffiths, editors, *Proc. User Interfaces to Data Intensive Systems, UIDIS 99*, pages 64–74, Edinburgh, Scotland, September 1999. IEEE Computer Society Press.

Distributed Scientific Databases

Query Processing with Description Logic Ontologies Over Object-Wrapped Databases

Martin Peim, Enrico Franconi, Norman W. Paton and Carole A. Goble
Department of Computer Science, University of Manchester
Oxford Rd, Manchester M13 9PL, UK
lastname@cs.man.ac.uk

Abstract

This paper presents an approach to answering queries over an ontology modelled using a description logic. The ontology acts as a global schema, providing a declarative description of the concepts of the domain, the instances of which are stored in (potentially many) object-wrapped sources. Queries are expressed using terms from the rich vocabulary of the ontology, and are translated into an equivalent calculus expression, which references only the objects available in the source databases. The query is then optimized on the basis of information from the ontology and the source databases. Distinctive features of the approach include: the use of the expressive $\mathcal{ALCQI}$ description logic, which supports both ontology definition and query expression; the adoption of a global-as-view approach to relating the ontology to the sources; and the use of the ontology to direct semantic optimization of queries phrased over specific sources. The approach is being developed in, and is illustrated using examples from, bioinformatics.

1 Introduction

Bioinformatics involves the storage and analysis of experimental and observational data from biology. Bioinformatics researchers have developed a wide range of biological databases and tools; for example, the January "Databases Issue" of *Nucleic Acids Research* in 2001 contained 96 papers describing bioinformatics databases. For the most part, bioinformatics databases are developed and maintained in isolation from each other, and thus manifest classical syntactic and semantic heterogeneities. As a result, there has been considerable attention devoted to the development of proposals that ease the task of exploring or accessing heterogeneous bioinformatics resources (e.g. [10, 7, 8]). However, few such proposals have provided truly declarative query interfaces or high levels of transparency.

The work described in this paper follows on from the TAMBIS project [13], in which queries written over the GRAIL description logic (DL) were mapped into execution plans in CPL/Kleisli [8]. Limitations of the online TAMBIS query processor (http://img.cs.man.ac.uk/tambis/) include: (i) the ontology is represented using a relatively inexpressive DL, in which certain features of the biological domain are difficult to express; (ii) the mapping between concepts in the ontology and collections in the sources is quite restrictive – for example, TAMBIS did not allow multiple sources for the same kind of data (e.g. both SwissProt and PIR as protein sources); and (iii) although queries are optimized, as described in [17], there is no semantic query optimization making use of axioms from the ontology. All of these limitations are addressed in the proposal described in this paper. Furthermore, the adoption of an object-oriented wrapper layer makes the proposal compatible with mainstream middleware proposals such as that standardised by the OMG, which in turn is associated with an important standardisation activity in bioinformatics (http://www.omg.org/homepages/lsr/).

In a wider technical context, the proposal presented in this paper is part of a collection of results on knowledge based query processing in distributed information systems. Such proposals can be classified as *global as view* (e.g., SIMS [1], OBSERVER [16]), where terms in the global schema are defined by views over the source schemas, or *local as view* (e.g. Information Manifold [15], DWQ [2], Picsel [12]), where the content of sources is defined in terms of views over the global schema [19]. This paper presents results in a global-as-view setting, but extends earlier proposals in several ways. Firstly, the language used to describe the global conceptual model and to express queries, namely $\mathcal{ALCQI}$, is considerably more powerful than in earlier global-as view proposals. This allows domains to be described more precisely in the ontology and allows more precise questions to be asked. Secondly, the reasoning services of the DL are used extensively during query processing to support semantic query optimization based on axioms within the ontology. This allows relationships between source and global model constructs to be described in a declarative manner, which can then be exploited by the op-

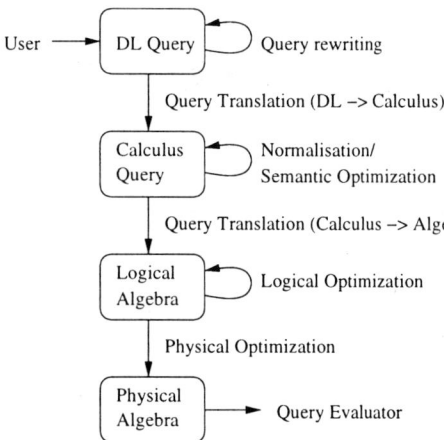

Figure 1. Query processor architecture.

timizer to detect redundant subqueries. Thirdly, the sources are wrapped using an object model, and not represented as relations, as in most other proposals. This is an important practical consideration, as many existing resources are more naturally wrapped using an object model than as tables.

The remainder of the paper is structured as follows. Section 2 provides an overview of the approach, introducing the models and languages used at different points during query processing. Section 3 describes the $\mathcal{ALCQI}$ description logic and illustrates its use for conceptual modelling and query expression. Section 4 introduces the object model and the calculus over that model that together represent the API of the object-wrapped sources. Section 5 describes the translation of $\mathcal{ALCQI}$ queries from the DL into the calculus of the wrapper layer, and Section 6 describes how (semantic) optimization can be applied over the calculus representation of the queries. Section 7 concludes the paper.

2 Overview

This section gives an overview of the components that participate in the query processing framework described in more detail later in the paper, which are illustrated in Figure 1.

The input to the process is a description logic (DL) query over an ontology. The ontology consists of a set of integrity constraints in the form of logical axioms. Efficient inference engines exist [14] for use with DL ontologies. In particular, we are interested in the consistency and containment reasoning tasks, which enable certain semantic query optimizations to be made. The ontology can also be used to guide the query formulation process, in a way similar to that used in the TAMBIS system [13].

The *query evaluator* executes queries over a collection of object wrapped sources. Sources descriptions consist of a set of class declarations described using the ODMG's object definition language ODL—that is, an ODL *schema*—plus information needed for physical query optimization, such as

access paths (existence of indices, for example) and cost information. Extent names in the wrapper layer are associated with certain terms (which are referred to as *ground* terms) in the ontology by an *Ontology-to-Object-Model mapping*.

Queries are formulated over the ontology in a query language which is itself a DL. The system translates these high-level user queries into queries over the definitions at the wrapper layer. The translation process can be broken down into several phases.

The *query rewriting* phase comprises a simple expansion algorithm to rewrite the query as an expression containing only ground terms; this is based on a global-as-view approach. Any ground term which appears in the rewritten query must be mapped to source constructs in the Ontology-to-Object-Model (OOM) mapping.

The *query translation* phase translates the rewritten query into an expression in Fegaras' Monoid Comprehension Calculus [11]. The calculus proviudes a state-of-the-art semantics for OODBs together with well-founded tools for optimization and evaluation of queries. We have enriched the calculus with a *match*() operator to perform *object fusion*. The *match*() operator supports the reconstruction of a unique object by gathering sparse information coming from one or more sources. The translation process works in a compositional fashion, using the OOM mapping to translate DL terms, and handling logical connectives using a set of rules given in Section 5.

The *semantic optimization* phase takes the (possibly deeply nested) calculus expression generated by the query translation algorithm and performs simplifying transformations on it. The ontology is used at this stage to improve the calculus expression by identifying redundant generators (potential iterations) over source extents.

The *calculus-to-algebra translation* phase. From this point on, the translation system is an adaptation of Fegaras' OQL optimizer [11]. The calculus expression is translated into an expression in a logical query algebra based on the nested-relational algebra. The translation rules are quite general and do not depend on source-specific information.

The *logical and physical optimization* phases are essentially those presented in [11]. Firstly, a calculus expression is translated to a corresponding expression in a logical algebra. This logical algebra is then subject to heuristic optimization by a logical optimizer, and then the logical algebra operators are replaced by operators in a physical algebra during physical optimization. For lack of space, the logical and physical optimizers are not described further in this paper.

3 The Conceptual Model and Query Language

At the top level of the system, a unified and user-centred view of the source data is abstracted in a global conceptual schema: the ontology. Queries are formulated with respect to this i.e., the user may use any term from the ontology in the query, without knowing where the data actually is, how it is structured, and how it should be merged and reconciled to fit the global schema. In this section the formalisms for representing the ontology and the top level queries are introduced.

3.1 The $\mathcal{ALCQI}$ Description Logic

The conceptual schema consists of an ontology expressed in the DL $\mathcal{ALCQI}$ [5]. The basic types of a DL are *concepts* and *roles*. A concept expression is a description gathering the common properties among a collection of individuals; from a logical point of view it is a unary predicate ranging over the domain of individuals. Inter-relationships between these individuals are represented by means of role expressions (which are interpreted as binary relations over the domain of individuals). We present here a only a brief summary of the syntax and semantics of $\mathcal{ALCQI}$.

$\mathcal{ALCQI}$ has a rich combination of constructors, including full boolean operators, qualified number restrictions, inverse roles and inclusion assertions of a general form. Its syntax rules in Figure 2 define valid concept and role expressions. C and R denote concept and role expressions, A is an atomic concept name and P is an atomic role name.

The semantics of $\mathcal{ALCQI}$ can be given in terms of an interpretation $\mathcal{I} = (\Delta^{\mathcal{I}}, \cdot^{\mathcal{I}})$ consisting of a non-empty set $\Delta^{\mathcal{I}}$ of individuals (the *domain* of $\mathcal{I}$) and a function $\cdot^{\mathcal{I}}$ (the *interpretation function* of $\mathcal{I}$) satisfying suitable axioms. Every concept is interpreted as a subset of $\Delta^{\mathcal{I}}$ and every role as a subset of $\Delta^{\mathcal{I}} \times \Delta^{\mathcal{I}}$. An ontology (or knowledge base) is a finite set Σ of axioms of the form $C \sqsubseteq D$, involving concept expressions C, D; we write $C \doteq D$ as a shortcut for both $C \sqsubseteq D$ and $D \sqsubseteq C$. An interpretation $\mathcal{I}$ satisfies $C \sqsubseteq D$ if and only if the interpretation of C is included in the interpretation of D, i.e., $C^{\mathcal{I}} \subseteq D^{\mathcal{I}}$; it is said that C is subsumed (or contained) by D. Details can be found in [4]

3.2 Example Biological Ontology

Ontologies have recently established a significant profile in bioinformatics, for example, for describing the biological functions of genes identified in genome sequencing projects [18]. Figure 3 contains a small example ontology that will be used throughout the paper to illustrate query processing. The sources for **protein** information are simplifications of SwissProt (see http://www.ebi.ac.uk/swissprot/) and PIR (see http://www-nbrf.georgetown.edu/pirwww/). Most of the axioms are self-explanatory. For example, the first states that any enzyme is a protein which catalyses some reaction. A protein as described in the SwissProt database is a protein with exactly one accession number and possibly some species (axioms 4 and 7). The role **cited-in** represents the relationship between a protein—as described in the PIR database—and journal articles in which it is cited (axiom 8). The concept **top-journal** represents (names of) those journals

$$\text{enzyme} \sqsubseteq \text{protein} \sqcap \exists\text{catalyses.reaction} \tag{1}$$

$$\text{enzyme} \doteq \exists\text{enz-protein}^{-}.\top \tag{2}$$

$$\exists\text{catalyses}.\top \sqsubseteq \text{protein} \tag{3}$$

$$\text{sp-protein} \sqsubseteq \text{protein} \tag{4}$$

$$\text{pir-protein} \sqsubseteq \text{protein} \tag{5}$$

$$\text{protein} \sqsubseteq (\exists\text{has-sequence}.\top) \sqcap (\exists^{\leq 1}\text{has-sequence}.\top) \tag{6}$$

$$\text{sp-protein} \sqsubseteq (\exists\text{sp-acc}.\top) \sqcap (\exists^{\leq 1}\text{sp-acc}.\top) \tag{7}$$
$$\sqcap (\forall\text{has-species.species})$$

$$\text{pir-protein} \sqsubseteq (\exists\text{pir-acc}.\top) \sqcap (\exists^{\leq 1}\text{pir-acc}.\top) \tag{8}$$
$$\sqcap (\forall\text{cited-in.reference})$$

$$\text{species} \sqsubseteq (\exists\text{common-name}.\top) \tag{9}$$
$$\sqcap (\exists^{\leq 1}\text{common-name}.\top)$$
$$\sqcap (\exists\text{latin-name}.\top) \sqcap (\exists^{\leq 1}\text{latin-name}.\top)$$

$$\text{mammal} \sqsubseteq \text{species} \tag{10}$$

$$\text{reference} \sqsubseteq (\exists\text{has-author}.\top) \tag{11}$$
$$\sqcap (\exists\text{has-journal}.\top) \sqcap (\exists^{\leq 1}\text{has-journal}.\top)$$
$$\sqcap (\exists\text{has-year}.\top) \sqcap (\exists^{\leq 1}\text{has-year}.\top)$$

$$\text{enz-entry} \sqsubseteq (\exists\text{enz-protein}.\top) \tag{12}$$
$$\sqcap (\forall\text{enz-protein.sp-protein})$$
$$\sqcap (\exists\text{enz-reaction}.\top)$$

$$\text{journal} \doteq \exists\text{has-journal}^{-}.\top \tag{13}$$

$$\text{top-journal} \sqsubseteq \text{journal} \tag{14}$$

Figure 3. The example ontology.

which are considered important, and corresponds to an extent top-journals consisting of a set of strings (axioms 13 and 14).

Given an ontology, such as that in Figure 3, a concept description can be taken as a query to retrieve all instances of the concept. For example, the following query asks for all proteins found in mammals:

$$\text{protein} \sqcap \exists\text{has-species.mammal} \tag{15}$$

The following query, which is revisited later, asks for all proteins referred to in important journals.

$$\text{protein} \sqcap \exists\text{cited-in}.(\exists\text{has-journal.top-journal}) \tag{16}$$

3.3 Query Rewriting

The query rewriting task can be phrased in general terms as follows. Given a query Q, an ontology, and a set of view definitions that characterise the actual source data, reformulate the query into an expression, the rewriting, that refers only to the views, and provides the answer to Q. In our case, the view definitions correspond to the classes and relationships of the object wrapped sources.

In this paper we will consider only the simplistic case of the logical (source) vocabulary being a subset of the

$$C, D \rightarrow A \mid \top \mid \bot \mid \neg C \mid C \sqcap D \mid C \sqcup D \mid \forall R.C \mid \exists R.C \mid \exists^{\geq n} R.C \mid \exists^{\leq n} R.C$$
$$R \rightarrow P \mid R^{-}$$

Figure 2. $\mathcal{ALCQI}$ **concept and role expressions**

conceptual DL vocabulary. We call these concepts and roles *ground*. Moreover, we restrict our attention to top level *class queries*, which are themselves DL concept expressions. Concept expressions which are to be treated as queries are rewritten into equivalent concept expressions involving only the ground concept names.

The simplest rewriting system we can use is a stratified system with a straightforward expansion algorithm. In a stratified system, a distinguished subset of the DL axioms, called the *expansion axioms*, are used for query rewriting. The expansion axioms are all of the form $A \doteq C$, where A is an atomic concept (a concept name) and C is a concept expression. The set of definitions in the expansion axioms is required to be acyclic, or *stratified*. Furthermore, each concept name which can appear in a query and is not represented by a source database (i.e., is not ground) must be defined in an expansion axiom. Also, the names used on the right-hand sides of definitions must all be queryable. The rewriting process is then a simple matter of expanding definitions of non-ground concepts until only ground ones are left. This expansion process is guaranteed to terminate and to produce a grounded query equivalent to the original one.

As an example, suppose the ontology contains the definition mammalian-enz $\doteq$ enzyme $\sqcap$ $\forall$has-species.mammal, for enzymes which only occur in mammals. Many such concept definitions could be added to the ontology for the convenience of users. We have not done so here to conserve space. Then the query mammalian-enz would expand into

$$(\exists \text{enz-protein}^{-}.\top) \sqcap \forall \text{has-species.mammal} \quad (17)$$

Once a query has been rewritten into an equivalent query containing only ground terms, it is important to check that the query is *safe* [9]. Essentially, a query (or concept) is considered safe if answering that query does not involve looking up information not referred to in the query. This is crucial to restrict the scope of a query. Our translation scheme in Section 5 only produces translations for safe queries. For example, the query $\forall R.C$ is unsafe because answering it involves, among other things, finding all individuals with no R fillers, and this information is not available from extents for R and C.

Let Q be a concept expression which has been expanded (so it contains only ground concept names) and rewritten into *negation normal form*[1]. Then Q is safe if it has the form

$\bot$, A (where A is a ground concept), $\exists R.C$ or $\exists^{\geq n} R.C$ (provided $n \geq 1$). It is unsafe if it has the form $\top$, $\neg A$, $\forall R.C$ or $\exists^{\leq n} R.C$. A conjunction is safe if and only if at least one of its conjuncts is safe. A disjunction is safe if and only if all of its disjuncts are safe. Note that, under this definition, a concept expression is safe if and only if its negation is unsafe.

Note that a more sophisticated rewriting system without the above restrictions, and allowing for a more expressive query language such as non-recursive datalog, could be used without affecting the rest of the proposal [19, 3].

4 The Object Model and Calculus Queries

The query translator takes a DL query and translates it into a calculus expression over an object model. The object model is described in Section 4.1, and the calculus in Section 4.3.

4.1 The Object Model

The source databases are presented to the rest of the system by software wrappers in such a way that they can be seen as forming an object database, conforming to the ODMG data model [6]. From an implementation point of view, this is a practical choice because of its compatibility with CORBA and the fact the model is associated with well understood query processing techniques [11]. The structure of the objects returned by the wrappers is given by a schema in the ODMG's Object Definition Language (ODL).

Each ground concept in the ontology is viewed as a named persistent set of database objects. This set may be an extent over an ODL class or a set of values of some simple type like String. For the sake of brevity, we will refer to all such named collections as *source extents*. So that we can use the DL reasoner to assist in query optimization (as described in Section 6), we require that each source name be represented by a name in the knowledge base, and we record any information about containment between sources in the ontology.

Figure 4 shows wrapper class definitions for the domain represented by the ontology in Figure 3. For example, the interface Protein represents protein data from the sources SwissProt and PIR, which are represented by the classes SP_Protein and PIR_Protein.

The attribute SPAccessionNumber in the source class SP_Protein corresponds to the accession number of a SwissProt entry. It is a unique identifier for the entry.

---

[1]By pushing negations inwards in the usual way, one can rewrite any $\mathcal{ALCQI}$ concept expression into an equivalent expression in *negation normal form* or NNF, where negations only appear in front of concept names.

```
interface Protein {
  attribute String sequence;
}

class SP_Protein (extent sp_proteins)
  extends Protein {
    attribute String SPAccessionNumber;
    attribute Set<Species> species;
    attribute String sequence;
}

class PIR_Protein (extent pir_proteins)
  extends Protein {
    attribute String PirAccessionNumber;
    attribute Set<Reference> references;
    attribute String sequence;
}

class Species (extent species) {
  attribute String common_name;
  attribute String latin_name;
}

class Reference (extent references) {
  attribute Set<String> authors;
  attribute String title;
  attribute String journal;
  attribute String year;
}

class EnzEntry (extent enz_entries) {
  attribute String enz_id;
  attribute Set<SP_Protein> enz_proteins;
  attribute Set<String> reactions;
  attribute Set<String> cofactors;
}

class Enz_catalyses_class (ex-
tent enz_catalyses) {
  attribute SP_Protein base;
  attribute String filler;
}

Set<Species> mammals
Set<String> top_journals
```

Figure 4. Declarations of source classes.

Concept	Source extents
protein	sp_proteins pir_proteins
sp-protein	sp_proteins
pir-protein	pir_proteins
enz-entry	enz_entries
species	species
reference	references
top-journal	top_journals
mammal	mammals

Table 1. Concept to source mapping for biological example

Role	Attribute	Cardinality
sp-acc	SPAccessionNumber	Single
pir-acc	PirAccessionNumber	Single
has-species	species	Multiple
has-sequence	sequence	Single
cited-in	references	Multiple
common-name	common_name	Single
latin-name	latin_name	Single
has-author	authors	Multiple
has-title	title	Single
has-journal	journal	Single
has-year	year	Single
enz-id	enz_id	Single
enz-protein	enz_proteins	Multiple
enz-reaction	reactions	Multiple
cofactor	cofactors	Multiple

Table 2. Attribute role mappings for biological example

The species attribute contains the set of species in which the protein can be found. Finally, sequence is a string representation of the protein's amino acid sequence.

Like SwissProt, PIR also identifies its protein entries with an accession number (see PIR_Protein). Note that PIR accession numbers are not the same as SwissProt accession numbers, so the attributes must be given different names and correspond to different roles at the DL level. The references attribute is a set of descriptions of references to the protein in the scientific literature. As with SP_Entry, the string sequence represents the amino acid sequence.

4.2 The Ontology-to-Object-Model Mapping

The OOM Mapping describes how ground DL concepts and roles relate to object model classes and relationships. For the example application, Table 1 gives the mapping between DL concepts and source extents, and Table 2 shows the mapping of DL roles to class attributes.

DL roles are divided into *enumerated roles*, which are represented by the OOM mapping directly as sets of pairs, and *attribute roles*, which are represented as attributes of the base objects. An enumerated role R is represented by a set of source extents. Each of these is an extent e over a class of the form

```
class C (extent e){
    attribute T_1 base;
    attribute T_2 filler;
}
```

The attribute names *base* and *filler* are used to refer to the first and second components of the binary relation represented by R (see Section 3.1). We say that e has type $T_1 \times T_2$.

The model contains a single enumerated role, catalyses, which represents the relationship between an enzyme and the reaction it catalyses. This role is mapped to the extent enz_catalyses of Figure 4, which is implemented as part of the wrapper on the Enzyme database.

An attribute role R is represented by an attribute name a_R. The attribute may be defined in several classes, and may have a different value type in each. For example, the fillers for a role like has-name may be simple strings in most classes, but structured objects (for example, botanical names of plants) in others. An attribute role can be either *single-valued* or *multiple-valued*. We require that a single-valued attribute role be represented by an attribute whose value type is a simple class name in each class which supports it, and that a multiple-valued attribute role be represented by an attribute whose value type is $\mathsf{Set}(T)$ for some class name T.

4.3 The Monoid Comprehension Calculus

The target language for the first stage of query translation is the Monoid Comprehension Calculus of Fegaras [11]. The calculus provides a uniform notation for collection types such as lists, bags and sets, based on the observation that the operations of set and bag union and list concatenation are *monoid* operations (that is, they are associative and have an identity element). Monoids for collection types are known as *collection monoids*. Operations like conjunctions and disjunctions on booleans and integer addition over collections can also be expressed in terms of so-called *primitive monoids*. A monoid comprehension has the form

$$\otimes\{e \mid q_1, \ldots, q_n\}. \qquad (18)$$

The symbol $\otimes$ is a monoid operator, and determines the type of the comprehension. The expression e is called the *head* of the comprehension. Each q_i is a *qualifier*, which can either be a *generator* of the form $v \leftarrow e'$, where v is a variable and e' is a collection-valued expression, or a *filter* of the form p, where p is a predicate (a boolean-valued expression). Each variable v is assigned a type T, and the corresponding collection monoid must have an element type which is a subtype of T. We will usually omit the variable type from our notation, except where it needs to be emphasised. If q_i is a filter, then the head expression e and the q_j for $j > i$ can contain free occurrences of the variable v. The identity, or zero, element of the monoid whose operation is $\otimes$ is denoted by $\mathcal{Z}_\otimes$.

The primitive monoids used in examples below are: (i) The logical-and monoid $\wedge$. This is a simple monoid whose underlying type is boolean. The monoid operation is boolean conjunction and $\mathcal{Z}_\wedge = \mathbf{true}$. (ii) The plus-monoid $+$. This is also a simple monoid whose underlying type is integer. The monoid operation is integer addition and $\mathcal{Z}_+ = 0$. The plus-monoid is used in examples in the form $+\{1 \mid \bar{s}\}$, to compute cardinalities of sets. The result of this expression is the number of distinct assignments to the generator variables in $\bar{s}$ which satisfy all the filters.

4.3.1 The Match-Union Monoid

A single individual belonging to the extension of a DL concept or query may be represented by several database instances (with distinct OIDs), coming from different sources.

For example, protein instances may be represented in both the SwissProt and PIR sources. Thus, a query may have alternative answers if more than one choice of database instance is available for some of the relevant individuals. In order to support the reconstruction of a unique individual object from the sparse information coming from the same or different sources, we introduce a new boolean-valued operator $match(x, y)$ which returns **true** if the database instances x and y represent the same individual.

This $match(\_, \_)$ operator is really a collection of operators $match_{S_1, S_2}(\_, \_)$, one for each pair of source databases S_1, S_2 (we can also divide it according to different object types returned by each source). We require that $match()$ defines an equivalence relation—that is, it must be reflexive, symmetric and transitive. We also assume that distinct elements from the same source are intended to represent distinct individuals, so that if x and y come from the same source S, $match_{S,S}(x, y)$ reduces to $x = y$. The $match()$ operator extends to tuples of objects in the obvious way. It can be interpreted as a simple equality test for domain values like integers and strings.

At the physical level, the implementation of $match()$ for any given pair of sources may consist of a function which performs a comparison between certain key attributes of the objects concerned. Alternatively, if the two source classes do not have a common key, it might be necessary to use a binary table to associate the corresponding elements.

An answer to a query, then, is a set S of object references, such that

$$\text{for all } x, y \in S, \; match(x, y) = \mathbf{false} \text{ unless } x = y. \quad (19)$$

Such sets are referred to as *match-sets*—they are still sets (rather than bags) even if we regard $match()$ as equality.

In order to capture query answers as match-sets, comprehensions are written in terms of a collection monoid whose merge operation $\oplus$ (read as *match-union*) is like the set union operation but preserves the uniqueness condition (19). So, if S_1 and S_2 satisfy (19) then $S_1 \oplus S_2$ may be any set $W \subseteq S_1 \cup S_2$ of object references, such that for each $x \in S_1 \cup S_2$ there is precisely one $w \in W$ such that $match(x, w) = \mathbf{true}$. For those elements of S_1 which match some element of S_2, we can choose which element to include in $S_1 \oplus S_2$. This choice can be made by the system on the basis of user preference or cost estimation or, if we have no preference, by taking the representative for S_1 whenever possible.

5 Translating Queries to Monoid Comprehensions

The rules given in this section show how to translate a safe $\mathcal{ALCQI}$ concept expression C into an expression E in the monoid comprehension calculus. To save space, we only consider expressions whose subexpressions are all safe. The implemented system also deals with cases where this is not so. The modifications to the translation rules for these cases are indicated below.

The rules constitute a compositional syntax-directed translation scheme. The expression E is a collection monoid comprehension using the monoid operation $\oplus$ described in Section 4.3.1. If the element type of this comprehension (the type of its head expression) is T we will say that E is a translation of C having type T.

The Empty Concept. The unsatisfiable concept $\perp$ is translated by the empty $\oplus$-monoid $\mathcal{Z}_\oplus$.

Atomic Terms. To translate a ground atomic concept (a concept name) A we consult the OOM mapping to find the set of database extent names which represent A. Each extent name has a type (a class name) T_i and refers to a set S_i of object references of type T_i. Let T be the most specific superclass of the T_i. Then $\bigoplus_i S_i$ is a translation of A having type T.

For instance, in our example application the concept protein is mapped to the source extents sp_proteins and pir_proteins, and so it has the translation sp_proteins $\oplus$ pir_proteins (of type Protein).

Conjunctions. If C and D are safe concepts with translations C' and D' of type $T_{C'}$ and $T_{D'}$ then $C \sqcap D$ can be translated to either of the following, with types $T_{C'}$ and $T_{D'}$ respectively:

$$\oplus\{c \mid c \leftarrow C', d \leftarrow D', match(c, d)\} \quad (20)$$

$$\oplus\{d \mid c \leftarrow C', d \leftarrow D', match(c, d)\} \quad (21)$$

If we don't want to commit to choosing all our answers from one of C' and D' and we have a choice function $choose(x, y)$ which selects one of x and y according to some unspecified criteria, we can make a third translation with the type, T which is the most specific superclass of T_1 and T_2:

$$\oplus\{choose(c, d) \mid c : T \leftarrow C', d : T \leftarrow D', match(c, d)\} \quad (22)$$

Note the use of type specifiers for the variables c and d to emphasise that we are assigning subclass references to superclass reference variables. That is, we just take the $\oplus$ merge of C' and D' but interpret the references as having type T.

In the case where C is safe but D is unsafe, we must filter the instances of C for non-membership (up to $machtop()$) of the safe concept $\neg D$ rather than for membership of D.

Disjunctions. If C, D are safe concepts with translations C' and D' of type $T_{C'}$ and $T_{D'}$, let T be the most specific superclass of $T_{C'}$ and $T_{D'}$. Then

$$(\oplus\{c \mid c : T \leftarrow C'\}) \oplus (\oplus\{d \mid d : T \leftarrow D'\}) \quad (23)$$

is a translation of $C \sqcup D$ of type T.

Existentially Quantified and At-least Formulae. Existentially quantified formulae and at-least formulae are closely related, since $\exists R.C$ is equivalent to $\exists^{\geq 1} R.C$; they are handled together here.

- If R is an *enumerated role*, let $\{R_i'\}$ be the set of source tables for R and let the type of R_i' be $T_{i1} \times T_{i2}$. Let T_1 be the most specific superclass of the T_{i1}, let T_2 be the most specific superclass of the T_{i2}, and let

$$R' = \bigoplus_i (\oplus\{r \mid r : \langle \texttt{base} : T_1, \texttt{filler} : T_2 \rangle \leftarrow R_i'\}). \quad (24)$$

Suppose C is safe and has a translation C'. Table 3 gives translations for enumerated roles.

Inverses of enumerated roles (for example, in $\exists R^-.C$) can be handled similarly, by exchanging the roles of base and filler (and of T_1 and T_2).

- If R is a *single-valued attribute role*, let $\{D_i\}$ be the set of domains for R. Let T_{i1} be the type of D_i and let T_{i2} be the value type of the attribute a_R in D_i. Let T_1 be the most specific superclass of the T_{i1} and let T_2 be the most specific superclass of the T_{i2}. Let

$$D_R = \bigoplus_i \oplus\{d \mid d : T_1 \leftarrow D_i\} \quad (25)$$

Then D_R represents the (potential) domain of the relation R.

Suppose C is safe and has a translation C'. Table 4 gives translations for single-valued roles. The translation of $\exists^{\geq n} R.C$ is equivalent to $\exists R.C$ if $n = 1$ and is empty if $n > 1$.

As an example of the translation of a single-valued attribute role, consider the translation of $\exists$has-journal.top-journal. According to the OOM mapping in Table 2, the role has-journal is a single-valued attribute role. It is mapped to the attribute journal, which is supported by the class extent references. The concept top-journal is mapped to the extent top_journals, as described in Table 1. Using the translation for $\exists R.C$ in Table 4, with renaming of variables, the translation is

$$\oplus\{r_1 \mid r_1 \leftarrow \texttt{references}, t \leftarrow \texttt{top\_journals},$$
$$match(r_1.\texttt{journal}, t)\} \quad (26)$$

- If R is a *multiple-valued attribute role*, let $\{D_i\}$, T_{i1}, T_{i2} T_1 and T_2 be defined as for single-valued attributes above, except that the value type of a_R in T_{i1} is now $\text{Set}(T_{i2})$

Suppose C is safe and has a translation C'. Table 5 gives translations for multiple-valued roles.

Concept	Translation	Type
$\exists R.C$	$\oplus\{r.\mathtt{base} \mid r \leftarrow R', c \leftarrow C', match(c, r.\mathtt{filler})\}$	T_1
$\exists^{\geq n} R.C$	$\oplus\{r.\mathtt{base} \mid r \leftarrow R', +\{1 \mid s \leftarrow R', c \leftarrow C', match(s.\mathtt{base}, r.\mathtt{base}) \wedge match(s.\mathtt{filler}, c)\} \geq n\}$	T_1

Table 3. Translation of enumerated roles.

Concept	Translation	Type
$\exists R.C$	$\oplus\{d \mid d \leftarrow D_R, c \leftarrow C', match(d.a_R, c)\}$	T_1
$\exists R^-.C$	$\oplus\{d \mid d \leftarrow D_R, c \leftarrow C', match(d.a_R, c)\}$	T_2
$\exists^{\geq n} R^-.C$	$\oplus\{d.a_R \mid d \leftarrow D_R, +\{1 \mid c \leftarrow C', match(c, d)\} \geq n\}$	T_2

Table 4. Translation of single-valued roles.

An expression such as $\exists R.C$ or $\exists^{\leq n} R.C$ where C is unsafe is translated in a similar manner to the above, except that, instead of checking that objects related to a given instance of the domain of R are $match()$-equivalent to instances of C we must check that they are not equivalent to any instance of the safe concept $\neg C$.

5.1 Example Translation

As an example of the query translation process, consider query (16), which asks for all proteins referred to in important journals. To translate this query, we first translate the subquery:

$$\exists \mathtt{has\text{-}journal.top\text{-}journal}. \quad (27)$$

This gives rise to the calculus expression (26) described above. Proceeding in this way we obtain the monoid comprehension:

$$
\begin{aligned}
\oplus\{p_2 \mid\ & p_2 \leftarrow \mathtt{sp\_proteins} \oplus \mathtt{pir\_proteins}, \\
& p_3 \leftarrow \oplus\{p_1 \mid p_1 \leftarrow \mathtt{pir\_proteins}, \\
& \qquad r_2 \leftarrow p_1.\mathtt{references}, \\
& \qquad r_3 \leftarrow \oplus\{r_1 \mid r_1 \leftarrow \mathtt{references}, \\
& \qquad\qquad t \leftarrow \mathtt{top\_journals}, \\
& \qquad\qquad match(r_1.\mathtt{journal}, t)\}, \\
& \qquad match(r_2, r_3)\}, \\
& match(p_2, p_3)\},
\end{aligned}
$$

$$(28)$$

supposing that the system decides to use both sources for the concept protein. This rather unwieldy form can be considerably simplified by the methods of Section 6.

6 Optimization

For the most part, the optimization of the queries that result from the translation process described in Section 5 follows that of Fegaras' optimizer [11]. Following translation, a normalisation algorithm rewrites the comprehension

into a normal form which is more amenable to further optimization. We have extended Fegaras' normalisation algorithm so that certain semantic optimizations are made during the normalisation process, to remove unnecessary iterations or generators. The normalisation rules are given in Section 6.1. The optimization rules are given in Section 6.2.

6.1 Normalisation

The first stage of Fegaras' optimizer [11] is a normalisation process which does some unnesting of nested comprehensions. The process results in a canonical form which is (in our case) a $\oplus$-merge of comprehensions of the form

$$\oplus\{e \mid v_1 \leftarrow path_1, \ldots, v_n \leftarrow path_n, pred\}, \quad (36)$$

where each $path_i$ is a database extent name or an expression of the form $v.a_1 \ldots a_n$, where v is a bound variable and the a_i are attribute names. Note that the head e and the predicate $pred$ may still contain nested comprehensions, though these will also be in canonical form. The normalisation rules needed to convert the comprehension expressions produced by the algorithm in Section 5 are given in Figure 5. In the figure, $\otimes$ and $\oslash$ may be any of the monoid operations $\oplus, \vee, \wedge$ or $+$, $*$ may be $\vee, \wedge$ or $+$, and $\odot$ may be $\oplus$, $\wedge$ or $\vee$. The notation $e[e'/v]$ denotes the result of substituting e' for the free occurrences of v in e. Further details on the normalisation of the monoid calculus can be obtained from [11].

6.2 Semantic Optimization

The translations given in Section 5 are applicable to any concept expression. However, in certain circumstances more efficient translations can be produced by exploiting knowledge about the types returned by translations of subexpressions and the containment relationships stored within the ontology.

For example, the concept $C \sqcap \exists R.D$ (where R is a multiple-valued attribute role, C is translated as C' and D as D') is translated to a monoid comprehension which, after

34

Concept	Translation	Type
$\exists R.C$	$\oplus\{d \mid d \leftarrow D_R, f \leftarrow d.a_R, c \leftarrow C', match(f,c)\}$	T_1
$\exists R^-.C$	$\oplus\{f \mid d \leftarrow D_R, f \leftarrow d.a_R, c \leftarrow C', match(c,d)\}$	T_2
$\exists^{\geq n} R.C$	$\oplus\{d \mid d \leftarrow D_R, +\{1 \mid f \leftarrow d.a_R, c \leftarrow C', match(f,c)\} \geq n\}$	T_1
$\exists^{\geq n} R^-.C$	$\oplus\{f \mid d \leftarrow D_R, f \leftarrow d.a_R, +\{1 \mid e \leftarrow D_R, g \leftarrow e.a_R, c \leftarrow C', match(g,f) \wedge match(c,e)\} \geq n\}$	T_2

Table 5. Translation of multiple-valued roles.

$$\langle A_1 = e_1, \ldots, A_n = e_n \rangle.A_i \longrightarrow e_i \tag{29}$$

$$\otimes\{e \mid \bar{q}, v \leftarrow \mathcal{Z}_\oslash, \bar{s}\} \longrightarrow \mathcal{Z}_\otimes \tag{30}$$

$$\odot\{e \mid \bar{q}, v \leftarrow e_1 \oplus e_2, \bar{s}\} \longrightarrow (\odot\{e \mid \bar{q}, v \leftarrow e_1, \bar{s}\}) \odot (\odot\{e \mid \bar{q}, v \leftarrow e_2, \bar{s}\}) \tag{31}$$

$$+\{e \mid \bar{q}, v \leftarrow e_1 \oplus e_2, \bar{s}\} \longrightarrow (+\{e \mid \bar{q}, v \leftarrow e_1, \bar{s}\})$$
$$+ (+\{e \mid \bar{q}, v \leftarrow e_2, \wedge\{\neg match(v,w) \mid w \leftarrow e_1\}, \bar{s}\}) \tag{32}$$

$$\otimes\{e \mid \bar{q}, \vee\{pred \mid \bar{r}\}, \bar{s}\} \longrightarrow \otimes\{e \mid \bar{q}, \bar{r}, pred, \bar{s}\} \tag{33}$$

$$\otimes\{e \mid \bar{q}, v \leftarrow \oplus\{e' \mid \bar{r}\}, \bar{s}\} \longrightarrow \otimes\{e[e'/v] \mid \bar{q}, \bar{r}, \bar{s}[e'/v]\} \tag{34}$$

$$*\{*\{e \mid \bar{r}\} \mid \bar{s}\} \longrightarrow *\{e \mid \bar{s}, \bar{r}\}, \tag{35}$$

Figure 5. Normalisation rules for the calculus.

normalisation (ignoring for the moment the fact that C' and D_R may be $\oplus$-unions), looks like

$$\oplus\{c \mid c \leftarrow C', x \leftarrow D_R, f \leftarrow x.a_R, d \leftarrow D',$$
$$match(c,x), match(f,d)\}. \tag{41}$$

However, if the type of C' is such that the elements can be guaranteed to be in D_R, the iteration over x can be dispensed with. In this case, each instantiation of the variable c has its own set $c.a_R$ of R-fillers and the query can be answered by the comprehension

$$\oplus\{c \mid c \leftarrow C', f \leftarrow c.a_R, d \leftarrow D', match(f,d)\} \tag{42}$$

The optimization rules (37) and (38) in Figure 6 achieve the required simplification.

Similarly, if a comprehension has two generators whose variables are supposed to match but whose domains are known to be incompatible from the ontology, then the comprehension is empty. This is captured by rule (39) in Figure 6.

Another optimization can be applied to a comprehension C that contains a generator of the form $x \leftarrow X \oplus Z$. If the predicate of C implies matches between x and other variables, then it may be that the intersection of the ranges of those variables is contained in X. In that case we can restrict x to range only over X. The formalisation of this rule (rule (40) in Figure 6) refers to the predicate *can-restrict* defined as follows. Let x be a variable, X a union of extents, p a predicate and $\bar{r}$ a sequence of generators. Let $\{y_i\}$ be the set of expressions which are related to x by $match()$ conjuncts in p, not including x itself. The y_i are the elements (excluding x) of the connected component of the graph de-

fined by the $match()$ conjuncts in p. Each y_i is either a variable or a path expression of the form $z.a_1 \ldots a_n$ where z is a variable and the a_j are attributes. So each y_i has a type which corresponds to a DL concept Y_i. Let X' be the DL concept corresponding to X. Then *can-restrict*$(x, X, p, \bar{q})$ is true if $(\prod_i Y_i) \sqsubseteq X'$ (which we can find out from the DL classifier) and false otherwise.

6.3 Simplification Example

As an example of query simplification, we can consider the translation (28) of query (16) from Section 5. This form immediately admits a simplification by rule (40), since the type of p_3 is PIR_Protein which corresponds to the concept pir_protein so that the sp_proteins summand in the domain of p_2 can be eliminated. Normalisation then yields

$$\oplus\{p_2 \mid p_2 \leftarrow \texttt{pir\_proteins}, p_1 \leftarrow \texttt{pir\_proteins},$$
$$r_2 \leftarrow p_1.\texttt{references}, r_1 \leftarrow \texttt{references},$$
$$t \leftarrow \texttt{top\_journals}, match(r_1.\texttt{journal}, t) \wedge$$
$$match(r_2, r_1) \wedge match(p_2, p_1)\}. \tag{43}$$

Two applications of rule (37) then eliminate the variables p_1 and r_1, leaving the form

$$\oplus\{p_2 \mid p_2 \leftarrow \texttt{pir\_proteins}, r_2 \leftarrow p_2.\texttt{references},$$
$$t \leftarrow \texttt{top\_journals}, match(r_2.\texttt{journal}, t)\}. \tag{44}$$

Note that the $match()$ comparison which remains is between values of type String and so it will be evaluated by a simple equality test.

$$\oplus\{e \mid \bar{q}, v \leftarrow X, \bar{r}, w \leftarrow Y, \bar{s}, match(v,w) \wedge p\} \longrightarrow \oplus\{e[v/w] \mid \bar{q}, v \leftarrow X, \bar{r}, \bar{s}[v/w], p[v/w]\}, \quad \text{if } X \sqsubseteq Y \quad (37)$$

$$\oplus\{e \mid \bar{q}, v \leftarrow X, \bar{r}, w \leftarrow Y, \bar{s}, match(v,w) \wedge p\} \longrightarrow \oplus\{e[v/w] \mid \bar{q}, v \leftarrow Y, \bar{r}, \bar{s}[v/w], p[v/w]\}, \quad \text{if } Y \sqsubseteq X \quad (38)$$

$$\oplus\{e \mid \bar{q}, v \leftarrow X, \bar{r}, w \leftarrow Y, \bar{s}, match(v,w) \wedge p\} \longrightarrow \emptyset, \quad \text{if } X \sqcap Y \doteq \bot. \quad (39)$$

$$\oplus\{e \mid \bar{q}, x \leftarrow X \oplus Z, \bar{r}, p\} \longrightarrow \oplus\{e \mid \bar{q}, x \leftarrow X, \bar{r}, p\}, \quad \text{if } can\text{-}restrict(x, X, p, concat(q,r)) \quad (40)$$

Figure 6. Semantic optimization rules for the calculus.

7 Conclusions

The provision of knowledge-based information integration systems has been a focus of research activity for a considerable period, as it holds out the hope that high-level, declarative representations of resources can be used for schema reconciliation and query answering. This paper seeks to contribute to this line of research by bringing together recent results on expressive description logics and object database query processing to provide more expressive modelling and query processing to global-as-view query systems. The paper not only shows how queries over an $\mathcal{ALCQI}$ ontology can be mapped to the monoid calculus for evaluation, but has also demonstrated how information from the ontology can be used to simplify the resulting calculus expression.

The motivation for this work has come from bioinformatics, in which query-oriented access to heterogeneous sources is seen important, but in which existing sources are both numerous and of complex structure. The results described in the paper have been implemented for validation purposes, and work is underway to couple the query optimizer to an evaluator running over object-wrapped bioinformatics sources.

Acknowledgements: This work has been funded by the UK Engineering and Physical Sciences Research Council, whose support we are pleased to acknowledge. We would also like to thank Ian Horrocks and Robert Stevens for useful discussions during the development of the techniques described in this paper.

References

[1] Y. Arens, C.A. Knoblock, and W-M. Shen. Query Reformulation for Dynamic Information Integration. *J. Intelligent Information Systems*, 6(2/3):99–130, 1996.

[2] D. Calvanese, G. De Giacomo, M. Lenzerini, D. Nardi, and R. Rosati. Information integration: Conceptual modeling and reasoning support. In *Proc. of the 6th Int. Conf. on Cooperative Information Systems (CoopIS'98)*, pages 280–291, 1998.

[3] D. Calvanese, G. De Giacomo, M. Lenzerini, and Moshe Y. Vardi. View-based query processing and constraint satisfaction. In *Proc. of the 15th IEEE Sym. on Logic in Computer Science (LICS 2000)*, 2000.

[4] D. Calvanese, M. Lenzerini, and D. Nardi. Description logics for conceptual data modeling. In J. Chomicki and G. Saake, editors, *Logics for Databases and Information Systems*, pages 229–263. Kluwer, 1998.

[5] D. Calvanese, M. Lenzerini, and D. Nardi. Unifying class-based representation formalisms. *J. of Artificial Intelligence Research*, 11:199–240, 1999.

[6] R. G. G. Cattell, D. K. Barry, M. Berler, J. Eastman, D. Jordan, C. Russell, O. Schadow, T. Stanienda, and F. Velez, editors. *The Object Data Standard: ODMG 3.0*. Morgan Kaufmann, 2000.

[7] I-Min A. Chen, A.S. Kosky, V.M. Markowitz, and E. Szeto. Constructing and Maintaining Scientific Database Views in the Framework of the Object Protocol Model. In *Proc. SSDBM*. IEEE Press, 1997.

[8] S.B. Davidson, C. Overton, V. Tannen, and L. Wong. BioKleisli: A Digital Library for Biomedical Researchers. *Journal of Digital Libraries*, 1(1):36–53, Nov 1997.

[9] P. T. Devanbu. Translating description logics to information server queries. In *CProc. of the Second International Conference on Information and Knowledge Management (CIKM'93)*, pages 256–263, 1993.

[10] T. Etzold and A. Ulyanov. SRS: information retrieval system for molecular biology data banks. *Methods Enzymol.*, 266:114–128, 1996.

[11] L. Fegaras and D. Maier. Optimizing object queries using an effective calculus. *ACM Transactions on Database Systems*, 2001. (to appear).

[12] F. Goasdoue, V. Lattes, and M-C. Rousset. The use of CARIN language and algorithms for information integration: the picsel system. *International Journal on Cooperative Information Systems*, 2000.

[13] C.A. Goble, R. Stevens, G. Ng, S. Bechhof er, N.W. Paton, P.G. Baker, M. Peim, and A. Brass. Transparent access to multiple bioinformatics info rmation sources. *IBM Systems Journal*, 40(2):534–551, 2001.

[14] I. Horrocks, U. Sattler, and S. Tobies. Practical reasoning for expressive description logics. In *Proc. of the 6th International Conference on Logic for Programming and Automated Reasoning (LPAR'99)*, pages 161–180, 1999.

[15] A.Y. Levy, D. Srivastava, and T. Kirk. Data Model and Query Evaluation in Global Information System s. *J. Intelligent Information Systems*, 5:121–143, 1995.

[16] E. Mena, A. Illarramendi, V. Kashyap, and A.P Sheth. OBSERVER: An approach for query processing in global information systems based on interoperation across pre-existing ontologies. *Distributed and Parallel Databases*, 8(2):223–271, 2000.

[17] N. W. Paton, R. Stevens, P. Baker, C. A. Goble, S. Bechhofer, and A. Brass. Query Processing in the TAMBIS Bioinformatics Source Integration System. In *Proc. SSDBM*, pages 138–147. IEEE Press, 1999.

[18] R. Stevens, C.A. Goble, and S. Bechhofer. Ontology-based knowledge representation for bioinformatics. *Briefings in Bioinformatics*, 1(4):398–414, 2000.

[19] J. D. Ullman. Information integration using logical views. In *Proc. of the 6th International Conference on Database theory (ICDT'97)*, volume 1186, pages 19–40. Springer-Verlag, 1997.

Chimera: A Virtual Data System
for Representing, Querying, and Automating Data Derivation

Ian Foster[1,2]   Jens Vöckler[2]   Michael Wilde[1]   Yong Zhao[2]

[1] Mathematics and Computer Science Division, Argonne National Laboratory, Argonne, IL 60439, USA

[2] Department of Computer Science, University of Chicago, Chicago, IL 60637, USA

{foster,wilde}@mcs.anl.gov, {voeckler,yongzh}@cs.uchicago.edu

Abstract

Much scientific data is not obtained from measurements but rather derived from other data by the application of computational procedures. We hypothesize that explicit representation of these procedures can enable documentation of data provenance, discovery of available methods, and on-demand data generation (so-called "virtual data"). To explore this idea, we have developed the Chimera virtual data system, which combines a virtual data catalog, for representing data derivation procedures and derived data, with a virtual data language interpreter that translates user requests into data definition and query operations on the database. We couple the Chimera system with distributed "Data Grid" services to enable on-demand execution of computation schedules constructed from database queries. We have applied this system to two challenge problems, the reconstruction of simulated collision event data from a high-energy physics experiment, and the search of digital sky survey data for galactic clusters, with promising results.

1   Introduction

In many scientific disciplines, the analysis of "data" (whether obtained from scientific instruments, such as telescopes, colliders, or climate sensors, or from numerical simulations) is a significant community activity. As a result of this activity, communities construct, in a collaborative fashion, collections of derived data (e.g., flat files, relational tables, persistent object structures) with relationships between data objects corresponding to the computational procedures used to derive one from another (Figure 1). Recording and discovering these relationships can be important for many reasons, as illustrated by the following vignettes.

"I've come across some interesting data, but I need to understand the nature of the corrections applied when it was constructed before I can trust it for my purposes."

"I want to search an astronomical database for galaxies with certain characteristics. If a program that performs this analysis exists, I won't have to write one from scratch."

"I want to apply an astronomical analysis program to millions of objects. If the program has already been run and the results stored, I'll save weeks of computation."

"I've detected a calibration error in an instrument and want to know which derived data to recompute."

"I want to find those results that I computed last month, and the details of how I generated them."

More generally, we want to be able to track how data products are derived—with sufficient precision that one can create and/or re-create data products from this knowledge. One can then explain definitively how data products are created, something that is often not feasible even in carefully curated databases. One can also implement a new class of "virtual data management" operations that, for example, "re-materialize" data products that were deleted, generate data products that were defined but never created, regenerate data when data dependencies or transformation programs change, and/or create replicas of data products at remote locations when re-creation is more efficient than data transfer.

In order to explore the benefits of data derivation tracking and virtual data management, we have designed, prototyped, and experimented with a virtual data system called *Chimera*. A *virtual data catalog* (based on a relational *virtual data schema*) provides a compact and expressive representation of the computational procedures used to derive data, as well as invocations of those procedures and the datasets produced by those invocations. A *virtual data language interpreter* executes requests for constructing and querying database entries.

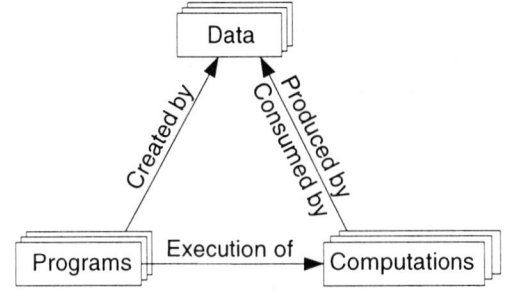

Figure 1: Relationships among programs, computations, and data

We couple Chimera with other Data Grid services [1, 11, 18, 28] to enable the creation of new data by executing computation schedules obtained from database queries, and the distributed management of resulting data.

We have applied the Chimera system successfully to two challenging physics data analysis computations, one involving the generation and reconstruction of simulated high-energy physics collision event data from the Compact Muon Solenoid (CMS) experiment at CERN [20, 25], and the other the detection of galactic clusters in Sloan Digital Sky Survey (SDSS) data [4, 29]. Our results demonstrate our ability to track data derivations and to schedule large distributed computations in response to user virtual data queries. Others have successfully used some of these techniques in the analysis of data from the LIGO gravitational wave observatory [14].

The importance of being able to document provenance is well known [30]. Our work builds on preliminary explorations within the GriPhyN project [5, 15, 16]. There are also relationships to work in database systems [9, 10, 31] and versioning [8, 26]. Cui and Widom [12, 13] record the relational queries used to construct materialized views in a data warehouse, and then exploit this information to explain lineage. Our work can leverage these techniques, but differs in two respects: first, data is not necessarily stored in databases and the operations used to derive data items may be arbitrary computations; second, we address issues relating to the automated generation and scheduling of the computations required to instantiate data products.

Early work on *conceptual schemas* [21] introduced virtual attributes and classes, with a simple constrained model for the re-calculation of attributes in a relational context. Subsequent work produced an integrated system for scientific data management called ZOO [22], based on a special-purpose ODBMS that allowed for the definition of "derived" relationships between classes of objects. In ZOO, derivations can be generated automatically based on these relationships, using either ODBMS queries or external transformation programs. Chimera is more specifically oriented to capturing the transformations performed by external programs, and does not depend on a structured data storage paradigm or on fine-grained knowledge of individual objects that could be obtained only from an integrated ODBMS.

We can also draw parallels drawn between Chimera and workflow [23, 27] and knowledge management systems that allow for the definition, discovery, and execution of (computational) procedures.

The rest of this article is as follows. We first introduce the Chimera virtual data system and describe its virtual data schema and language (Sections 2-4). Then, we discuss the integration of Chimera with Data Grids (Section 5), our experiences applying the system to challenge problems (Section 6), and future directions.

2   Chimera Architecture

The architecture of the Chimera virtual data system is depicted in Figure 2. In brief, it comprises two principal components: a *virtual data catalog* (VDC; this implements the Chimera *virtual data schema*) and the *virtual data language interpreter*, which implements a variety of tasks in terms of calls to virtual data catalog operations.

Applications access Chimera functions via a standard *virtual data language* (VDL), which supports both *data definition* statements, used for populating a Chimera database (and for deleting and updating virtual data definitions), and *query* statements, used to retrieve information from the database. One important form of query returns (as a directed acyclic graph, or DAG) a representation of the tasks that, when executed on a Data Grid, create a specified data product. Thus, VDL serves as a lingua franca for the Chimera virtual data grid, allowing components to determine virtual data relationships, to pass this knowledge to other components, and to populate and query the virtual data catalog without having to depend on the (potentially evolving) catalog schema.

Chimera functions can be used to implement a variety of applications. For example, a *virtual data browser* might support interactive exploration of VDC contents, while a *virtual data planner* might combine VDC and other information to develop plans for computations required to materialize missing data (Section 5).

The Chimera virtual data schema defines a set of relations used to capture and formalize descriptions of how a program can be invoked, and to record its potential and/or actual invocations. The entities of interest—transformations, derivations, and data objects—are as follows; we describe the schema in more detail below.

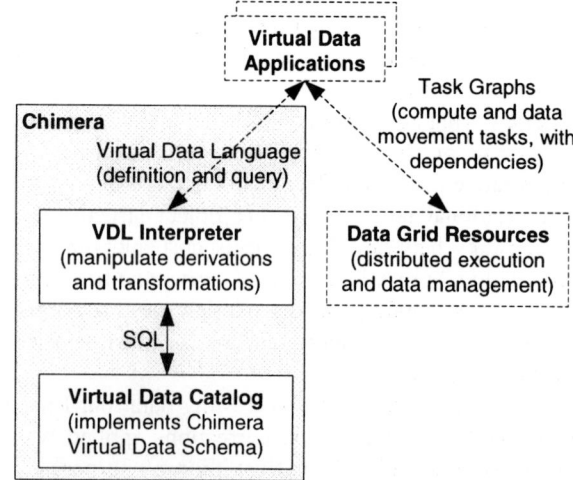

Figure 2: Schematic of the Chimera architecture

- A *transformation* is an executable program. Associated with a transformation is information that might be used to characterize and locate it (e.g., author, version, cost) and information needed to invoke it (e.g., executable name, location, arguments, environment).

- A *derivation* represents an execution of a transformation. Associated with a derivation is the name of the associated transformation, the names of data objects to which the transformation is applied, and other derivation-specific information (e.g., values for parameters, time executed, execution time). While transformation arguments are *formal* parameters, the arguments to a derivation are *actual* parameters.

- A *data object* is a named entity that may be consumed or produced by a derivation. In the applications considered to date, a data object is always a *logical file*, named by a *logical file name* (LFN); a separate *replica catalog* or *replica location service* is used to map from logical file names to physical location(s) for replicas [2, 11]. However, data objects could also be relations or objects. Associated with a data object is information about that object: what is typically referred to as metadata.

We do not address here the question of how the data dependency information maintained within the Chimera system is produced. Information about transformations and derivations can potentially be declared explicitly by the user, extracted automatically from a job control language, produced by higher-level job creation interfaces such as portals, and/or created by monitoring job execution facilities and file accesses.

Information can be recorded in the virtual data system at various times and for various purposes. Transformation entries generated before invocation can be used to locate transformations and guide execution. Derivation entries generated before jobs are executed can provide information needed to generate a file. Entries generated *after* a job is executed record how to *re*generate a file. Consider, for example, transformation entries that define the four stages of a simulation pipeline. An initial query for the output of this pipeline returns a DAG that, when executed, generates files called f1 - f4. Subsequent deletion of file f3 followed by a retrieval request for that file results only in the re-execution of stage 3 of the pipeline.

3 Chimera Virtual Data Schema

We describe here the Chimera virtual data schema, shown in Figure 3.

A logical transformation is characterized by its identifying name, the namespace within which the name is unique, and a version number. The signature of the transformation includes input and output parameters, which need not be files. A transformation may have an arbitrary number of formal arguments. Thus the relationship between TRANSFORMATION and FORMALARG is 1:N.

A transformation may have more than one derivation, each supplying different values for the parameters. A derivation may be applicable to more than one transformation. Thus, versioning allows for a range of valid transformations to apply, increasing the degrees of freedom for schedulers to choose the most applicable one.

An ACTUALARG relates to a derivation. Its value captures either the LFN or the value of a non-file parameter. A FORMALARG may contain an optional default value, captured in a similar fashion by the same VALUE class. The VALUE class is an abstract base class for either a single value (SCALAR) or a list of similar values (LIST), which are collapsed union-fashion into a single table.

The relationships between a transformation and its formal parameters, on the one hand, and a dependent derivation and its actual parameters, on the other, are not independent of each other. Each instantiation of an actual parameter maps to exactly one formal parameter describing the entry. The binding is created using the argument name, not its position in the argument list.

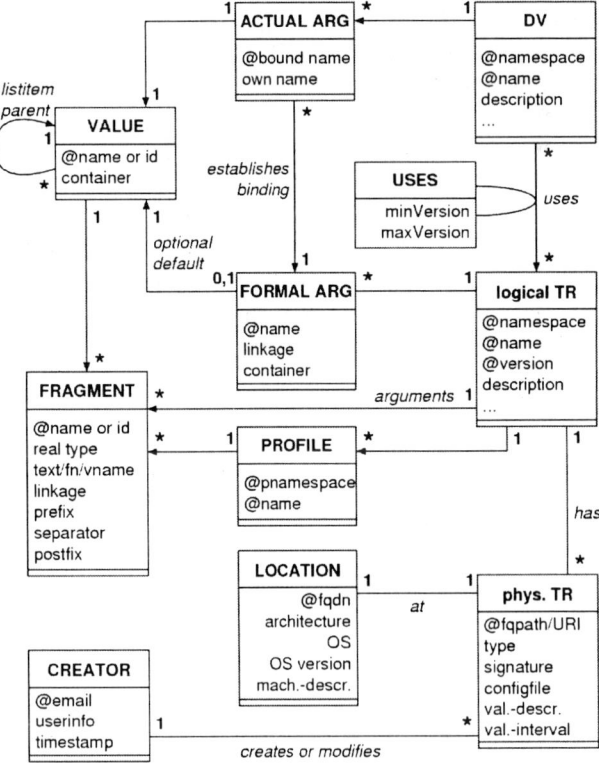

Figure 3: UML description of the Chimera schema

The arguments on the command line of a transformation are captured in multiple fragments, each either a reference to a formal argument, or a textual string.

Scheduler- and runtime environment-specific data is abstracted in the PROFILE table. For example, in the case of a Unix environment variable, the namespace is "env", the key within this namespace is the environment variable name, and the value is a list of fragments, either references to bound variables or textual strings.

The FRAGMENT table captures three child classes, either a textual string, a LFN or a reference to a bound variable. The three child classes are collapsed into a single table.

The lower portion of the diagram deals with the physical location of any given transformation [16].

Finally, we note that Chimera applications will typically also require a METADATA table, which maps from (key, value) attribute pairs to LFNs, and a REPLICA table, which maps from LFNs to physical file locations. However, these relations are frequently implemented via a separate metadata catalog [7] and/or replica catalog [11] and so they are not considered here.

4 Chimera Virtual Data Language

As noted above, the Chimera *virtual data language* (VDL) comprises both data definition and query statements. We first introduce two data derivation statements, TR and DV, and then discuss queries. While our implementation uses XML internally, we use a more readable syntax here.

4.1 The TR Data Definition Statement

A TR statement defines a transformation. When the VDL interpreter processes such a statement, it creates a transformation object within the virtual data catalog. For example, the following definition provides the information required to execute a program app3.

```
TR t1( output a2, input a1,
       none env="100000",
       none pa="500" ) {
  app vanilla = "/usr/bin/app3";
  arg parg = "-p "${none:pa};
  arg farg = "-f "${input:a1};
  arg xarg = "-x -y ";
  arg stdout = ${output:a2};
  profile env.MAXMEM = ${none:env};
}
```

This definition reads as follows. The first line assigns the transformation a name (t1) for use by derivation definitions, and declares that t1 reads one input file (formal parameter name a1) and produces one output file (formal parameter name a2). The parameters declared in

the TR header line are *transformation arguments* and can only be file names or textual arguments.

The APP statement specifies (potentially as an LFN) the executable that implements the execution.

The first three ARG statements describe how the *command line arguments* to app3 (as opposed to the *transformation arguments* to t1) are constructed. Each ARG statement comprises a name (here, parg, farg, and xarg) followed by a default value, which may refer to transformation arguments (e.g., a1) to be replaced at invocation time by their value. The special argument stdout (the fourth ARG statement in the example) is used to specify a filename into which the standard output of an application would be redirected.

Argument strings are concatenated in the order in which they appear in the TR statement to form the command line. The reason for introducing argument names is that these names can be used within DV statements to override the default argument values specified by the TR statement.

Finally, the PROFILE statement specifies a default value for a Unix environment variable (MAXMEM) to be added to the environment for the execution of app3.

4.2 The DV Data Definition Statement

A DV statement defines a derivation. When the VDL interpreter processes such a statement, it records a transformation invocation within the virtual data catalog. A DV statement supplies LFNs for the formal filename parameters declared in the transformation and thus specifies the *actual* logical files read and produced by that invocation. For example, the following statement records an invocation of transformation t1 defined above.

```
DV t1(
  a2=@{output:run1.exp15.T1932.summary},
  a1=@{input:run1.exp15.T1932.raw},
  env="20000", pa="600" );
```

The string immediately after the DV keyword names the transformation invoked by the derivation.

In contrast to transformations, derivations need not be named explicitly via VDL statements. They can be located in the catalog by searching for them via the logical filenames named in their IN and OUT declarations as well as by other attributes, as discussed below.

Actual parameters in a derivation and formal parameters in a transformation are associated by name. For example, the statements above result in parameter a1 of t1 receiving the value run1.exp15.T1932.raw and a2 the value run1.exp15.T1932.summary.

The example DV definition corresponds to the following invocation:

```
export MAXMEM=20000
/usr/bin/app3 -p 600 \
```

```
       -f run1.exp15.T1932.raw -x -y \
       > run1.exp15.T1932.summary
```

Filenames listed as IN and OUT in a transformation need not necessarily appear as command line arguments in a corresponding derivation. For example, if a filename was determined directly by the executable through an internal definition or is determined dynamically, it might not appear on the command line even though the file is read or written by the application. In such cases, applications that know or can detect what filenames were read and written by an application could, after the fact, create a derivation record to describe these dynamic data dependencies.

The filename insertion semantics described here support a wide variety of argument passing conventions. Executables with argument passing conventions that cannot be expressed in these terms must be executed by creating "wrapper" scripts or executables that adapt the VDL conventions to those expected by the executable. For example, applications that read the names of further input files from a "control" file can often be handled with a wrapper that accepts the filenames as command line arguments and places them in the control file before calling the actual application. In some cases, the creation of the control file itself could be described as a transformation that reads several files and produces the control file as its output. Then the real transformation can be described as having the control file as an input.

4.3 Tracking Derivation Dependencies

Chimera's VDL supports the tracking of data dependency chains among derivations. For example, the following statements define two derivations (as well as two transformations), such that the output of the first is the input to the second. Thus, we can conclude that (unless file2 or file3 exist) to generate file3 we must run trans1 before trans2, and use its output file (file2) as the input file for trans2.

```
TR trans1( output a2, input a1 ) {
    app vanilla = "/usr/bin/app1";
    arg stdin = ${input:a1};
    arg stdout = ${output:a2};
}
TR trans2( output a2, input a1 ) {
    arg vanilla = "/usr/bin/app2";
    arg stdin = ${input:a1};
    arg stdout = ${output:a2};
}
DV trans1( a2=@{output:file2},
    a1=@{input:file1} );
DV trans2( a2=@{output:file3},
    a1=@{input:file2} );
```

We can thus construct arbitrarily complex directed acyclic execution graphs (DAGs) automatically. For example, consider the transformations illustrated in Figure 4. Four logical files, named f.a, f.b, f.c, and f.d in the figure, are produced as a result of this computation. The output from a first node generate is stored into file f.a. Two processes findrange each operate on disjoint subsets of the input f.a, publishing their results in f.b and f.c, respectively. A final node, analyze, combines the two halves.

The following statements define the transformations generate, findrange, and analyze, and the derivations that produce files f.a, f.b, f.c, and f.d.

```
TR generate( output a ) {
    app vanilla = "generator.exe";
    arg stdout = ${output:a2};
}
TR findrange( output b, input a,
    none p="0.0" ) {
    app vanilla = "ranger.exe";
    arg arg = "-i "${:p};
    arg stdin = ${output:a};
    arg stdout = ${output:b};
}
TR analyze( input a[], output c ) {
    app vanilla = "analyze.exe";
    arg files = ${:a};
    arg stdout = ${output:a2};
}
DV generate( a=@{output:f.a} );
DV findrange( b=@{output:f.b},
    a=@{input:f.a}, p="0.5" );
DV findrange( b=@{output:f.c},
    a=@{input:f.a}, p="1.0" );
DV analyze( a=[ @{input:f.b},
    @{input:f.c} ], c=@{output:f.d} );
```

Notice that the transformation findrange is invoked twice, with different values for the command line argument -i specifying different search ranges.

4.4 Compound Transformations

A *compound transformation* describes the coordinated (perhaps concurrent) execution of multiple programs and the passing of files among them. It is described in the same manner as a simple transformation, with a single derivation statement. All internal transformation invocations within a compound transformation are tracked in the catalog, along with all files read and produced by the internal transformation steps. The system can thus remain fully cognizant of all data dependencies, and arbitrary files within those dependency chains can be deleted and later re-derived, based on stored knowledge.

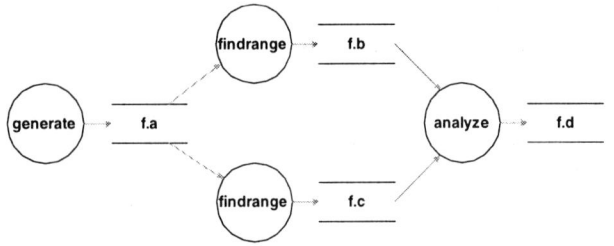

Figure 4: Example directed acyclic graph

A transformation is either a simple transformation or a compound transformation. A compound transformation is itself composed of references to one or more transformations, each of which in turn are either simple or compound. In all other respects, and in particular, from the point of view of its external interface and semantics, compound and simple transformations are indistinguishable. Thus, compound transformations can themselves contain compound transformations.

4.5 Queries

VDL provides various commands for extracting derivation and transformation definitions. Since VDL is implemented in SQL, this query set is readily extensible. Query output can (optionally) be returned in the same format as the commands that could be used to re-create the matching entries. Query commands can be used both by end-user query systems (e.g., a virtual data browser) and by automated Grid components such as a data analysis system.

In brief, VDL query commands allow one to search for *transformations* by specifying a transformation name, application name, input LFN(s), output LFN(s), argument matches, and/or other transformation metadata. One can search for *derivations* by specifying the associated transformation name, application name, input LFN(s), and/or output LFN(s). An important search criterion is whether derivation definitions exist that invoke a given transformation with specific arguments. From the results of such a query, a user can determine if desired data products already exist in the data grid, and can retrieve them if they do and create them if they do not.

Query output options specify, first of all, whether output should be recursive or non-recursive (recursive shows all the dependent derivations necessary to provide input files assuming no files exist) and second whether to present output in columnar summary, VDL format (for re-execution), or XML.

5 Chimera as a Data Grid Component

We discuss some of the issues that arise when the Chimera system is incorporated as a component within a larger Data Grid system. As illustrated in Figure 2, a virtual data "application" can combine information from both Chimera and other Data Grid components as it processes user requests for virtual data. For example, an application might combine information about the materialization status of a requested derivation with information about the physical location of replicas and the availability of computing resources to determine whether to access a remote copy or (re-)generate a data value.

One proposed Data Grid architecture [17] interposes a planner between an application and other components illustrated in Figure 2. The planner accepts *abstract DAGs* from the application, that is, DAGs that refer only to LFNs and not to specific physical instances of files, and that are thus not yet bound to specific Grid locations. For LFNs needed as input, an abstract DAG does not specify if these files already exist at a computation site, need to be copied there, or should be re-derived; for LFNs produced as output, the planner must determine where to place the newly created file. Location decisions must also be made recursively for any additional derivations needed.

The planner examines the abstract DAG, selects an execution site for each node, and then determines how to obtain and transport the data needed by each computation. The planner must also determine how to deal with the relocation of physical files *produced* by a job, if policies require that these files be relocated to specific physical file storage servers. The planner may evaluate several different execution plans, based for example on cost estimates for data movement vs. re-creation.

The output of the request planner is a *concrete* DAG that refers only to real physical file names and specifies the steps that must be followed to compute or transport any input data that does not yet exist at its execution site.

6 Experiences with the Chimera System

We describe application experiments with our Chimera prototype, conducted on the small-scale Data Grid shown in Figure 5. (Subsequent experiments will use the larger International Virtual Data Grid Laboratory [6].) This Grid used Globus Toolkit resource management and data transfer components [17], Condor schedulers and agents [19, 24], and the DAGman job submission agent to coordinate resources at four sites.

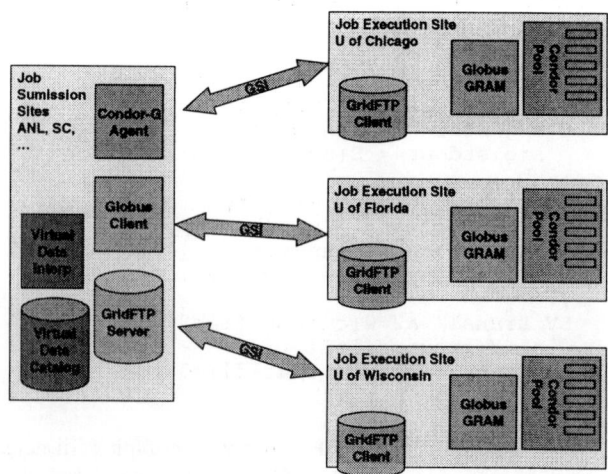

Figure 5: The Data Grid used in our experiments

In our experiments, we did not consider data replication. The persistent location of all data products was the site at which the VDL interpreter resided. All data files needed by executables were pulled to the executing sites, and all data files from successful executions were returned to the submission site upon completion of data derivations. All data transfers used the Grid-enabled data transfer tool GridFTP [2].

As we explain, the results demonstrate that Chimera can manage complex interdependencies among application invocations that occur in practice. We have also validated the capability of the Chimera system on a variety of larger and more complex artificial DAGs.

6.1 CMS Data Reconstruction

We used the Chimera prototype to assess the feasibility of using virtual data descriptions for data products involved in the production of Monte Carlo-based simulations of high-energy physics collision events in the CMS experiment [20].

Event simulation is critical to the design and operation of the complex detector that is at the heart of CMS, and is also used to test the scientific and data management software systems on which CMS will depend.

This complex physics data derivation process comprises the following four-stage pipeline of transformations (i.e., executables).

1. **pythia** using Monte Carlo techniques to determine randomly the physics attributes of a specific collision event.
2. **cmsim** determines how that event would affect the CMS detector.
3. **writehits** converts the information into a persistent object data structure in an object-oriented database system.
4. **writedigis** determines the digital signals that the detector would produce from the event.

The first two stages produce files as output, while the second two stages produce object-oriented databases that are contained in the files of an Objectivity federation.

In these experiments, we sidestepped many of the complications of managing data stored in object form by configuring simulations to produce just one event per file, instead of the usual hundreds. However, we did use the Objectivity database used in normal CMS simulation.

6.2 SDSS Galactic Structure Detection

Our second Chimera application concerns the analysis of data from the Sloan Digital Sky Survey (SDSS) [29]. As described at www.sdss.org, "...the Sloan Digital Sky Survey is the most ambitious astronomical survey project ever undertaken. The survey will map in detail one-quarter of the entire sky, determining the positions and absolute brightness of more than 100 million celestial objects. It will also measure the distances to more than a million galaxies and quasars." The project, which will survey the night sky at an unprecedented resolution, is currently in its third year of data taking. When complete in 2004 it will have collected around 40 TB of image and spectroscopic data, and 3TB of catalog metadata.

Working with collaborators at Fermilab, we applied the concept of virtual data to one scientific challenge problem on the SDSS project—that of locating galactic clusters in the image collection [3]. The goal of this application is to create a database of galaxy clusters for the entire survey. A highly simplified view of this problem is as follows.

The sky is tiled into a set of regular "fields." For each field, clusters are searched for in that field and in some set of neighboring fields, using the concepts of "brightest cluster galaxy" (BCG) and "brightest red galaxy" (BRG) to determine cluster candidates [3]. The resulting algorithm is a tree-structured pipeline (Figure 6) comprising the following five transformations.

1. **fieldPrep** extracts from the full data set required measurements on the galaxies of interest and produces new files containing this data. The new files are about 40 times smaller than the full data sets.
2. **brgSearch** calculates the unweighted BCG likelihood for each galaxy. The unweighted likelihood may be used to filter out unlikely candidates for the next stage.
3. **bcgSearch** calculates the weighted BCG likelihood for each galaxy. This is the heart of the algorithm, and the most expensive step.
4. **bcgCoalesce** determines whether a galaxy is the most likely galaxy in the neighborhood.
5. **getCatalog** removes extraneous data and stores the result in a compact format.

Further details of the algorithms and astrophysics mentioned here are provided elsewhere [3, 4, 32].

Figure 7 shows the actual dataflow found in the last three stages of a small computation in which 24 *brgSearch* transformations (the leaves) reduce 156 files down to the root, where the *getCatalog* transformation produces the cluster catalog for a single field of the sky.

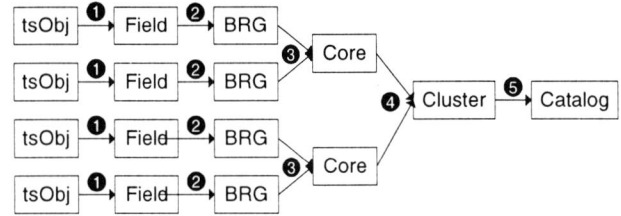

Figure 6: SDSS cluster identification workflow.

43

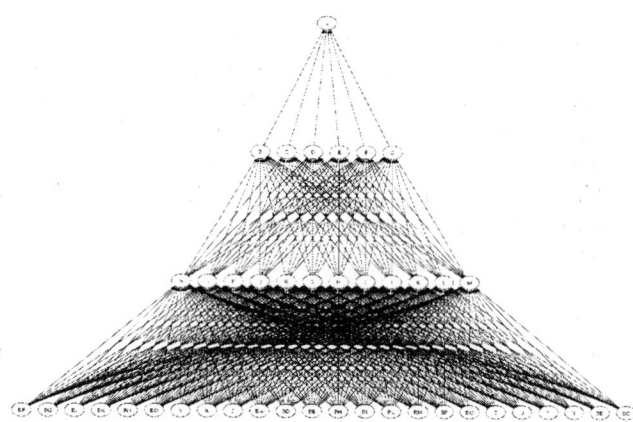

Figure 7: DAG for cluster identification workflow.

The derivation *getCatalog* now becomes a function that can invoke all the prior four dependent steps. To generate "virtual" results for the entire sky, we define one derivation of the *getCatalog* transformation for each field of the current survey.

This application motivated an important extension to the virtual data paradigm to permit the specification of transformations where the input and output file names are a function of one or more transformation arguments and thus cannot be known at the time when the transformation or even the derivation are defined. In the cluster finding mechanism, the celestial coordinates of the neighboring fields that will be required to locate the clusters within a given field of the survey are a function of that central field's coordinates—i.e., they are the neighbors of that central field, and their coordinates and hence their file names must be computed dynamically once the coordinates of the central field are known. Table 1 characterizes the five transformations used in the pipeline in terms of the number of static and dynamic file names.

Table 1: Input and output files for galactic cluster identification computation

Transform	Fixed files in	No. of varying input lists & files/list	Fixed files out	Varying output lists
fieldPrep	1	1 (13*12)		1 (13*12)
brgSearch	1	1 (13*12)		1 (13*12)
bcgSearch	1	2 (13*12,13*12)		1 (7*12)
bcgCoalesce	1	1 (7*12)		1 (12)
getCatalog	1	1 (12)	1	

Table 2: Numbers of files produced and consumed for galactic cluster finding on entire sky survey

Transform	# derivations	#files in	#files out
fieldPrep	(45x600x12)/10	45x600x12	45x600x12
brgSearch	(45x600x12)/10	45x600x12	45x600x12
bcgSearch	(45x600x12)/10	**2x**45x600x12	45x600x12
bcgCoalesce	(45x600x12)/10	45x600x12	45x600x12
getCatalog	(45x600)/10	45x600x12	45x600
Totals	**132,300**	**1,944,000**	**1,323,000**

The amount of computation and data access required to perform a complete analysis of all SDSS data is substantial. Table 2 summarizes the number of files that would be produced and consumed. At the time of writing, just over 2% of this data (one of the 45 eventual stripes of the sky) has been entered into the virtual data catalog as part of our experiments. The largest DAG executed to date for this application contained over 700 nodes. We have just started the larger computations required to complete the process, but it is already clear that the virtual data system simplifies the task tremendously. Without Chimera, the bookkeeping required to track a production effort of this magnitude would be considerable, and would involve a large amount of custom programming.

The virtual data mechanism can be thought of as a paradigm for the management of batch job production scripts. The Chimera system and its underlying Grid mechanisms automate all resource scheduling, synchronization, data movement, bookkeeping, and retry needed to manage the this large amount of work on a loosely coupled set of distributed resources.

The mechanism can also be thought of as a "makefile" for data production. For example, if the program *bcgCoalesce* is changed, a data administrator can request the re-creation of the final catalogs and Chimera will determine that 35,100 out of the total of 132,300 jobs need to be re-executed.

Similarly, if a few hundred of the 45x600x12 raw input files need to be revised (due to, say, errors discovered in the input capture system), then it is easy to determine what must be recomputed to update the entire output set.

It is also interesting to note that data production can be performed in parallel with interactive use by users who are requesting final output data products. In this mode, the virtual data grid acts much like a large-scale cache. If data products are produced through the batch process before they are needed interactively, then the system knows, at the time of the interactive request, that a data product is already available and no computations need be scheduled to produce it. If on the other hand a data product is requested before the batch process has produced it, the required derivations will be executed on demand to support the interactive need and will then be skipped when the batch process encounters a similar need at a later point in time.

7 Future Directions

We briefly discuss concepts that we plan to explore in future versions of the Chimera system.

Three data representation modes are dominant within GriPhyN experiments: files, relational tables (in RDBMSs) and persistent object structures (in ODBMSs). In addition, XML is proliferating as a universal and perhaps unifying underlying data representation. We believe that the mechanisms we have developed for file-based transformations and derivations extend naturally to encompass these other data modalities. If we can identify encapsulated data units, name these units, and specify a name and a pointer to transformations on the data, then our system can catalog both transformations and derivations in these data representation modes.

We define a unit of granularity that represents the entity that can be tracked: a file; an entire SQL table; or an entire object structure or "closure" that can be identified within the OODB and extracted, deleted, or regenerated as an atomic unit. Transformations described by Chimera can be executable programs, SQL command-level queries (including stored procedure invocations), ODBMS command-level queries or method invocations, or applications that access a SQL or OO database directly.

We believe that within this model it should be feasible to represent code transformations that freely exchange and transform data (within certain restrictive and well-defined limits) between these three modes of data storage. We plan to test this hypothesis on various challenge problems from the GriPhyN experiments.

We plan to augment the transformation description with information about the nature and state of the software and hardware environment in which a transformation executes [16]. This information can extend into the realm of configuration management systems.

We will also assess the utility of providing data type-based transformation templates—for example, to specify a transformation that translates "raw event data" files to "reconstructed event data" files, much as a makefile specifies a rule for translating a ".c" file into a ".o" file.

We will continue to explore the range of operations supported by the Chimera VDL, with the goal of validating our ability to realize the full spectrum of scenarios presented in the introduction.

Another significant research goal is to develop and test higher-level knowledge-based representations of domain-specific data, and to create databases and tools for representing and manipulating this knowledge. The question of how to support discovery of data, derivations, and transformations in a uniform fashion raises many challenging problems.

8 Conclusions

We have described Chimera, a virtual data tracking and generation system that can be used to audit and trace the lineage of derived data produced by computation and also to manage the automatic, on-demand (re)derivation of such data. This system comprises a relational database schema used to represent the various entities involved in data derivation, a virtual data language used to represent derivations and to manage the virtual data database, and a virtual data system used to manage the virtual data derivation process in large distributed data grid systems.

While the value of on-demand data derivation remains to be demonstrated in the general case, the value of auditing and tracing the lineage of scientific data in a large collaboration appears clear, as evidenced by the commitment of the four groundbreaking science experiments that comprise the Grid Physics Network. In general, we believe that virtual data techniques can significantly increase the usability of scientific data management systems by permitting science users to search for data based on application-level characteristics and automatically request the derivation of the data from pre-stored algorithm descriptions and derivation "recipes."

We have achieved positive results in our first tests of the Chimera system. These tests involved the automatic derivation of collider event simulation data in an application relating to the CMS high energy physics experiment, and automatic invocation of galactic cluster finding algorithms on SDSS data. We have also used various artificial problems to demonstrate our ability to manage more complex data derivation relationships.

These initial results encourage us that the Chimera design is viable and that it is feasible not only to represent complex data derivation relationships but also to integrate virtual data concepts into the operational procedures of large scientific collaborations. Further studies will provide additional insights into the utility of our techniques. We also plan to investigate the derivation of relational and object data, and the integration of higher-level techniques for representing ontologies.

Acknowledgements

We gratefully acknowledge helpful discussions with Rick Cavanaugh, Peter Couvares, Ewa Deelman, Greg Graham, Carl Kesselman, Miron Livny, and Alain Roy. This research was supported in part by the National Science Foundation under contract ITR-0086044 (GriPhyN), and by the Mathematical, Information, and Computational Sciences Division subprogram of the Office of Advanced Scientific Computing Research, U.S. Department of Energy, under Contract W-31-109-Eng-38 (Data Grid Toolkit).

References

1. The DataGrid Architecture. EU DataGrid Project document DataGrid-12-D12.4-333671-3-0, 2001, www.eu-datagrid.org.
2. Allcock, W., Bester, J., Bresnahan, J., Chervenak, A.L., Foster, I., Kesselman, C., Meder, S., Nefedova, V., Quesnel, D. and Tuecke, S., Secure, Efficient Data Transport and Replica Management for High-Performance Data-Intensive Computing. In *Mass Storage Conference*, (2001)
3. Annis, J., Kent, S., Castander, F., Eisenstein, D., Gunn, J., Kim, R., Lupton, R., Nichol, R., Postman, M. and Voges, W., The MaxBCG Technique for Finding Galaxy Clusters in SDSS Data. In *AAS 195th Meeting*, (2000)
4. Annis, J., Zhao, Y., Voeckler, J., Wilde, M., Kent, S. and Foster, I. Applying Chimera Virtual Data Concepts to Cluster Finding in the Sloan Sky Survey. Technical Report GriPhyN-2002-05, 2002, www.griphyn.org.
5. Avery, P. and Foster, I. The GriPhyN Project: Towards Petascale Virtual Data Grids. Technical Report GriPhyN-2001-15, 2001, www.griphyn.org.
6. Avery, P., Foster, I., Gardner, R., Newman, H. and Szalay, A. An International Virtual-Data Grid Laboratory for Data Intensive Science. Technical Report GriPhyN-2001-2, 2001, www.griphyn.org.
7. Baru, C., Moore, R., Rajasekar, A. and Wan, M., The SDSC Storage Resource Broker. In *Proc. CASCON'98 Conference*, (1998)
8. Buneman, P., Khanna, S., Tajima, K. and Tan, W.-C., Archiving Scientific Data. In *ACM SIGMOD International Conference on Management of Data*, (2002)
9. Buneman, P., Khanna, S. and Tan, W.-C., Why and Where: A Characterization of Data Provenance. In *International Conference on Database Theory*, (2001)
10. Chen, I.A., Kosky, A.S., Markowitz, V.M. and Szeto, E., Constructing and Maintaining Scientific Database Views. In *9th Conference on Scientific and Statistical Database Management*, (1997)
11. Chervenak, A., Foster, I., Kesselman, C., Salisbury, C. and Tuecke, S. The Data Grid: Towards an Architecture for the Distributed Management and Analysis of Large Scientific Data Sets. *J. Network and Computer Applications* (23). 187-200. 2001.
12. Cui, Y. and Widom, J., Practical Lineage Tracing in Data Warehouses. In *16th International Conference on Data Engineering*, (2000), 367–378
13. Cui, Y., Widom, J. and Wiener, J.L. Tracing the Lineage of View Data in a Warehousing Environment. *ACM Transactions on Database Systems*, 25 (2). 179–227. 2000.
14. Deelman, E., Blackburn, K., Ehrens, P., Kesselman, C., Koranda, S., Lazzarini, A., Mehta, G., Meshkat, L., Pearlman, L., Blackburn, K. and Williams., R., GriPhyN and LIGO, Building a Virtual Data Grid for Gravitational Wave Scientists. In *11th Intl Symposium on High Performance Distributed Computing*, (2002)
15. Deelman, E., Foster, I., Kesselman, C. and Livny, M. Representing Virtual Data: A Catalog Architecture for Location and Materialization Transparency. Technical Report GriPhyN-2001-14, 2001, www.griphyn.org.
16. Deelman, E., Kesselman, C. and Mehta, G. Transformation Catalog Design for GriPhyN. Technical Report GriPhyN-2001-17, 2001, www.griphyn.org.
17. Foster, I. and Kesselman, C. A Data Grid Reference Architecture. Technical Report GriPhyN-2001-12, 2001, www.griphyn.org.
18. Foster, I. and Kesselman, C. (eds.). *The Grid: Blueprint for a New Computing Infrastructure*. Morgan Kaufmann, 1999.
19. Frey, J., Tannenbaum, T., Foster, I., Livny, M. and Tuecke, S., Condor-G: A Computation Management Agent for Multi-Institutional Grids. In *10th International Symposium on High Performance Distributed Computing*, (2001), IEEE Press, 55-66
20. Innocente, V., Silvestris, L. and Stickland, D. CMS Software Architecture: Software Framework, Services, and Persistency in High-level Trigger, Reconstruction, and Analysis. *Computer Physics Communications*, 140. 31-44. 2001.
21. Ioannidis, Y.E. and Livny, M. Conceptual Schemas: Multi-faceted Tools for Desktop Scientific Experiment Management. *International Journal of Cooperative Information Systems*, 1 (3). 451-474. 1992.
22. Ioannidis, Y.E., Livny, M., Gupta, S. and Ponnekanti, N., ZOO : A Desktop Experiment Management Environment. In *22th International Conference on Very Large Data Bases*, (1996), Morgan Kaufmann, 274-285
23. Leymann, F. and Altenhuber, W. Managing Business Processes as an Information Resource. *IBM Systems Journal*, 33 (2). 326–348. 1994.
24. Litzkow, M., Livny, M. and Mutka, M. Condor - A Hunter of Idle Workstations. In *Proc. 8th Intl Conf. on Distributed Computing Systems*, 1988, 104-111.
25. M. Della Negra, S., The CMS Experiment,. A Compact Muon Solenoid. cmsinfo.cern.ch/Welcome.html.
26. Marian, A., Abiteboul, S., Cobena, G. and Mignet, L., Change-Centric Management of Versions in an XML Warehouse. In *27th International Conference of Very Large Data Bases*, (2001)
27. Mohan, C., Alonso, G., Gunthor, R. and Kamath, M. Exotica: A Research Perspective on Workflow Management Systems. *Data Engineering Bulletin*, 18 (1). 19-26. 1995.
28. Stockinger, H., Samar, A., Allcock, W., Foster, I., Holtman, K. and Tierney, B., File and Object Replication in Data Grids. In *10th IEEE Intl. Symp. on High Performance Distributed Computing*, (2001), IEEE Press, 76-86
29. Szalay, A. and Gray, J. The World-Wide Telescope. *Science*, 293. 2037-2040. 2001.
30. Williams, R., Bunn, J., Moore, R. and Pool, J. Interfaces to Scientific Data Archives. Center for Advanced Computing Research, California Institute of Technology, Technical Report CACR-160, 1998.
31. Woodruff, A. and Stonebraker, M. Supporting Fine-Grained Data Lineage in a Database Visualization Environment. Computer Science Division, University of California Berkeley, Report UCB/CSD-97-932, 1997.
32. Zhao, Y. Virtual Galaxy Clusters: An Application of the GriPhyN Virtual Data Toolkit to Sloan Digital Sky Survey Data. MS thesis, University of Chicago, Technical Report GriPhyN-2002-06, 2002, www.griphyn.org.

An Architecture for Managing Distributed Scientific Resources

Maria Cláudia Cavalcanti[1], Marta Mattoso[1], Maria Luiza Campos[2], Eric Simon[3], François Llirbat[3]

[1]*COPPE Sistemas – UFRJ – Rio de Janeiro, Brazil*
[2]*DCC/IM – UFRJ - Rio de Janeiro, Brazil*
[3]*INRIA Rocquencourt – Paris, France*

{yoko, marta}@cos.ufrj.br, mluiza@nce.ufrj.br, {eric.simon, francois.llirbat}@inria.fr

Abstract

There are many examples where cooperation among scientists takes place through exchanging scientific resources, such as data, programs and mathematical models. This is particularly true for environmental applications. Finding the right resource to apply in an environmental problem is a difficult task. Usually, this decision is based on previous experience. Scientists have to cooperate in order to solve such problems. To facilitate the exchange, reuse and dissemination of information we propose an architecture for managing distributed scientific resources. Our proposal combines a mediation-based heterogeneous distributed database system and an enhanced metadata support system for effective management of distributed scientific models and data.

1. Introduction

Environmental applications are characterized by the cooperation among scientists from different disciplines and organizations, bringing them together on scientific experiments. Scientists typically work with experiments based on models, which are simplified representations of real phenomena. In particular, decisions over environmental problems are usually based on such experiments.

Scientific experiments have traditionally been developed in isolation, i.e., scientists from different disciplines work with their own sets of models. As technology improved, scientific data from different sources became largely available in digital media. Scientists can take advantage of such data availability by using them to enhance their experiments. Moreover, scientists can exchange not only data but also scientific models and their implementations (programs). Several standard tools can be used for this exchange, such as Web services [19] protocols and languages (SOAP and WSDL), as well as platforms for Grid computing [12]. In this scenario, the main difficulty is to find the right model and correspondent resources for each experiment.

A scientific experiment can be viewed as a flow of data transformations that starts from raw data and finally produces data with added scientific value. An experiment may begin when a scientist selects models and relevant input data for the problem to be studied, determining or developing an adequate flow of programs that can process the selected input data. Many of these programs are implemented for some specific platform, such as high performance and parallel machines.

Usually, it is the scientist's previous experience that guides the choice of a model for a new experiment. To take advantage of such knowledge, the scientist has to access documentation on previous experiments. This documentation is not always available and may not be described within a common framework. Specially when dealing with empirical models, the scientist has to analyze contextual details of such experiments, verifying similarities with the problem in hand. Therefore, models descriptions and a catalog system may be of great help to scientists.

In order for distributed scientific resources to become part of a large information system on the Internet, they must be located, understood and efficiently accessed over the network. Sharing scientific data requires identifying not only data but also what model and model implementations (programs) are useful, where these programs are located and when (in which order) should programs be executed. It also requires enabling remote data access and their load management for execution at program locations. A step toward this direction includes providing program and model descriptions to facilitate the selection of the appropriate model and consequently the appropriate program. However, this is not a simple task. First of all, model developers come from many different areas, dealing with different kinds of models and description standards. Also, each model description should include its use conditions, i.e., contextual and operational constraints, which are difficult to formalize.

Now, suppose a user understands a model and selects it, finds the correspondent program and runs it. Feedback information on such model usage may be valuable, as other users may need to investigate previous case studies involving that model to fully understand it.

This paper proposes an architecture to facilitate the exchange, reuse and dissemination of distributed scientific resources, where scientists may share data, programs and models. To accomplish this goal, we need to provide solutions to three main problems: (i) how to deal with the distribution and heterogeneity of data and program sources; (ii) how to describe models; and (iii) how to monitor the actual distributed usage of models, programs and data. The Le Select extended architecture addresses these three problems, by the combination of five modules: Navigation, Publication, Execution, Resource Operation, and Resource Description Modules. The Publication and Resource Description Modules address item (ii), while the Execution and Navigation Modules address item (iii). These components are coupled with a fifth module, the Resource Operation Module to address item (i), that can be a middleware architecture such as Le Select [16].

This work is organized into four sections. Section 2 discusses some of the current similar architectures. The third section presents our architecture, describing its components in more detail. Finally, we conclude the paper with comments on some future enhancement ideas.

2. Current Approaches to Distributed Resources Management

Mediator-based Heterogeneous Distributed Database Systems (HDDS) are some of the current approaches to manage data resources [1][8][18]. Most of these systems usually focus on well-behaved data and do not address scientific resources. Grid computing based systems typically use the Globus infrastructure [9] to distribute the execution of scientific programs. However data management is still not well developed with respect to metadata. Specially for scientific resources, metadata issues are even more specific and mandatory. Nevertheless, we found in the literature some very interesting architecture proposals that focus specifically on distributed scientific resources management.

The DataFoundry architecture, developed at the Lawrence Livermore National Laboratory (LLNL), has a very similar approach to our proposal in what it concerns to scientific data management. In [9] the authors present a mediator-based architecture specially conceived to address genomics community. They focus on a metamodel for describing genomics data, embedded in a data warehouse metadata repository. Mediator modules are responsible for translating data based on these metadata, and for storing them into the data warehouse. Their metamodel include four concepts: *abstractions, transformations, schema, and mappings.* However, these concepts are related only to data description (data focused), as no description is provided for scientific programs and the models behind them. Another interesting work, also partially developed at LLNL, uses models to reduce data volume [15]. However, they do not consider describing these models as part of the scientific resources.

An interesting initiative nurtured by NASA is the Federation of Earth Science Information partners – ESIP. It also follows a data focused approach, i.e., it does not consider program description. ESIP brings together government agencies, universities, non-profit organizations, and businesses in an effort to make Earth Science information available to a broader community. Two dozens projects were selected by NASA in 1997. The Earth System Science Workbench (ESSW) project [7], for instance, proposes a non-intrusive data management infrastructure for researchers who desire to publish large data sets derived from environmental models and global satellite imagery. Another ESIP project is the Earth Science Partners' Private Network (ESP2Net) [6] that focuses on problems of dealing with scientific datasets, such as cost/performance, security, and reliability, in a collaborative scientific computing environment.

Some approaches go beyond scientific data in their architecture proposals. For instance, the Poseidon project [17] involves the development of sharable ontologies and metadata for measured/simulated marine data and for modeling software. In their proposal, they use existing metadata standards such as Dublin Core and FGDC, embedded on the Warwick metadata framework. On top of it, they also identify the need to use an ontology to enhance a metadata searching tool. Furthermore, in the Poseidon System Architecture, they include a model management system with enhanced functionality, supporting the design of workflows, and the validation and management of their instances execution.

The Arion [14] and Poseidon architectures are very related to each other. Both of them are developing a distributed data and programs architecture to locate, retrieve and utilize (in simulations and analyses) scientific resources. In the same way as our proposal, these architectures include some metadata support to describe programs and data. However, in our work, we add an extra abstraction layer, where we propose a generic metamodel to describe such resources and their relationship. This metamodel is used as a central reference, facilitating the development of interface tools.

Another important similar approach that should be mentioned is the Grid architecture, where geographically distributed, heterogeneous collections of computing resources are accessed through a single point of contact.

Many projects within Grid computing are also concerned with data, not only programs [13][4]. They focus on scale up the computational power and access to multi-Petabyte data, considering the metadata problem an important but not well-addressed issue. A good example of a scientific resource Grid is the Virtual Sky Project leaded by Jim Gray. On a recent talk [11], he has declared metadata as a challenge for his project. Using his own words *"it is hard to publish, find and understand data and programs"*.

Developed at INRIA, Le Select [16] acts as a mediator-based HDDS. However, differently from other HDDS, Le Select was specially developed to support environmental applications, offering unique features to share both data and programs, while maintaining the general principles of mediator/wrapper architectures. The main objective of Le Select is to allow resource owners to easily publish their resources (data and programs) to the community, to give a uniform and integrated view of distributed published resources to potential users, and to let these users manipulate available resources through a high-level language. However, we believe that finding the right program means first to find the right model, and Le Select does not consider describing models or keeping track of their use. Therefore, Le Select should be enhanced towards a better metadata support, in order to provide an adequate solution to environmental applications.

In our approach, we propose to extend Le Select architecture with a metadata support layer where an embedded metamodel differentiates theoretic models from their implementations, i.e., programs. This is particularly useful because it allows users to retrieve and consider similar programs as alternatives to solve the same problem. We believe that the existence of a metamodel is an essential requirement for the effective distributed management of scientific resources.

3. Le Select Extended Architecture

The main goal of this proposal focuses on scientific resources available throughout the Web. The target user is not the scientific modeler, but the scientific application user. Scientists should be able to publish models for direct real case usage. However, an architecture is required to make these resources really useful. This architecture should include mechanisms for description and management of such resources. We propose an extension of Le Select's architecture where these mechanisms are provided. Figure 1 shows that the extended architecture has two layers, one for Web services and the other for scientific resources management. There are two main modules to manage scientific resources: the Resource Operation Module and the Resource Description Module. Le Select plays the role of the Resource Operation Module, which deals with data and programs. The

Resource Description Module is a metadata repository manager, dealing with data and program descriptions, and also with model, experiment and workflow descriptions.

The Web services layer is composed of three other modules: Publication, Execution and Navigation. The Navigation Module allows scientists to browse scientific resources and their correspondent descriptions. After browsing models and data, the user chooses a program and specific data to be used as input. According to the user choices, the Navigation Module works with the Execution Module, which verifies program constraints by querying the Resource Description Module. The Execution Module interacts back with the Navigation Module, helping the user on the selection process. Then, if the choice is validated, the Execution Module interacts with Le Select (Resource Operation Module) by issuing a job execution command. After Le Select starts the program execution with the specified input data, the Execution Module can keep track of the ongoing experiment, by issuing job query commands. Since the execution of an experiment may take days to finish, it is very useful to have a job monitoring interface. In summary, the Execution Module guides the user on the correct use of the available models, providing an on-the-fly interface for executing them. Finally, the Execution Module should be able to publish the finished experiment by interacting with the Publication Module.

The Publication Module is responsible for publishing scientific resources. When a publisher enters some resource descriptions, the module checks these inputs by interacting with both the Resource Operation and Resource Description Modules. Once validated, the Resource Description Module stores these inputs.

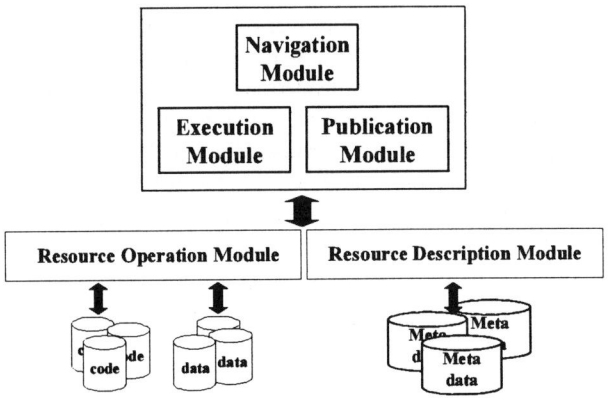

Figure 1: Le Select Extended Architecture

The following sections describe each module of the Le Select Extended Architecture in more detail.

3.1. Resource Description Module

The Resource Description Module manages metadata of scientific resources. The Scientific Publication Model (SPM) is the metamodel (schema) behind the Resource Description Module, and is described in more detail in [2]. Each resource instance is validated and stored in accordance with the SPM.

Considering some of the requirements raised before, XML seems to be the most adequate language to express scientific resources descriptions. Therefore, the SPM is expressed as an XML Schema. The idea of having a separate metadata repository manager came from the need to store semi-structured descriptions expressed in XML. XML-enabled RDBMS and native XML databases are both alternatives for XML storage. However, due to the similarity between the object oriented model and the XML model, we have decided to store XML documents in an ODBMS, in our architecture.

The Goa System [10] is an ODBMS prototype currently under development at COPPE, in the Federal University of Rio de Janeiro. The Resource Description Module (Figure 2) was implemented as a Goa Client and uses the Goa System as the database server. More than just a Goa Client, the Resource Description Module includes the Goa XML enabler (*Goaxe*) facility. With *Goaxe*, the Goa System is now able to understand and store XML documents. *Goaxe* manipulates XML documents by creating a Goa XML Schema that reflects the W3C DOM API classes. Goaxe takes an XML document instance, reads it and breaks it down into DOM class instances. Then, each of these instances is translated into a Goa XML Schema instance. Therefore, Goa Server can be viewed as a generic XML repository.

A further enhancement of Goaxe would be to provide a specific object schema to an XML document, based on the SPM XML schema. Considering the following XML example, the current Goaxe version instantiates three objects of the `element` class (specified at the W3C DOM API): "model", "author", and "description". Using the specific schema policy, Goaxe would instantiate only one object from the `model` class that would have two attributes, `author` and `description`, which would assume the string values "Kuznetsova" and "H_2S Reduction Bacteria presence", respectively.

```
<model>
    <author>Kuznetsova</author>
    <description> H₂S Reduction
Bacteria presence</description>
</model>
```

This approach is more expressive than the first one, narrowing the distance between the real object and its representation.

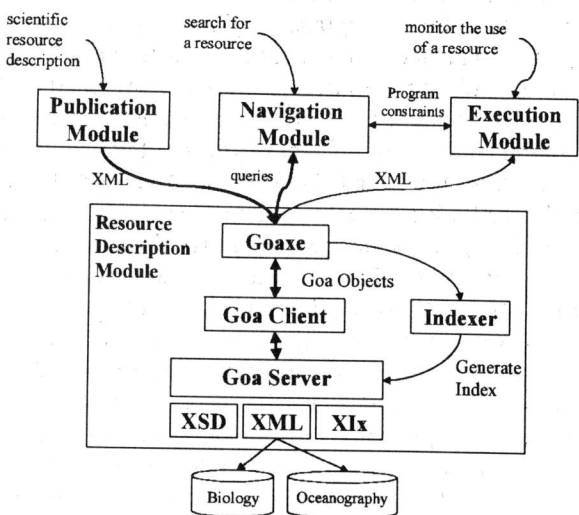

Figure 2: Resource Description Module

In the Resource Description Module, Goa Server maintains three main databases: XML, XSD, and XIx. The XML database stores all XML documents with metadata about any scientific resources area, e.g. Biology, Oceanography, etc. The XSD database stores XML Schemas used for XML documents validation. In this architecture, XSD contains the SPM Schema, which specifies how scientific resources such as data category, transformation, data, code and experiment should be described. Finally, the XIx database stores an index over the XML documents to provide direct access to these documents, facilitating keyword searches. The Indexer builds the XIx database by indexing XML documents stored in Goa Server.

The Publication Module feeds the Resource Description Module with XML documents. Section 3.3 discusses how the Publication Module works in more detail. The Navigation Module interacts with the Resource Description Module by issuing XML queries [20]. Goaxe is responsible for retrieving and returning XML documents that satisfy such queries. The Execution Module interacts with the Resource Description Module when it needs to validate a request for running a program.

The Resource Description Module was built on top of the Goa System. Its current version includes a client API (Goa Client API) implemented in Java, which connects to the Goa Server running on any Internet site. Goa Server is implemented in C++. Goaxe and Goa Client API were encapsulated as Java servlets. Goaxe manipulates XML documents using the Apache Group DOM API, which was implemented based on the W3C DOM API specification. The Publication Module is currently under implementation as a Java Servlet, interfacing with any Web browser through HTML pages and forms. The idea is to deliver an XML valid document to the Resource Description Module, ready to be stored by the Goa Server.

3.2. Resource Operation Module

Le Select is a mediator/wrapper architecture that offers unique features: it allows data and code providers to share their resources with the scientific community. Users publish data in their original format and location. There is no need for transformation or replication of these data. Similarly, programs remain installed in their original configuration and computer platform.

To publish data or code, the publisher needs to install a Le Select server at some Internet server site. Le Select's publishing policy requires that the user provides the wrapper to access his code or data, or that he configures an existing wrapper. Then, the wrapper should be registered within Le Select server. Data wrappers provide a uniform representation of published resources as relational tables and transform SQL queries issued by Le Select server into the particular language of the data source. To facilitate data publishing, Le Select provides built-in data wrappers (e.g. table, XML and JDBC wrappers), which can be easily configured by the publisher. For instance, water sample measurement data may be published at a petroleum station called Cabiúnas, via a table wrapper that exports the table `WSample(date, tankTime, calcium, magnesium, sodium, potassium)`.

Published data can be persistent or just virtual, which means that they are only generated on demand by some computing means offered by the data source (e.g., a C program or a CGI script). Published codes are data processing programs that take a set of relational tables as input, a set of parameters as arguments, and return a set of relational tables as output.

All Le Select's published resources have universal names that are based on the URL of their wrappers. Data resources can be manipulated by the standard high-level language SQL. "*Select-from-where*" statements are used to query tables exported by multiple distributed wrappers within a single query. Code resources are manipulated through a specific Le Select's language that includes a "*job execute*" statement. This statement starts the asynchronous execution of a code. In a "job execute" statement, each input relational table is specified by means of SQL *Select* statements, and arguments are passed by value. Code execution takes place at the site where it is published, and its wrapper is responsible for getting its operand data from possibly remote Le Select servers. Then the wrapper invokes the code and makes its result available as relational tables through a table wrapper. For instance, scientists in Salvador city may want to prevent H_2S propagation by computing water sample data that reside in Cabiúnas city through a program that implements the Kuznetsova model, which resides in Rio de Janeiro. In Le Select, this is possible by issuing the following statement:

```
Job execute
//cenpes.petrobras.br/leselect/Kuzn
etsova.exe

input data set is
Select * from
//cabiunas.petrobras.br/leselect/WS
ample/am345
```

Le Select's servers are distributed as publication sites over the Internet. Each site is capable of publishing local or remote resources (views to other site published resources), as well as processing (optimizing and executing) SQL queries. Schemas of data and signatures of Le Select programs are only known to the wrappers that publish them. There is no notion of global catalog and integrated schema shared by all Le Select servers. Le Select server architecture is presented in [5].

3.3. Publication Module

The Publication Module is responsible for scientific resource description input. According to our SPM there are four types of scientific resources to be described: Model, Program, ModelDC and ProgramDC. Figure 3 shows a class diagram (a subset of the complete SPM) where concepts behind these types of scientific resources are represented. A scientific *program* is the implementation of a theoretic *model* and both are described separately. Although they belong to different usage levels, both concepts have many characteristics in common. To take advantage of such similarity, either *model* or *program* can be viewed as a *transformation*. A *transformation* is a description of a data transformation process that produces some output data and requires input data. Therefore, a transformation should be associated to at least one input and one output, and each one of such *I/O data* refers to a *data category*. A *data category* can be associated to a model (ModelDC) or to a program (ProgramDC). Real code and data are represented as resources. *Code* and *data resources* refer to *program* and *programDC* descriptions, respectively. Both resources contain a URL address.

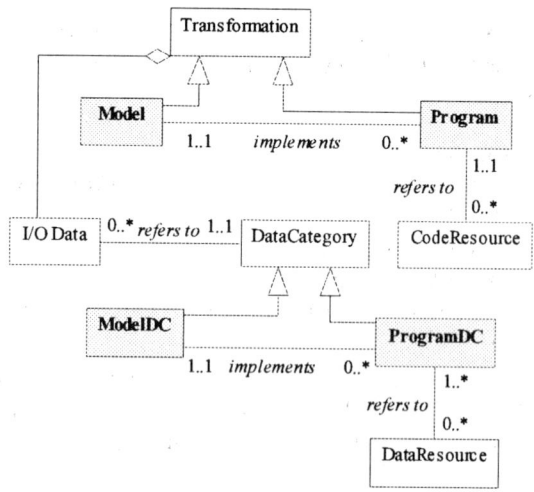

Figure 3: SPM Subset

For each of these four metadata resources, the Publication Module provides a different entry form (Figure 4). ProgramDC and ModelDC are description forms for programs and models data categories, respectively. Each form is based on the SPM XML Schema stored in the XSD database. When the user submits one of these forms to the Form Handler facility, the submitted data are translated into XML format and validated against the SPM Schema, which is obtained by querying the XSD database at the Goa Server.

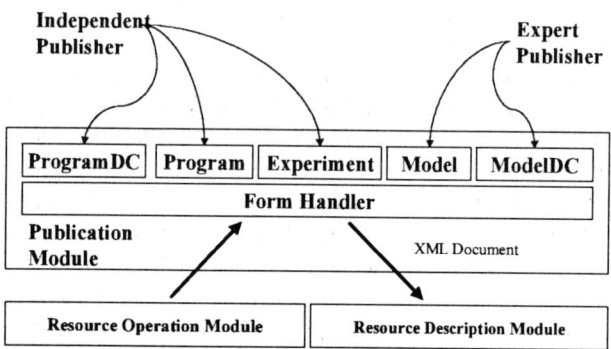

Figure 4: Publication Module

There are two main publication scenarios. In the Expert scenario, the publisher is usually a scientist that has been publishing theoretic model resources within a scientific community. The expert publisher starts publishing *models* and *model data categories*, by filling up the correspondent forms. Then, if that is the case, the expert may also publish *program* and *program data categories* related to these resources, playing the role of an independent publisher.

In the Independent scenario, the publisher is either a *data* provider or a *code* provider. To describe each data

set, the publisher has to fill up a program data category form (ProgramDC). Also, the data provider gives the URL, which that data description refers to. After the form submission, the Publishing Module generates an XML document, validates it and stores it through the Resource Description Module.

Assuming this data resource is available through the Resource Operation Module (Le Select), the Publication Module provides the mapping between the data description and the relational data description. On the other hand, if the data resource is not yet published through the Resource Operation Module, the publisher should follow Le Select's publishing policy, which is to provide a data wrapper for his data resource or to build a Le Select configuration file (wd file) for an existing wrapper.

Now, suppose the publisher finds an existing program data category that is compliant to his data resource. In this case, the publication model should aid the user in choosing the correct existing program data category, and map it to the new data resource URL.

Likewise, the code publisher has to fill up a Program form. However a code publisher should also fill up ProgramDC forms to describe his I/O data types. He may either use existing program data categories that are compliant to his I/O needs or create new ones. Furthermore, these program data categories may be implementations of model data categories. In this case, the Publication Module helps the publisher map the program to the corresponding model, as well as its correspondent data categories.

Differently from the data publisher, the code publisher cannot count on an existing program wrapper. In this case, the program publisher should build a specific wrapper for his program, and should also provide access to such program. In order to access (i.e. execute) a Le Select program, the user needs to provide a wrapper configuration file. The Execution Module described in section 3.5, is responsible for executing a program.

An experiment is another scientific resource that describes the use of a program, i.e., which values were assigned to which parameters, which data resource was used as input to run which code resource, and which data resource was generated as output. To publish an experiment, the user may use the experiment form at the Publication module. The Execution module may also publish an experiment automatically.

3.4. Navigation Module

A quick glance at the most usual queries suggests that the scientific user searches for different data characteristics, such as substance names (e.g. Calcium), magnitudes (e.g. concentration) and units (e.g. mg/l). The Resource Description Module usually answers these queries. However, due to the diversity of scientific users, there is not a pre-defined way to present such queries. Therefore, the need for a keyword-based search facility is clearly identified. Also, a guided navigation is required; based on the results of a keyword search request. Dynamically configured interface pages should guide the user through the selected available resources.

The Navigation Module is responsible for handling scientific resource user queries and navigation. Therefore, the Navigation Module includes facilities for querying XML documents stored by Goa Server and for handling query results and preparing them for presentation in a Web browser (Figure 5). Depending on the request of the user, the Query Interface facility may submit it either to the Keyword Search Handler or to the Query Handler.

The Keyword Search Handler facility takes advantage from the built-in index (XIx) stored by Goa, to retrieve a set of possible XML document references, which are formatted by the Page Builder facility and then sent back to the user browser.

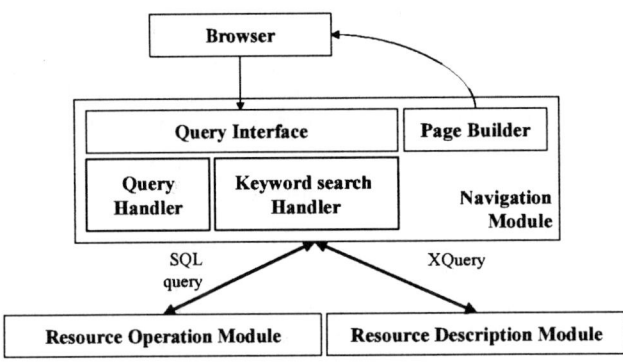

Figure 5: Navigation Module

The Navigation Module also provides some pages through which the user can navigate and ask for more specific information on available resources. These user requests are translated into XQueries by the Query Handler facility and then submitted to the Resource Description Module. The result of such queries may either be a set of references or a single XML document. Both are returned to the Page Builder facility.

Some of these queries may refer to specific data values. These queries require interaction with Le Select (Resource Operation Module). In this case, SQL queries are issued to the Le Select server that manages the specified data.

3.5. Execution Module

The Execution Module provides three facilities (Figure 6). This module interacts mainly with the Navigation and Publication Modules. When the user chooses the program he needs to execute, the Navigation Module starts interacting with the Execution Module, which uses the Job

Validator facility to check program constraints. These interactions may occur many times during a user resource navigation/selection process, and the Job Validator facility will provide different solutions according to the type of constraint to be checked.

So far, we have identified three basic types of constraints that express use conditions for models and programs: contextual, input data and operational. Contextual constraints are usually defined over a model parameter that represents a model calibration variable. For instance, suppose a model is constrained to specific oil storage types. In this case, the model expert publisher previously defined a parameter to represent the variable StorageType and also defined a constraint on this parameter that would accept values not different from "Iron", "Steel" and "Aluminum".

Based on such constraint definitions provided by the Execution Module, the Navigation Module dynamically builds a user interface where the user may configure the program according to the problem in hands.

After configuring the program, it is time to select some data resources to be used as program input, from a list of data resource sites. When the user chooses some data resource from this list, the Navigation Module calls the Execution Module that checks input data constraints, using the Job Validator facility. For instance, suppose a program input data attribute named Chloride should have values constrained to the interval (0, 140000). The program constraint definition would be expressed as:

```
Chloride > 0 and < 140.000
```

To verify these constraints, the Job Validator issues SQL queries to the Resource Operation Module, asking for invalid data. If the query result is empty, then it means that the data resource is all valid and can be used. However, if there are result tuples, then the user should agree on adding such constraint clause to the job execution command, in order to make that data resource a valid input for that program.

Operational constraints are related to program operational issues, such as precision and accuracy. If these constraints refer to data attribute values then the checking process is similar to input data constraints. If these constraints refer to program running options, then they should be defined on parameter values and, similarly to contextual constraints, the Execution Module provides the proper dynamically built user interface.

Finally, after checking all constraints, the job is ready to be executed. The Job Executor facility issues a "job execute" command to the Resource Operation Module. Then, the Job Monitor facility keeps track of the job execution by issuing "job query" commands to the Resource Operation Module. When the job is finished, then the Job Monitor facility interacts with the Publication Module to publish this experiment.

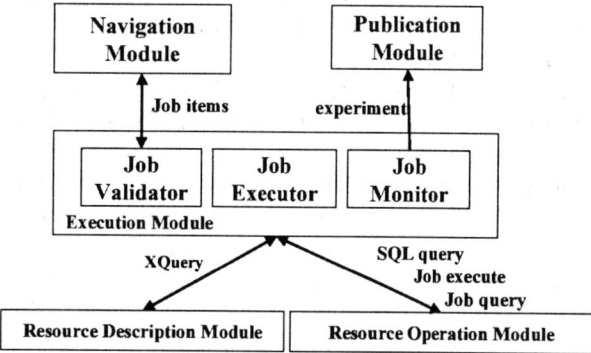

Figure 6: Execution Module

4. Conclusion

Our architecture focuses on overcoming difficulties of using scientific resources in environmental applications. The goal is to provide better metadata support for managing distributed scientific resources, i.e., programs and data. Currently we are using Le Select. By enhancing its functionality, the Extended Le Select Architecture addresses all three problems identified. Le Select itself deals with the distribution and heterogeneity of data and program sources. The extension specifically addresses the other two problems by providing mechanisms to capture model descriptions and to monitor the actual distributed usage of models, programs and data.

The SPM metamodel embedded in our architecture is now being enhanced to capture more semantic details about scientific resources. SPM main contribution is the explicit semantic representation of scientific models and a more comprehensive description of programs, constraints and data sources. The distinction between program and model provides the representation of models at both theoretic and operational levels. We have been validating this metamodel with scientists from Petrobras Petroleum Company, describing models used in different stages of oil extraction and production.

Portability, interoperability and flexibility are some of the main benefits when managing distributed scientific resources with our prototype under development. It may reside in any Web server that serves Java servlets. Other applications may interact with it by issuing XML queries and handling XML documents resulting from such queries. Finally, even though our architecture is based on a specific metamodel, XML schemas provide mechanisms that allow schema extensions to accommodate different XML documents.

5. Acknowledgements

This work was partially sponsored by CNPq and INRIA through KIWI and SIMBIO-CTPetro projects. We also would like to thank Faperj and NCE/UFRJ for supporting software development.

6. References

[1] Carey, M. J. et al **Towards Heterogeneous Multimedia Information Systems: The Garlic Approach**. Tech. Report, IBM Almaden Research Center, USA, 1995.

[2] Cavalcanti, M., Mattoso, M., Campos, M, Llirbat, F., Simon, E. Sharing Scientific Models in Environment Applications. In **Proc. of ACM Symposium on Applied Computing**, pp. 453-457, Madrid, Spain, March 2002.

[3] Critchlow, T., Musick, R., Slezak, T. Experiences Applying Meta-Data to Bioinformatics. In: **Information Sciences**, v.139 (1-2), Elsevier Science Inc., November 2001.

[4] DataGrid Project – Available from <http://www.gridcomputing.com/>.

[5] Ecobase project members – The Ecobase Project: Database and Web Technologies for Environmental Information Systems. In: ACM Sigmod Record, September 2001.

[6] ESP2Net Project – Available from <http://dml.cs.ucla.edu/projects/dml_esip>.

[7] ESSW Project – Available from <http://essw.bren.ucsb.edu/>.

[8] Garcia-Molina, H., Papakonstantinou, Y., Quass, D. et al. The TSIMMIS Approach to Mediation: Data Models and Languages. In: **Journal of Intelligent Information Systems**, 1997. Available from <http://www-db.stanford.edu/tsimmis/publications.html>.

[9] Globus – Available from <http://www.globus.org>.

[10] Goa System – Available from <http://www.cos.ufrj.br/~goa>.

[11] Gray, J. **Mining the Sky – The World Wide Telescope.** Invited talk at the Workshop on Distributed Data and Structures - WDAS-2002, University Paris 9, March, 2002. Available from <http://www.research.microsoft.com/~Gray/talks/Databases_Meet_Astronomy_WDAS.ppt>.

[12] Grid Computing Info Centre – Available from <http://www.gridcomputing.com/>.

[13] Grid Datafarm Project – Available from <http://datafarm.apgrid.org/overview.en.html>.

[14] Houstis, C., Lalis, S. ARION: A Scalable Architecture for a Digital Library of Scientific Collections. In: **8th Panhellenic Conference on Information**, November 2001.

[15] Lee, B., Snapp, R., Musick, R., Critchlow, T. Ad hoc Query Support for Very Large Simulation Mesh Data: the Metadata Approach. In: **Proc. of Brazilian Symposium on**

Databases, pp.199-212, Rio de Janeiro, Brazil, October 2001.

[16] Le Select – Available from <http://caravel.inria.fr/ Fprototype_LeSelect.html>.

[17] Patrikalakis, N., Abrams, S., Bellingham, J., Cho, W., Mihanetzis,K., Robinson, A., Schmidt, H., Wariyapola, P. The Digital Ocean. Invited paper in **Proc. of Computer Graphics International, GCI'2000**, pp. 45-53, Geneva, Switzerland, June 2000. IEEE Computer Society Press. Los Alamitos, CA: IEEE, 2000.

[18] Tomasic, A., Rachid, L., Valduriez, P. A Data Model and Query Processing Techniques for Scaling Access to distributed Heterogeneous Databases in Disco. In: **IEEE Transactions on Knowledge and Data Engineering,** v.10 (4), July 1998.

[19] Web Services – Available from <http://www.w3.org/2002/ws/>.

[20] XML Query – Available from <http://www.w3.org/XML/Query>.

Scientific Metadata Management

Annotating Scientific Images: A Concept-based Approach

Michael Gertz[1]   Kai-Uwe Sattler[1,5]   Fredric Gorin[3,4]   Michael Hogarth[2]   Jim Stone[4]

[1]Department of Computer Science, University of California, Davis
{gertz|sattler}@cs.ucdavis.edu

[2]Department of Pathology and Internal Medicine, University of California, Davis
mahogarth@ucdavis.edu

[3]Department of Neurology, [4]Center for Neuroscience, University of California, Davis
{fagorin|jmstone}@ucdavis.edu

[5]Permanent Address: Department of Computer Science, University of Magdeburg, Germany
kus@iti.cs.uni-magdeburg.de

Abstract

Data annotations are an important kind of metadata that occur in the form of externally assigned descriptions of particular features in Web accessible documents. Such metadata are eventually used in data retrieval tasks on heterogeneous, possible distributed Web-accessible documents.

In this paper, we present the model and realization of an annotation framework that scientists can employ to semantically enrich differerent types of documents, primarily scientific images made availabe through an image respository. Although we employ ontology like structures, called concepts, for metadata schemes used in annotations, our primary focus is on how concepts are actually used to annotate images and regions of interest, respectively, that exhibit features of interest to a researcher. It turns out that the combined consideration of domain specific concepts and annotated regions in images provides interesting means to analyze the usage of metadata regarding certain correctness and plausibility criteria. We detail our annotation management framework in the context of the Human Brain Project in which Neuroscientists record their observations on specific brain structures, and share and exchange information through concept-based annotations associated with images.

1. Introduction

Since the establishment of the Human Brain Project by the US National Institute of Mental Health of the National Institute of Health in 1993, there have been tremendous advancements in brain and behavioral research [12, 16, 3]. A major factor contributing to these advancements are recent developments in imaging and visualization techniques and tools (e.g., [20, 1]). Through these, neuroscientists are now able to investigate experiments and neurobiological phenomena at an unprecedented level of detail and precision. Another contributing factor to these advancements is the current image repository technology which allows researchers to manage and query images generated at different sites in a logically centralized fashion (see, e.g., [25] for an overview). Sophisticated data retrieval methods atop such image repositories, however, are still in their infancy. Pattern recognition and feature extraction methods that operate on diverse types of images and extract certain content descriptive, text-based metadata from such images are only of limited help. There are several reasons for this. First, many features are hard to describe and are often only discovered "manually" by researchers who investigate and interpret a given image in a specific research context. This is in particular the case where the classification of features in an image, e.g., neurons, or nuclei, is based on functional or biochemical properties of these features which are not explicit in the image. Second, many features are not yet known but are discovered while an image is investigated and interpreted by a researcher for a different reason.

In general, what is needed is an extensible model that allows neuroscientists to (1) define semantic rich metadata schemes specifying features of interest for a particular application domain, (2) use such metadata schemes to describe instances of features discovered in images at different levels of granularity, and (3) use the metadata associated with such instances in different data retrieval tasks on an image repository.

In this paper, we present the components of such a model and their realization in a database framework. The core idea underlying this model is to employ domain specific concepts as metadata schemes for the description of features of interest in images. Upon the identification of a features in an image, a researchers chooses a concept providing a metadata template for the feature and then instantiates the text-based metadata for a region of interest (ROI) in the image through a *data annotation process*. Data annotations thus can be understood as well-defined, typed links between schema like metadata structures and ROIs and can easily be employed in data retrieval tasks. There are several advantages of the model we propose. First, annotations are kept separately from images and thus several users can annotate the same image using perhaps different concepts. Second, regions of interest in an image are specified as spatial structures and thus allow fine grained data annotations instead of just whole images. Third, the underlying model allows for various text-based data retrieval scenarios. A major novelty of our model is that it supports checking for the compatibility of annotations and underlying concepts.

In the following section, we present our annotation model and its realization in a specific research project of the Human Brain Project conducted at the University of California at Davis [24]. Our primary focus is on annotating images of neuroanatomical structures of the human brain. In Section 3, we present different mechanisms we realized atop of the model to check for the compatibility of annotations in images and usage of underlying concepts. A prototype application of our approach including some basic usage scenarios for annotating images is presented in Section 4. After a review of related work in Section 5, we draw some conclusions and outline future work in Section 6.

2. Representing and Managing Conceptualized Annotations

In the following, we present the model underlying our approach to annotate regions of interest in images using domain specific concepts.

2.1. Requirements and Assumptions

An annotation of an image basically can be understood as a typed link between a spatial object (so-called *region of interest* or *ROI*) in the image and a domain specific concept representing a metadata template. Associated with the annotation are values for properties that describe the feature according to the concept. In order to specify, represent, and in particular query concepts, annotations and images in a uniform fashion,

these types of information need to be not only modeled appropriately, but a respective model should also be easy to implement and use in different data management and retrieval tasks.

The model should be extensible with regard to different conceptual structures adopted as metadata schemes. Conceptual structures can include simple standard vocabulary like structures, such as Neuronames [15] or UMLS [4], as well as complex (bio)ontology like structures [2], provided such structures support the notion of uniquely identifiable concepts. Since such conceptual structures are developed in a collaborative fashion and represent various domain specific aspects, not only different views on such structures and thus annotations need to be supported, but a respective infrastructure needs to be in place to negotiate concept specifications such as the naming or properties of anatomical structures. In Section 3, we will describe some mechanisms that can be employed for realizing such infrastructure.

We assume that images are managed by an image repository which can be used by individual researchers and research groups in a collaborative fashion. Images can be registered and Dublin Core like creational metadata are associated with images and describe authorship, experiment and a very basic content description. Such metadata, different from the concepts used to annotate images, provide researchers an entry point to image data of interest. Such images are further investigated and interpreted and perhaps annotated using concepts. We assume high resolution image data that cover a wide variety of neuroanatomical and biological phenomena of the human brain, ranging from photographed slices of sections of the brain up to images of individual cells, cell structures, and nuclei, perhaps in different stages of function and/or behavior.

2.2. Annotation Graph Model

In the following, we detail an *annotation graph model* that addresses the above requirements in terms of expressiveness, extensibility, and ease in implementation. In this model, concepts, annotations, and images are represented as different types of nodes. Edges between nodes describe respective relationships such as how concepts are related and images are annotated using concepts.

Assume a set $T = \{String, Int, Date, \ldots\}$ of simple data types. Let $\mathrm{dom}(T)$ denotes the domain, i.e., the set of all possible values for T. In the annotation graph model, a property of a node is defined by an identifier and a type $PDef = String \times T$. An instantiation of a property consists of an identifier and a value $PVal = String \times \mathrm{dom}(T)$.

As outlined in the previous sections, concepts provide templates for annotations that are associated with ROIs in images. In our model, concept are represented by a simple yet extensible form of *concept nodes*. We assume that each concept node has the following components: (1) a concept identifier *cid*, (2) a natural language definition that associates an agreed upon, well-defined meaning with the concept (*def*), (3) a set *terms* of phrases or words that are typically used to name the concept (e.g., synonyms), (4) and a set *pdefs* $\subset \mathbb{P}\,PDef$ of property definitions. A concept thus is similar to a class definition used in the context of object-oriented modeling. In the following, we denote the set of all concepts by $\mathcal{C}$.

The second type of node in our graph model represents *images*, which are assumed to be Web accessible, either through a direct URL or a query against the image repository. Images thus are identified by a URI (Uniform Resource Identifier). The set of all image nodes is denoted by $\mathcal{I}$. Finally, *annotation nodes* provide the basis to specify links between concepts and ROIs in images. An annotation node has an identifier and a set *PVal* of property instantiations induced by the concept underlying the annotation. The set of all annotation nodes in a graph is denoted as $\mathcal{A}$.

With each of the above nodes, further creational properties are associated, including author information, date of creation etc, and are not specified explicitly. Note that from an operational point of view creational properties of image nodes can be provided by the image repository.

The set of all nodes $\mathcal{V}$ in an annotation graph is defined as the union of the component sets $\mathcal{A}$, $\mathcal{C}$ and $\mathcal{I}$: $\mathcal{V} = \mathcal{A} \cup \mathcal{C} \cup \mathcal{I}$. Links between nodes are represented as directed, typed edges to which optional property instantiations are assigned. The types of edges are drawn from concepts (see below) and property instantiations are determined by the concepts underlying the edges. Finally, the set $\mathcal{E}$ of all edges is define as $\mathcal{E} \subseteq \mathcal{V} \times \mathcal{V} \times \mathcal{C} \times \mathbb{P}\,PVal$. The meaning of the components of an edge $e = (\textit{from, to, type, pvals}) \in \mathcal{E}$, with *from*, *to*, and *type* being nodes (or rather node identifiers), is as follows. The edge e connects the node with id *from* with the node *to* (in this direction). With the edge e the concept with the id *type* is associated, and *pvals* is a set property instantiations induced by the concept *type*.

Based on these definitions, our annotation graph model comprises both metadata template components (concepts) and metadata instance components (annotations). An instance of the model defined by one or more users is then represented by a graph $\mathcal{G} = (\mathcal{V}, \mathcal{E})$.

Naturally, nodes can be connected via edges of ar-

bitrary types (concepts). However, in most cases only edges of certain types are reasonable. In order to deal with the specific meaning of the different kinds of nodes and how they can be connected via edges, we introduce the following *default concepts*, which are assumed to be contained in any specification of a collection of concepts:

- *annotates* is used to represent edges from annotations to images. Since a basic requirement in our model is to allow for fine-grained annotations, that is, regions of interest in an image, we assume a set of *ROI descriptions* as properties of this default concept. An ROI description comprises information about the region in an image in form of a spatial object. Currently, our model supports polygons, rectangles, circles, as well as single points as spatial objects.[1]

- *annotatedBy* is a concept for representing the inverse of *annotates*, thus supporting edges from images to annotations.

- The concept *ofConcept* represents the fact that an annotation is based on a certain concept, i.e., it assigns the concept to the annotated image and instantiates the properties specified by this concept. Each annotation $a \in \mathcal{A}$ must be related to a concept, i.e., $\forall a \in \mathcal{A} : \exists c \in \mathcal{C} \wedge (a, c, \textit{ofConcept}) \in \mathcal{E}$.

- *hasAnnotation* describes the inverse relationship of *ofConcept*.

It is important to note that besides these default *relationship type concepts*, other such concepts can be introduced to specify relationships among base concepts. In the context of our current research, these include in particular concepts that define spatial (e.g., *contains* and the inverse relationship *containedIn*) and type-based is-a relationships (*isA/hasSubtype*). Type-based is-a relationships naturally involved the inheritance of property definitions among concepts related through such a relationship, i.e., $\forall e \in \mathcal{E} : e.type = isA \rightarrow e.from, e.to \in \mathcal{C} \wedge e.from.pdefs \supseteq e.to.pdefs$. In should also be noted that the concept part of an annotation graph as it typically occurs in our project basically consists of a collection of classification hierarchies. Individual concepts can occur in one or more such hierarchies, depending on whether the focus of the classification is based on functional, biochemical, physiological etc. aspect. Figure 1 illustrates a typical subgraph of an annotation graph. It represents

---

[1]It should be noted that the same principles can be applied to text documents where a document is viewed as a tree like structure. In this case, XPath expressions can be used as ROI descriptions, assuming that documents are represented in X(HT)ML.

a hierarchy of base concepts related through a spatial containment type relationship concept (left side) and a type-based hierarchy (right side). Among certain base concepts in these two hierarchies, there is another relationship type concept, here representing the fact that cells of the type A typically occur in the ephithalamus.

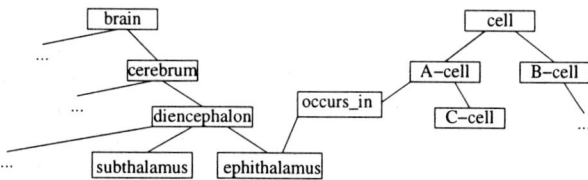

Figure 1. Concept Classification Hierarchies

In our current application for annotating images showing neuroanatomical phenomena, specifications of base and relationship type concepts as part of an annotation graph actually turn out to be very similar to cross-linked Yahoo-like hierarchies, thus providing users with an intuitive and easy to employ entry point to annotated images.

2.3. Querying an Annotation Graph

Querying and navigating an annotation graph is supported by two kinds of operations: selection and path traversal. The input for a selection operation is either one of the basic sets $\mathcal{A}$, $\mathcal{C}$ or $\mathcal{I}$ (but not a union of them) or a derived set resulting from a prior operation. Let S be one of the sets $\mathcal{A}$, $\mathcal{C}$ or $\mathcal{I}$, and $P(s)$ a predicate on $s \in S$. Then a selection operation σ_P is defined as

$$\sigma_P(S) = \{s \mid s \in S \land P(s)\}$$

A predicate P is a boolean expression made up of a number of clauses like *prop* $<op>$ *value* which can be connected by logical connectives. In addition, path expressions of the form $rel_1.rel_2 \ldots rel_n.prop$ indicating the traversal of edges of concepts rel_1, rel_2 etc. from the current node to the property *prop* of the target node are allowed as long the result is a single-valued expression. Accessing non-existing properties always evaluates to false.

Path traversal enables following links between nodes of the graph. Given a start node v_s and a relationship type concept rel, the operation ϕ_{rel} returns the set of target nodes based on respective edges:

$$\phi_{rel}(v_s) = \{v_t \mid (v_s, v_t, rel) \in \mathcal{E}\}$$

Since we mainly have to deal with sets of nodes in query expressions, this operation is easily extended to set of nodes $V \in \mathcal{V}$ as $\Phi_{rel}(V) = \{v_t \mid \exists v_s \in V : (v_s, v_t, rel) \in \mathcal{E}\}$. A special kind of the path traversal operation is the computation of the transitive closure. It extends ϕ_{rel} by traversing the path indicated by the relationship as long as edges can be found that have not already been visited. The result set of nodes visited during the traversal thus is

$$\phi_{rel}^+(v_s) = \{v_t \mid (v_s, v_t, rel) \in \mathcal{E} \lor \exists v_i \in \phi_{rel}^+(v_s) : (v_i, v_t, rel) \in \mathcal{E}\}$$

As for ϕ_{rel}, this operation is defined on a set of nodes:

$$\Phi_{rel}^+(V) = \{v_t \mid \exists v_s \in V : (v_s, v_t, rel) \in \mathcal{E} \lor \exists v_i \in \Phi_{rel}^+(v_s) : (v_i, v_t, rel) \in \mathcal{E}\}$$

Using these operations, query expressions containing node selections and edge traversal can easily be formulated. The initial set of nodes for a traversal always has to be obtained by applying a selection on one of the basic sets $\mathcal{A}$, $\mathcal{C}$ or $\mathcal{I}$. Then, following edges specified by special relationship-type concepts *ofConcept*, *annotates* etc. allows to go to another type of node.

In order to provide for easy specification and implementation of services on top of the model, we have developed a simple language in the spirit of XPath. In this language, the sets $\mathcal{A}$ (*annotation*), $\mathcal{C}$ (*concept*), and $\mathcal{I}$ (*images*) are valid root elements. If views as filters on these sets are defined, they can be used as root elements as well (see also Section 2.4). Selections on nodes are formulated by appending a [*condition*] clause to a term. In *condition*, the properties of the nodes can be accessed and – in combination with the standard logical connectives – used for formulating predicates. The Φ-operator is expressed by appending /*relship* to the term. *relship* denotes a relationship type concept that has to be used for following the links. The optional + indicates that the transitive closure has to be computed.

In the following example, we start from an image with a given URI and then retrieve the annotations associated with that image. If now the *ofConcept* relationship type concept is followed, we are able to obtain the concepts underlying these annotations, and by traversing to the annotations and images, we obtain "similar" images (ROIs), i.e., images that are annotated using the same concepts. This query can be extended further by considering a relationship type concepts, say *is-of-cell-type*. We are then able to find images that have been annotated based on more general concepts:

image[uri=. . .]/annotatedBy/ofConcept/
   is-of-cell-type+/hasAnnotation/annotates

As another example, the query below returns the image(s) and ROI(s), respectively, that is/are linked to a given image by an annotation of a certain concept C:

$$image[uri=\ldots]/annotatedBy[ofConcept.cname='C']/$$
$$annotates$$

In Section 4, we will outline how queries expressed in this language are managed and evaluated against a database storing an instance of an annotation graph.

2.4. View Mechanism

In many emerging areas of the biosciences and in particular in Neuroscience, new domain concepts and knowledge are acquired almost every day and thus general, fully agreed upon conceptual structures among research communities in form of, e.g., standard vocabularies, do not exist. Typically, such structures and vocabularies are developed over time in individual research projects and later homogenized and made available to specific research communities. Thus, for associating concept-based metadata with images as proposed in this work, there is a strong need to support different vocabularies or conceptual structures as the basis for metadata schemes.

We support these requirements by providing *view mechanisms* on annotation graphs. On each of the base component sets, a view can be defined. More precisely, a view specifies a (virtual) sub-graph $\mathcal{G}'$ of the annotation graph: $\mathcal{G}' = (\mathcal{V}', \mathcal{E})$ with $\mathcal{V}' \subset \mathcal{V}$. $\mathcal{V}'$ is specified by formulating queries using operations presented in Section 2.3 and which restrict the set of annotations, concepts, and images to be considered in selection and graph traversal.

For example, if we want to provide a view containing only (1) concepts that have been introduced by a certain author and (2) annotations made this year, we could define this as follows[2]

```
define view my_view as
    annotation := annotation[created>= '01/01/02']
    concept    := concept[author='Jim Smith']
```

If a base component set is not involved in the view definition, e.g., the set $\mathcal{I}$ of images as in the above case, the complete base set is used by default. For restricting queries in views, any valid query expression is allowed as long as it returns a proper subset of a base set, for example, a set $\mathcal{C}' \subset \mathcal{C}$ for the concept set. Since the set $\mathcal{C}$ of concepts contains default relationship type concepts such as *annotates*, *ofConcept* etc, this set is

handled in a special way, guaranteeing the inclusion of such default concepts in each view.

Views are used in queries by simply giving the name of the view as an additional parameter of the query service invocation. For example, the invocation

```
query("image['uri=...']/annotatedBy",
    my_view)
```

is evaluated based on the annotation sub-graph defined by my_view.

It should be noted that from a practical perspective, views are a viable approach for protecting researchers from concept or annotation "overload". When a single image annotation service is used by researchers from different domains, it will probably contain large portions of concepts and annotations that are not of interest for all researchers in all these domains.
)

3. Analysis and Synthesis of Annotations and Concepts

In order for concepts and annotations created by different researcher to be useful in data retrieval tasks, mechanisms need to be devised that ensure a certain degree of compatibility among concepts and annotations. In this section, we present the basic principles underlying the realization of such mechanisms in the context of annotating images in the Neuroscience.

3.1. Overview

As indicated in the introduction, in order for metadata to be useful, it is essential to devise mechanisms that guard against inconsistent or incompatible metadata and metadata schemes. There have been major advancements in the development and usage of metadata schemes for Web-accessible data, but there has been little attention given to the correctness and plausibility of metadata associated with data. The problem obviously is that it is hard to precisely define what consistent metadata are and thus to develop respective mechanisms.

In the context of annotating images using concepts specific to a Neuroscience application domain, we have devised such mechanisms. The core idea behind these mechanisms is to (1) exploit region information associated with annotations, and (2) investigate relationships between the concepts underlying these annotations. Depending on what spatial properties such regions have and what relationships exist among the annotations describing ROIs, the user can be provided with feedback about possible incompatibilities. It should be noted

---

[2]Currently, we do not provide a view definition language. Instead, a view is defined in context of a dedicated service.

that no precise definition of consistent annotations and concepts is possible in this context since annotations and concept specifications typically are based on a specific perception and interpretation a neuroscientist has regarding a real-world concept or image representing some data specific to an application domain. A respective framework thus has to provide the user with mechanisms that allow her to specify (1) what is considered to be possibly incompatible and (2) how to react to an incompatibility. The latter aspect necessitates certain annotation policies the user can specify and which describe actions to be performed in case incompatibilities have been discovered.

In the following, we discuss a framework in which mechanisms checking for the compatibility of annotations and concepts has been devised. In order to have a workable but still useful setting, our mechanisms are based on two common types of concept classification hierarchies, namely those based on spatial containment (i.e., the spatial containment of one brain structure in another brain structure), and the type-based classification of brain structures (with a particular focus on cells and nuclei). An excerpt of two such hierarchies is shown in Figure 1 where the left hierarchy is based on spatial containment and the right hierarchy is based on type-based cell classification.

3.2. Annotation-level Mechanisms

Assume the scenario where a user interprets an image $i \in \mathcal{I}$ and that she chooses concept c underlying the new annotation a of a particular region r in i. Based on this information, the task of annotation level mechanisms is to determine other annotations in i that might be incompatible with the new annotation a. Let $A_i = \{a_1, \ldots, a_n\}$ denote all annotations that already exist for the image i (under the user specified view). For each annotation $a_k \in A_i$, let c_k be the concept underlying a_k, and let r_k and $pval_k$ be the region information and properties, respectively, associated with a_k. This information can easily be obtained through querying the annotation graph instance.

Based on A_i, there is one procedure that partitions A_i into three disjoint sets A_{same}, A_{over} and A_{disj}, based on the spatial relationships the existing annotations have with respect to the new annotation a. The set $A_{same} \subseteq A_i$ contains all annotations that employ the same region as a. The sets A_{over} and A_{disj} are defined similarly for overlapping and disjoint regions associated with annotations.

A subsequent set of procedures then checks for each annotation in such a set how the concept underlying the annotation is related to the new annotation a. Before we detail these procedures, we first describe the

partitioning of the set A_i.

Since regions in an image can be free drawings (circles, rectangles, or polygons), there is no precise and general definition of what "same region" actually means. In our framework, we thus employ user-defined predicates, which are typically detailed by a group of researchers and are based on some agreed upon criteria. For our annotation level mechanisms, we provide the user with three types of predicates, which are defined on pairs of regions and determine the spatial relationship among these regions (see also Figure 2).

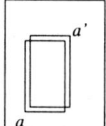 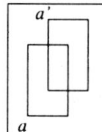 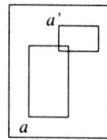

Figure 2. Spatial relationships among regions (ROIs) underlying annotations in an image

$same\_region(r_1, r_2, sr\_threshold)$ evaluates to true if there is an overlap among the two regions r_1 and r_2 of more than $sr\_threshold$ percent. In this case, the two regions are considered to be equal. The predicate $overlap\_region(r_1, r_2, or\_threshold, sr\_threshold)$ evaluates to true if there the two regions r_1 and r_2 overlap more than $or\_threshold$ percent but less than $sr\_threshold$ percent. The predicate $disj\_region(r_1, r_2, dr\_threshold, or\_threshold)$ used to determine whether two regions are disjoint is devised similarly. Checking these predicates for each annotation $a_k \in A_i$ and the new annotation a results in a partitioning of A_i into A_{same}, A_{over}, and A_{disj}. For each set, now individual mechanisms are applied that check for possible incompatibilities among the concepts underlying the annotations. Our main focus will be on the case where two annotations are based on the same regions. Mechanisms for cases where two regions are overlapping or disjoint can be devised in an analogous fashion and are only outlined.

Same Region. Assume an annotation $a_k \in A_{same}$ based on concept c_k and the new annotation a based on concept c. There are three cases to consider:

- $c = c_i$: Both annotation are based on the same concept. If they also have the same properties ($pval$, see Section 2), then the annotation a is redundant (case 1). The equivalence of properties is checked based on another function $match\_properties: A \times A \to [0, 1]$ in order to account for similarities among values users choose to describe instances for concepts. If the value determined by the function is below a certain user-specified threshold, the properties are considered to be different and thus a *data conflict* is determined (case 2). In this case, the mechanism triggers a respective action, e.g., a negotiation process with the user who specified the annotation a_k.

- $c \neq c_i$: The two annotations are based on different concepts. We refer to such a situation as *concept reference conflict* whose handling will be detailed in the following.

As indicated in Section 3.1, our main focus is on concept classification hierarchies that are based on spatial containment and sub-type/super-type relationships. Assume two annotations a and a_k, based on concepts c and c_k, respectively. If there is no (direct) relationship between c and c_k, then the two annotations are likely to reflect different views on the same ROI. The user annotating the image i then can initiate respective actions through the specification of annotation policies, as indicated in Section 3.1. The more interesting cases are when c and c_k belong to the same classification hierarchy.

Assume a hierarchy based on spatial containment. If c is a (direct) sub-concept of c_i, then the annotation a is either to fine-grained or the annotation a_k is to coarse-grained. That is, for either annotation a different region needs to be specified. The analogous case holds if c is a (direct) super-concept of c_k. In both cases, a review process is triggered by the annotation level mechanisms which then allow the two users who made the annotations a and a_k to review and negotiate a correct annotation. A similar scenario holds when both concepts belong to the same classification hierarchy but there is no super/sub-concept relationship between the two concepts. This case indicates that either annotation is based on a misclassification of the feature described by the annotations. The above scenario is adopted in an equivalent fashion where the two concepts c and c_k belong to the same type-based classification hierarchy.

Overlapping and Disjoint Regions. Mechanisms that are applied to the two sets A_{over} and A_{disj} are based on the same principles adopted for annotations that are associated with the same ROI. In particular, cases where the two concepts are specified in a concept hierarchy based on spatial containment can be handled in exactly the same fashion. For example, if two concepts c and c_k are associated with two disjoint regions r and r_k and there is a spatial containment among the two concepts in terms of a super/sub-concept relationship, then this clearly indicates a possible misclassification. A simple example, based on Figure 1, is when a user associated a region with the concept diencephalon

and another user associates an overlapping (or disjoint) region with the concept cerebrum. In order for the two annotations to be compatible, there must be no overlap among the two regions, but containment since the concept diencephalon is a super-concept of ephithalamus in the spatial containment hierarchy.

3.3. Concept-level Mechanisms

Concept-level mechanisms are invoked whenever a user creates or modifies a concept or introduces a new relationship type concept. Achieving the desired functionality of such mechanisms is much more critical for the overall approach since concepts provide the basis for annotations and thus require a high degree of compatibility in terms of consistency and non-redundancy.

For the specification of a concept, we employ a function that determines the similarity among a new and existing concepts and which is automatically invoked by the mechanism. Basis for this function, named *similar_concept*, are two components that individually check the similarity among terms and properties between the new and an existing concept using user-defined weights. Formally, the function is defined as

$$similar\_concept(c_1, c_2) := t * sim\_terms(c_1, c_2) + p * sim\_prop(c_1, c_2) \in \mathcal{R}[0, 1]$$

with t, p being weights such that $t + p = 1$. The functions *sim_terms* and *sim_prop* determine the similarity among individual components of two concepts c_1 and c_2 using word and phrase similarity measures as they are used in, e.g., schema matching approaches in database integration [19], natural language processing techniques [14], or approaches in consolidating clinical terminology [17]. Each function returns a value between 0 and 1, which is then passed to the function *similar_concept*. Upon the creation of a concept, the mechanism realizing the above function provides the user with a list of similar concepts, ranked based on their computed similarity value. Besides comparing the new concept to the existing concept, our approach in particular provides the user with means to investigate how similar existing concepts have been used in annotating images. In that respect, our framework provides the user with more functionality than just simply checking similarity among terms and properties used for specifying concepts.

The latter aspect is also of particular concern for mechanisms that check for possible incompatibilities among concepts *after* concepts have been specified and used for annotating images. The basic idea for this is that for pairs of concepts the similarity value is recorded. This is only done for pairs of concepts that are not be considered similar but whose similarity value is above a certain threshold. For these pairs of concepts, at user specified times, automated mechanisms check how these concepts have been used to annotate data. The invocation of the mechanism can either occur on a regular basis (e.g., weekly) or based on how many annotations have been made using these concepts. If it turns out that the two concepts have been used to basically annotate the image (or rather ROIs), then these two concepts are likely to be similar. The realization of such checks again utilizes the notion of *same_region* as introduced in the previous section for analyzing the compatibility of annotations.

4. Prototype Application

In this section, we briefly describe the application of the presented framework as part of a collaborative environment for annotating images in the context of the Human Brain Project. In this project, brain slices are digitally photographed under microscope and utilized by researchers who mark specific regions (e.g. cell structures) and assign concepts (e.g., a certain cell type) to these regions. The annotation graph model is used to represent concepts, annotations and images as well as their relationships. Furthermore, the query operations allow to formulate declarative queries for retrieving elements and traversing the graph.

The implemented annotation system follows the typical client/server paradigm. Basic services for defining concepts and assigning annotations as well as formulating queries are provided by an annotation server, whose architecture is shown in Figure 3.

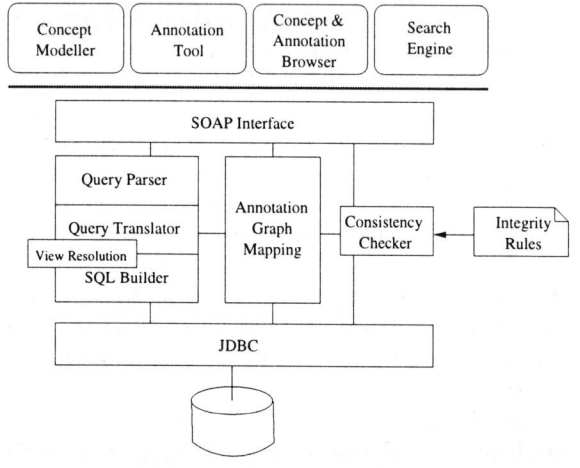

Figure 3. System architecture

The main component is the graph mapping module, which represents an annotation graph by mapping

its nodes and edges to relations stored in a relational database. Tightly related to this is the query component consisting of a parser for the language described in Section 2.3, the translator which transforms a query into a relational algebra expression by applying a set of transformation rules [9] and the SQL builder for deriving SQL queries from such expressions. The translator also implements the view mechanism by substituting all references to the basic sets (concept, annotation, images) in a query by the restricting expressions of the according view definition. Finally, the SQL query is sent to the DBMS for evaluation. For an efficient evaluation of similarity predicates both as part of queries as well as for consistency checking we are currently investigating the usage of DBMS cartridges for text and spatial data.

The components of the system are implemented in Java using JDBC for accessing the DBMS. The interface to the services of the system is realized using SOAP. In this way, the annotation server can be used as a Web Service by different (possibly Web-based) tools as shown in Figure 3.

A screenshot with two of these tools is given in Figure 4. On the lower left side, the concept browser is shown, which is used for querying and browsing the concepts. It offers different views, e.g., a simple tabular presentation of query results as well as tree presentations, where the primary relationship can be chosen by the user (for example, *contains* for browsing containment hierarchies and *hasSubtype* for specialization hierarchies). Beside visualizing the concept space, a second function of this browser is to select a concept for creating a new annotation. The latter step can be performed with the help of the annotation tool shown behind the browser. It allows to mark regions of interest in an image and to assign an annotation based on the previously selected concept.

This requires to chose a certain concept from the available set or sometimes – if new structures are discovered – to define a new concept, possibly as a specialization of another one. In the latter case, the consistency/compatibility checking mechanisms (cf. Section 3) are involved to notify the user about possible conflicts or redundancies. In addition, concepts act as a kind of template for annotations by defining a set of properties which have to be instantiated, i.e., by specifying values for the annotation.

A collection of annotated images can later be used to visualize and explain various structures of the brain. For this purpose, images are shown together with their annotations, which not only give an explanation on important regions of the image, but allow also to follow links to the concept underlying the annotation, to re-

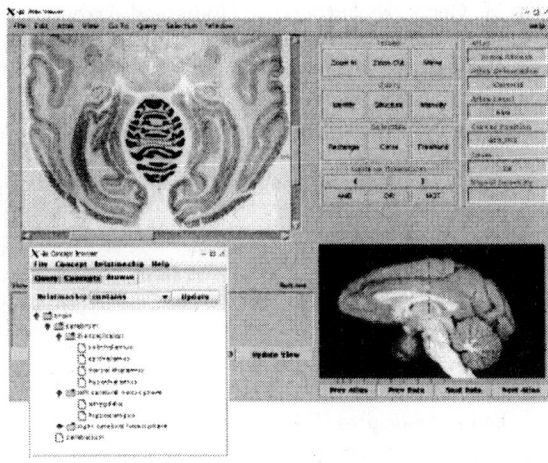

Figure 4. Tools from the annotation system

lated concepts and finally to images annotated with these concepts. In this way, an atlas of the human brain can be built, consisting of marked regions in images which are linked by concepts.

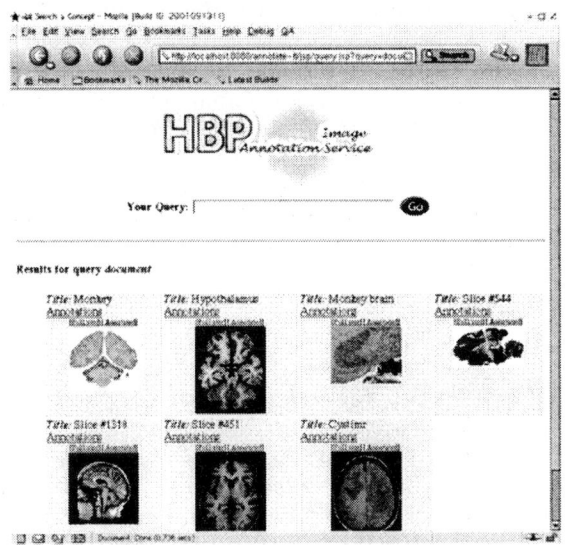

Figure 5. The search engine

Finally, Figure 5 shows a screenshot of the search engine for image annotations. It simply evaluates queries formulated in our query language using the query engine of the annotation service and displays the results. In addition, with each image the associated annotations and concepts are shown as links referring to a page with details about them including further links to related concepts, images etc.

5. Related Work

There is an increasing amount of work on models and methodologies to semantically enrich the Web (see www.semanticweb.org for an extensive overview). The major focus in these works is on building semantic rich and expressive ontology models that allow users to specify domain knowledge. The most prominent approaches in this area the Ontobroker project [6, 7], SHOE [10], the Topic Maps standard [23] as general ontology frameworks, and TAMBIS [21] and OIL [22] as specific ontology frameworks tailored to the biological domain. We consider these ontology-centric works as orthogonal to our annotation-centric approach. Also, most of these work do not put much emphasis on how remote Web documents or images can be annotated by different users at a fine-level of granularity. We consider the need for external and fine-grained annotations as essential and appropriately include these aspects in our model for annotating scientific images. Furthermore, whereas the above approaches concentrate on querying ontologies using, e.g., RDF-based languages, our focus is to have a simple, expressive, and easy to implement language that allows to query all three components, concepts, annotations, and Web accessible images in a uniform fashion.

At the other end of the spectrum, several systems have been proposed that provide users with means to annotate data. This includes the multivalent document approach [18], the SLIMPad approach [5], the Annotea project [13] as well as some commercial systems (see, e.g., [8, 11] for an overview). While none of these approaches supports a query framework for annotations, only [5] support the notion of concept like structures underlying annotations. Finally, to the best of our knowledge, there has been no work that considers the aspects of the consistent usage of metadata in annotating or enriching Web accessible documents.

6. Conclusions

In many computational sciences, the association of different types of metadata with heterogeneous and distributed collections of scientific data play a crucial role in order to facilitate data retrieval tasks in an integrated and uniform way. In this paper, we have presented an approach that allows researchers to associate well-defined metadata in form of concept instances with image data. Although our focus primarily is on image data as they typically occur in the Neurosciences, the underlying model of data annotations and concept-based metadata templates is applicable to a wide variety of different forms of scientific data. The model and its realization provide all features researchers in collaborative research environments deem necessary to enrich (possibly remote) data and thus to "semantically index" data that is not easy to classify or analyze otherwise. In particular, we have shown how properties of concepts and annotated regions in images can actually be used to investigate the compatibility or consistency of metadata associated with images.

While the usage of the first prototype of our system confirms this hypothesis, several new challenges come up. These include aspects of scalability of the system as well as efficiency and effectiveness of user interfaces. Due to the centralized graph storage, the presented architecture and its services are appropriate only for smaller communities. A distributed approach using multiple instances of an annotation graph and a distribution of data retrieval service alleviate such problems.

References

[1] I.N. Bankman (Editor-in-Chief): *Handbook of Medical Imaging – Processing and Analysis*, Academic Press, 2000.

[2] 3rd annual Bio-Ontolgies Workshop – Sharing Experiences and Spreading Best Practice. La Joalla, CA, www.ingenuity.com, August 2000.

[3] M. Chicurel: Databasing the brain. *Nature* 406:822-825, 2000.

[4] K.E. Campbell, D.E. Oliver, E.H. Shortliffe: The unified medical language system: towards a collaborative approach for solving terminology problems. *Journal of the American Medical Informatics Association*, Volume 8, 12–16, 1998.

[5] L.M. Delcambre, D. Maier, S. Bowers, M. Weaver, L. Deng, P. Gorman, J. Ash, M. Lavelle, J. Lyman: Bundles in Captivity: An Application of Superimposed Information. In *Proc. of the 17th International Conference on Data Engineering (ICDE 2001)*, IEEE Computer Society, 111-120, 2001.

[6] S. Decker, M. Erdmann, D. Fensel, R. Studer: Ontobroker: Ontology based Access to Distributed and Semi-Structured Information. In *Database Semantics - Semantic Issues in Multimedia Systems, IFIP TC2/WG2.6 Eighth Working Conference on Database Semantics (DS-8)*, 351–369. Kluwer, 1999.

[7] D. Fensel, J. Angele, S. Decker, M. Erdmann, H.-P. Schnurr, S. Staab, R. Studer, A. Witt. On2broker: Semantic-based access to information sources at the WWW, 1999. In *Proceedings of the World Conference on the WWW and Internet (WebNet 99)*, 1999.

[8] J. Garfunkel: Web Annotation Technologies. look.boston.ma.us/garf/webdev/annote/software.html

[9] M. Gertz, K. Sattler: A Model and Architecture for Conceptualized Data Annotations. Technical Report, Department of Computer Science, University of California, Davis, 2001.

[10] J. Heflin, J. Hendler: Dynamic Ontologies on the Web. In *Proc. of the 17th National Conference on Artificial Intelligence (AAAI 2000)*, 443–449, AAAI/MIT Press, 2000.

[11] R.M. Heck, S.M. Luebke, C.H. Obermark: A Survey of Web Annotation Systems, www.math.grin.edu/~luebke/Research/Summer1999/survey_paper.html.

[12] S. Koslow, M. Huerta (eds.): *Neuroinformatics: An Overview of the Human Brain Project*. Lawrence Erlbaum Associates, NJ, 1997.

[13] J. Kahan, M.-R. Koivunen, E. P. Hommeaux, R. R. Swick: Annotea: An Open RDF Infrastructure for Shared Web Annotations. In *Proceedings of the 10th International World Wide Web Conference (WWW10)*, 623–632, ACM, 2001.

[14] C.D. Manning, H. Schütze: *Foundations of Statistical Natural Language Processing*. MIT Press, 1999.

[15] braininfo.rprc.washington.edu/mainmenu.html, Neuroscience Division, Regional Primate Research Center, University of Washington.

[16] Neuroinformatics – The Human Brain Project. www.nimh.nih.gov/neuroinformatics/index.cfm.

[17] D.E. Oliver: Synchronization of Diverging Versions of a Controlled Medical Terminology. In *Proceedings of the 1998 AMIA Annual Fall Symposium*, 850–854, 1998.

[18] T. A. Phelps, R. Wilensky: Multivalent Annotations. In *Research and Advanced Technology for Digital Libraries – First European Conference*, 287–303, LNCS 1324, Springer, 1997.

[19] E. Rahm, P.A. Bernstein: A Survey of Approaches to Automatic Schema Matching. *VLDB Journal* 10(4):334–350, 2001.

[20] R.A. Robb: *Biomedical Imaging, Visualization, and Analysis*. Wiley-Liss, 2000.

[21] R. Stevens, P. Baker, S. Bechhofer, G. Ng, A. Jacoby, N.W. Paton, C.A. Goble, A. Brass: TAMBIS: Transparent Access to Multiple Bioinformatics Information Sources. *Bioinformatics* 16(2):184–186, 2000.

[22] R. Stevens, C. Goble, I. Harrocks, S. Bechhofer: Building a Bioinformatics Ontology using OIL. To appear in a special issue of IEEE Information Technology in Biomedicine on Bioinformatics, 2001.

[23] Topic Maps. www.topicmaps.org

[24] UC Davis/UC San Diego Human Brain Project Informatics of the Human and Monkey Brain, neuroscience.ucdavis.edu/HBP.

[25] A. Wong, S. Lou: Medical Image Archive and Retrieval. In *Handbook of Medical Imaging – Processing and Analysis*, Academic Press, 771–783, 2000.

Managing Heterogeneous Ecological Data Using Morpho

Daniel Higgins
Chad Berkley
Matthew B. Jones

{higgins, berkley, jones,}@nceas.ucsb.edu

National Center for Ecological Analysis and Synthesis (NCEAS)
University of California, Santa Barbara

Abstract

Ecological and environmental data cover a wide range of topics, from biodiversity surveys to measurements of trace gas fluxes, and are modeled using a tremendous variety of schemas. We have developed Morpho, a data management application designed to assist researchers in managing this heterogeneous collection of ecological data. Our goal in developing Morpho was to ease the burden of data management on scientists while improving access to and documentation for ecological data.

Morpho allows ecological researchers to describe their data using a comprehensive and flexible metadata specification, and to share their data publicly or to specific collaborators over the Knowledge Network for Biocomplexity (KNB). Morpho's main features include: (1) flexible metadata creation and editing using an XML syntax for metadata exchange; (2) a 'wizard' interface for collecting metadata; (3) automated metadata extraction while importing data; (4) an XML editor that is configurable using multiple XML DTDs; (5) compliance with the Ecological Metadata Language; (6) powerful metadata search on the network or locally; and, (7) comprehensive revision control for data and metadata.

1. Introduction

Ecological and environmental data cover a wide range of topics, from biodiversity surveys to measurements of trace gas fluxes. Information about this diverse data is described using a tremendous variety of schemas. This diversity makes it difficult to create simple, standardized methods for sharing such data. One attempt to help enable the sharing of ecological data is the Knowledge Network for Biocomplexity (KNB) [1].

The KNB is a national network intended to facilitate ecological and environmental research on biocomplexity. It enables the efficient discovery, access, interpretation, integration, and analysis of complex ecological data from a highly distributed set of field stations, laboratories, research sites, and individual researchers. Several software tools have been developed as part of the KNB project. One of these tools, a schema-independent XML database called *Metacat* was described previously [2,3]. This paper describes a companion tool, called *Morpho*. The overall system architecture for the KNB system is shown in Figure 1, with Morpho appearing in the upper left. Morpho allows ecological researchers to describe their data using a comprehensive and flexible metadata specification, and to share their data publicly or to specific collaborators over the KNB, using Metacat as a network based repository for data and metadata. Standard web communication protocols (http) are used for communication between Morpho and Metacat. See [2,3] for additional details.

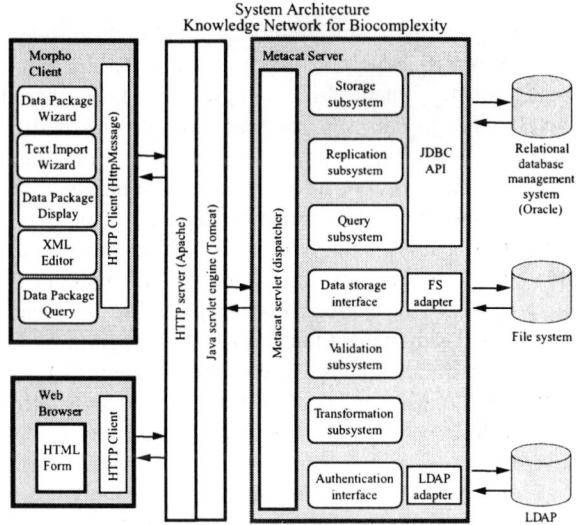

Figure 1 – System architecture for the Knowledge Network for Biocomplexity

Morpho is a cross-platform, desktop application that allows the ecologist to create and organize metadata describing data and to store that information either locally or on a shared network server. Metadata can then be searched and retrieved, either locally or from the network. Metadata is stored in an XML format; more specifically, it uses a specialized XML schema called Ecological

Metadata Language (EML) [4,5]. EML is a metadata standard developed by the ecology discipline and for the ecology discipline. It is based on prior work done by the Ecological Society of America and associated efforts [4]. EML is implemented as a series of XML document types that can by used in a modular and extensible manner to document ecological data [5]. Each EML module is designed to describe one logical part of the total metadata that should be included with any ecological dataset.

Although EML is the primary XML schema used by Morpho, it is recognized that other schemas may be of interest in the future. Morpho is thus designed to allow the use of other schemas with a change of configuration information. This allows new types of metadata to be added to the suite as needed by particular scientific users, and allows us to revise the metadata schemas without having to make programming changes to Morpho. This is critical in light of the rapid changes in metadata schemas that have occurred over the last 5 years.

Morpho operations generally fall into one of two categories: 1) the creation of data packages by the combination of metadata and data; and 2) the retrieval of previously created data packages. These are described in more detail in the following sections.

2 Describing Ecological Datasets with Ecological Metadata Language (EML)

Ecological data are traditionally stored in relational databases with schemas that are specific to the particular data being managed. Integrating the incredibly heterogeneous data sources found in ecology is difficult specifically because a universal schema for all ecological data is infeasible. We have taken an alternative approach by standardizing the metadata that describe the data instead of standardizing the data schemas themselves. In so doing, researchers are allowed to maintain their heterogeneous data models while providing the semantic and technical metadata needed for the integration of their data with other models. Thus, the Ecological Metadata Language (EML) is the result of a grassroots effort to standardize ecological metadata across the ecological community.

EML itself is divided into several modules, each representing one logical part of the total metadata that should be included with any dataset. The modules are connected through a flexible RDF-like [6] framework that allows for the arbitrary association of one module with another. This allows a single module to be reused in situations where it is appropriate, thus reducing the metadata input burden on the scientist. It also allows the data to be linked to the metadata in a cohesive unit that we call a *data package*. As used here, a data package is defined as a collection of data and related metadata. This is useful for software applications such as Morpho that need some way of associating the metadata with actual data for analytical and organizational purposes.

EML-Module	Description
eml-access	Access and authentication information.
eml-attribute	List of attributes within a table with relevant attribute level metadata.
eml-contraint	Relational constraint metadata. Used to describe a relational model.
eml-dataset	General dataset level metadata. Data collectors, originators, contacts, keywords, etc.
eml-entity	Entity level metadata. Table access information.
eml-literature	Citation metadata.
eml-physical	Physical file information. Such as format, encoding, compression, etc.
eml-project	Project level metadata. This file could describe more than one dataset.
eml-protocol	Research protocol metadata. Information about the research processes involved in the collection of this dataset.
eml-software	Software metadata. Any analytical engines, custom software, etc. used in the creation or processing of this dataset.

Table 1 – EML modules

The data package framework is implemented within the eml-dataset module (which is considered the head of each data package). The framework consists of a listing of triples that link two modules (or a module and a data file) and provide a definition of the linking. An example triple in XML format is:

```
<triple>
  <subject>DataFile1</subject>
  <relationship>isDataFileFor</relationship>
  <object>eml-datasetFile1</object>
</triple>
```

In the above example DataFile1 is linked to eml-datasetFile1 and the relationship shows that the subject (DataFile1) of the triple isDataFileFor the object eml-datasetFile1. In general, triples can be interpreted using the phrase "subject has some relationship to object".

Such triples can be used to create a data package with the relationships like those indicated in the diagram shown in Figure 2. This example shows how EML modules can be combined in a manner customized for the data being described. The data package concept can thus be used for widely varying types of data.

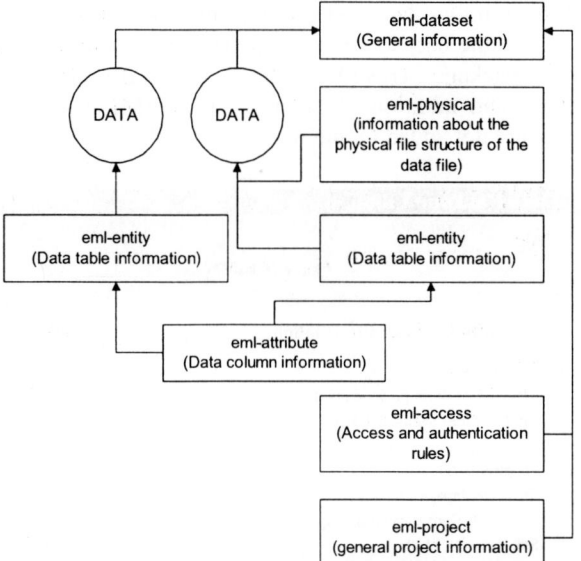

Figure 2 - A diagram of a data package that contains two data files. Each data file contains one table that has the same attributes as the other. Note that the attribute metadata document is reused instead of having redundant metadata. The eml-dataset module is the "head" of the data package.

Flexibility is emphasized because EML must be able to describe such a wide variety of ecological data. EML has the ability to represent data formats from tab-delimited text to Oracle relational databases. Morpho makes use of the flexibility of EML through its customizable interfaces and its extensible design.

3. Creating Data Packages with Morpho

Data Package Wizard

Morpho uses a wizard interface to help users construct new data packages. One of the initial screens in the Package Wizard is reproduced below in Figure 3. The process for creating a new data package is form-based. The user enters values for a set of fields. The fields are based on a subset of EML Required elements are shown on the forms (and indicated in red) and a few of the more common optional elements are also included.

The Data Package Wizard's content and layout are determined by text-based configuration files that can be edited independently of the Morpho source code. XML files are used to describe the content and layout of the forms. This configuration information is read at run time to dynamically create the wizard screens. This flexibility makes it easier to handle alternative metadata schemas that may well be needed due to the diversity of ecological data. This also allows scientists to customize the metadata acquisition process to meet their site based needs, which gives the scientists the freedom they need to describe their

heterogeneous datasets.

The metadata created by the wizard is relatively simple. Once a package is created, however, the metadata modules in the data package can be edited and expanded using the XML editor described in a later section of this paper.

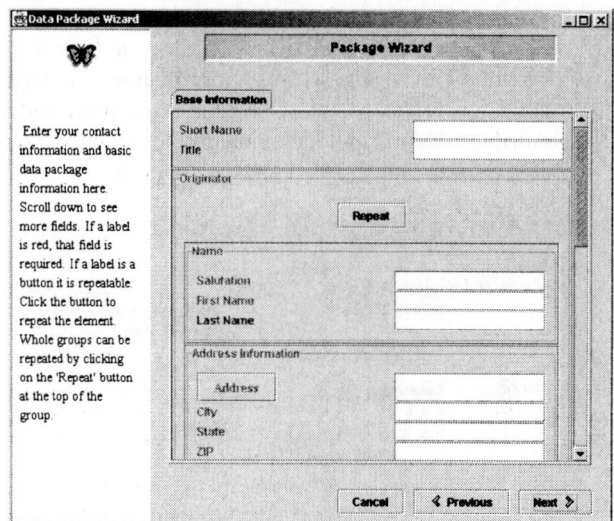

Figure 3 - Typical screen from the Package Wizard for creating a new data package

Automated Metadata extraction

Ecological data is often stored in text-based tables. Each column in the table represents a variable or attribute in the data. Metadata describing each of the attributes (i.e. each of the columns) is needed. In EML, this metadata is stored in the eml-attribute module and there are a number of metadata elements that are available for describing each attribute. Entering the metadata for every attribute in a large data table is thus necessary but can be quite tedious.

Data tables often have some embedded metadata that can be automatically extracted. For example, often the first line in a text-based table contains column names that are at least brief descriptions of the attribute. Attribute values can also be automatically scanned by heuristic algorithms in order to intelligently guess at the type of information in an attribute (e.g. is it a number, a date, or just text).

The Text Import Wizard parses a data file and extracts metadata. The Text Import Wizard takes a data set in text format (either from a file or pasted from the clipboard), determines the delimiters between columns of data, and creates a table as shown in Figure 4. Morpho scans the data associated with each attribute and guesses data types. It also displays unique items in the column, and minimum, maximum, and average values for numeric columns. The user can simply select each column, see the metadata created by Morpho for that column, and then correct/add information as needed. Some of the information (e.g. a list of unique items in each column) gathered by the Text Import Wizard also helps the user find mistakes in the input data.

When the Text Import Wizard is finished, it will create several EML documents containing metadata about the data table. They include: an eml-physical document, which has basic information about the data stream such as size, delimiters, and character encoding; a table-entity document with table level information including temporal, geographic, and taxonomic coverages; and an eml-attribute document which includes all the information about each of the attributes (i.e. each column).

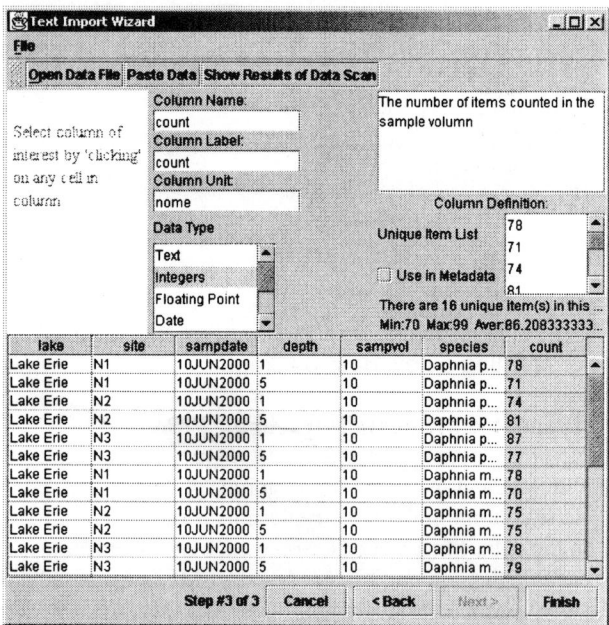

Figure 4 - One of the screens displayed by the Text Import Wizard for enabling the extraction of metadata from data

Data Package Summary

The Data Package and Text Import wizards in Morpho attempt to hide much of the complexity from users by only presenting a subset of all the possible metadata information that is contained in EML. However, EML is designed to be comprehensive with respect to the suite of metadata elements needed for long-term data preservation. We also recognize the burden involved in providing this information. Because of this, the simple wizard interfaces used to initially create a data package do not require the user to consider all of these possible metadata options. However, various other EML modules encoding varying levels of complexity can be added to a data package later by editing or adding various metadata elements in the data package.

The Data Package Editor (Figure 5) is the main

interface into a data package once the user has created it. It allows the user to view the general contact information in the package (as well as edit it) and it also allows the editing and addition of other types of EML documents that were not added by the wizard.

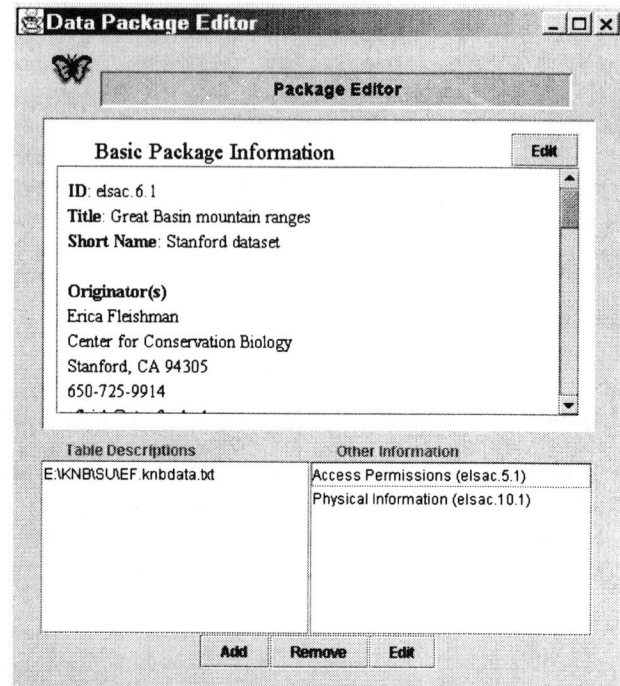

Figure 5 - Data Package Editor. The summary information is shown at the top. Any data tables associated with the data package are shown in the list on the lower left (only one in this example). Other metadata information, including an access control list, is summarized in the list on the right.

When the user edits a table description, Morpho displays the Table Editor (Figure 6). The Table Editor allows the user to edit the basic table information as well as view and edit any attributes associated with the table.

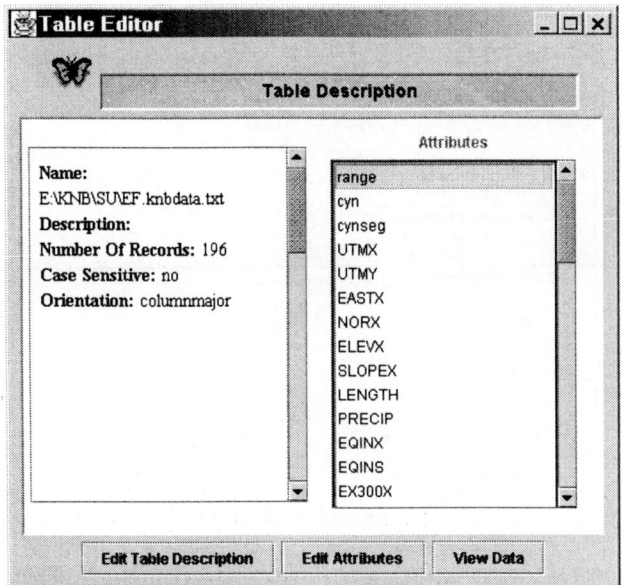

Figure 6 - Table entity and attribute summary display

When adding new types of EML metadata to the package, another wizard is executed to help the user through the process (Figure 7). The wizard allows the user to select the type of metadata that he wishes to insert, then it opens the XML editor (Figure 8) for the user to add the information.

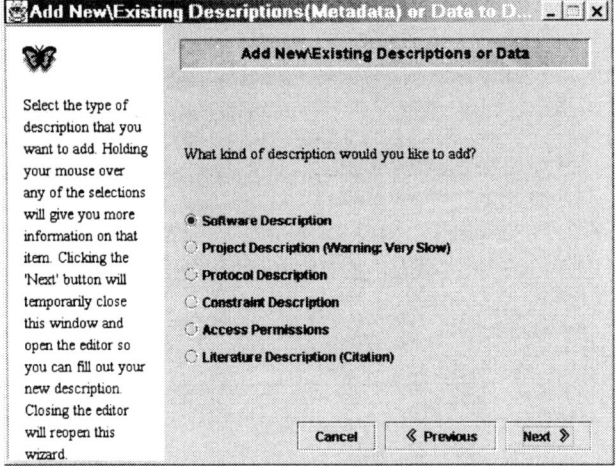

Figure 7 - Example screen showing other types on metadata that can be added to a data package

XML Editor

Morpho includes a comprehensive XML editor (Figure 8) tied to EML by the schemas contained in a set of EML DTDs. A tree showing the hierarchical structure of the XML document is shown on the left, while a form-representation is shown on the right., Metadata usage guidelines and term definitions are also displayed with each node.

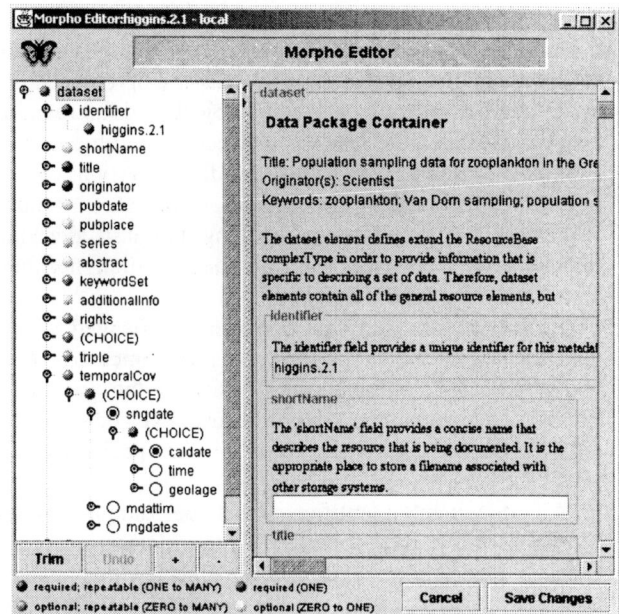

Figure 8 - Sample Morpho XML editor screen. An eml-dataset document is displayed.

The Morpho XML editor differs from many XML editors in that its default behavior is to display not just the content of the specific XML instance being edited, but also all the possible nodes that are defined by the schema (i.e. the DTD) for the document type. This display of possible nodes gives the user a quick view of all the metadata elements that they could enter, if appropriate.

The editor works by creating a tree from the nodes in the instance document. The DTD declared in that document is then used to create a second tree with all the possible nodes of the schema. These two trees are then merged (along with the node descriptions) to create the information displayed by the editor. The icon attached to each node in the tree view indicates cardinality and node requirements, as defined by the DTD.

Optional nodes (CHOICE nodes in the DTD) appear as the word "CHOICE" followed by one or more radio buttons. The user can choose one of several options by simply selecting the option of interest. Only the selected option is saved when the editor is closed. Similarly, all optional elements that do not contain any text are stripped from the output document created by the editor.

Although the sample screen shown here is a specific type of EML document, the editor is designed to handle any type of XML document. All it requires is a DTD defining the allowed content. Also, special widgets can be dynamically added for displaying/editing the content of certain nodes.

Revision control for data and metadata

Each document handled by Morpho has a unique identifier associated with it. This identifier is made up of three components separated by a delimiter ("."): a prefix, a serial number, and a revision number. The prefix is used to create human-understandable clusters of objects, the serial number indicates a specific object, and the revision number indicates the specific version of the object. For example, "higgins.1234.1" is the first revision of the object "higgins.1234". Morpho generates unique identifiers for the user by tracking the last identifier number used and synchronizing identification numbers with the KNB Metacat network.

Each time any metadata document is changed in any way, the last integer in the identifier is incremented. Thus, "higgins.1234.4" is the fourth revision of the document. Previous revisions of a document are never deleted, but only the most recent is shown in any data package result set.

All document identifiers must be unique. When Morpho sends a new document to the Metacat server, a check is made to detemine if a document with the same identifier already exists on the server. If it does, the document with the conflicting identifier cannot be sent to the server. Morpho has to increment the id and try submitting the document again with the new identifier.

4. Metadata Search

The search facility in Morpho (Figure 9) is designed to be convenient and intuitive to ecological scientists. Its default layout and settings reflect the types of queries we expect ecologists will commonly want to perform, while also allowing more comprehensive boolean queries to be constructed. Morpho can perform the same searches on data packages stored on the local computer and on the KNB Metacat network, allowing ecologists to easily manage their own data and locate data that is relevant to them from the KNB network.

On the first tab, users can search a variety of "Subject" related metadata such as "Title" or "Keywords" and specify the type of match (e.g. "contains" or "equals"). The "All" check box enables a complete search of all metadata elements. This option provides a great deal of flexibility to the system since it can be applied to any XML-based metadata schema, not just EML. Search terms can be combined using boolean operators for more advanced searches.

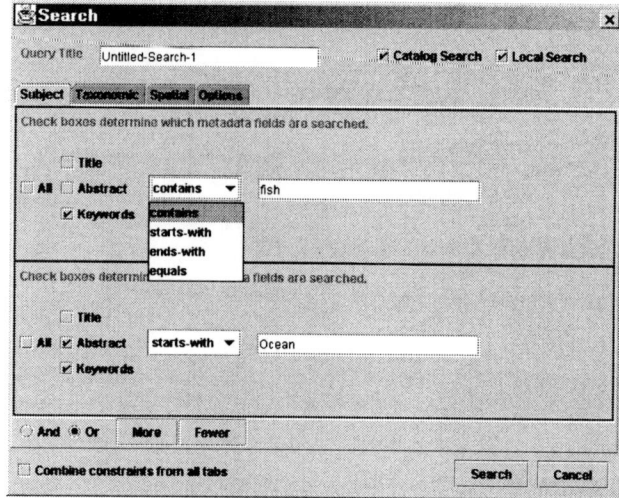

Figure 9 - Morpho search dialog (subject searches)

Other search types are also available as indicated by the tabs at the top of Figure 9. For example, the Taxonomic tab (Figure 10) allows users to constrain matches to data that relates to particular biological taxa and their synonyms. The synonyms are dynamically determined by querying the Integrated Taxonomic Information System (ITIS) [10]. This is a critical feature for ecologists because of the central role that taxonomy plays in ecological research. Search constraints across multiple tabs can be combined using controls at the bottom of the screen.

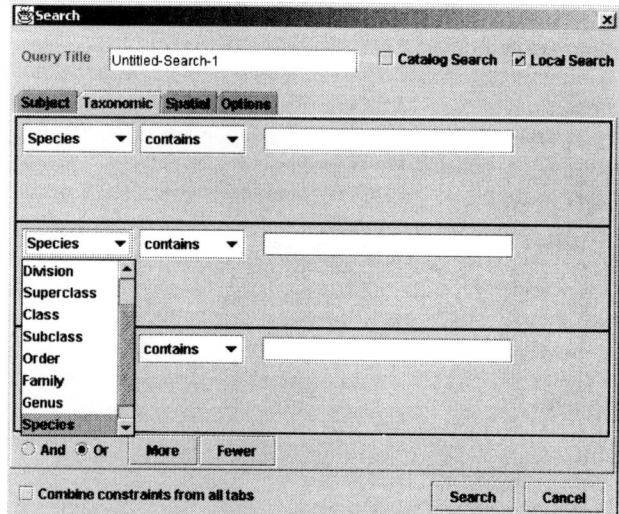

Figure 10 - Morpho search dialog (taxonomic searches)

The results of a search are displayed in a table that lists matching data packages (Figure 11). Each data package may be located on the local computer, on the KNB Metacat network, or both; the storage location is indicated by the icon in the left most column of the table. The data package in each row can then be examined, downloaded, and edited, depending on the access permissions available to the user.

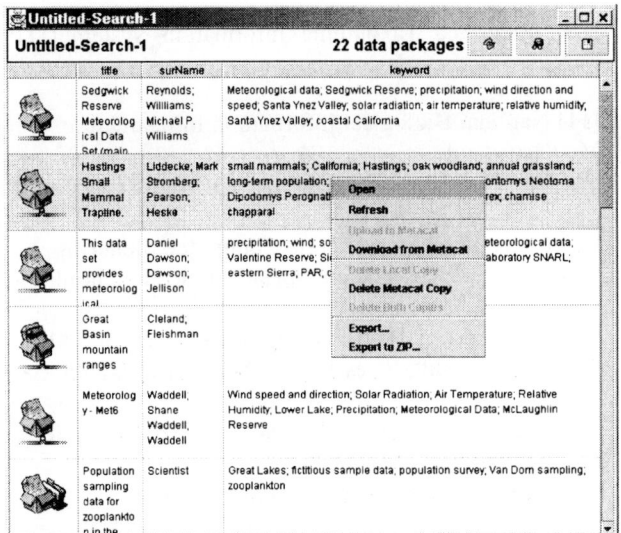

Figure 11 - Search results table

The query system is implemented using path-based searches against the XML document tree. A document matches a query request if it contains a node at the specified path in the tree and the text in that node matches the search criteria. We use XPath [7] to specify the search criteria, and an XML formatted query syntax that was described in [2,3] for combining these various path expressions using boolean operators. Locally, Morpho actually carries out the specified query using XPath and the Xalan XSLT processor [8], while on the Metacat server the search criteria are translated to SQL and carried out in a relational database (see [2,3]). We expect to transition to the XQuery specification as it matures from a working draft into a W3C recommendation [9].

An example of an XPath expression is:

```
/eml-dataset/originator/individualName
/surName[contains(text(),"Higgins")]
```

Documents would match this expression if they contain a surName node at the indicated path in the XML tree, and if the text of that node contained the string "Higgins".

5. Summary and Conclusions

Morpho is a software tool designed to help ecologists create and locate collections of data and metadata. However, Morpho is but one part of the Knowledge Network for Biocomplexity effort, which is intended to provide a world-wide network of archived and searchable ecological data.

Because ecological data is widely diverse, there is no single format that can be used by all ecologists. Thus we have taken the approach of characterizing any dataset using a collection of metadata documents in XML format into a single cohesive unit we call a data package

It has also been recognized that a single set of metadata elements is not appropriate for all types of ecological data. Data packages are thus not all identically organized. Morpho was designed so that it could help the ecologist create data packages appropriate for the data being stored and locate related packages, even if the packages are organized in different ways.

Morpho was therefore built to be quite flexible and to minimize assumptions about metadata organization and content. As a result, many of its features can be set at run-time with user modifiable configuration files. Morpho is designed to use specific metadata schemas defined in Ecological Metadata Language (EML). However, a change in EML or the use of some other schema is readily accommodated.

There are, of course, other software tools for creating and searching metadata. Several older tools can be found through the NBII or FGDC web sites [11],[12]. Some newer tools that are similar to Morpho are described in references [13] and [14]. Morpho differs from these other tools in several ways. First of all, many metadata tools are strongly tied to specific metadata schemas like Dublin Core or the FGDC standard for geospatial metadata. Morpho is tied to EML, but only weakly, so that details of the metadata created and queried can be changed with a few changes in configuration files. Secondly, Morpho combines metadata creation and query functions into one application and provides both stand-alone and network functionality.

A copy of Morpho can be downloaded from http://knb.ecoinformatics.org.

6. Acknowledgments

The authors wish to acknowledge the contributions of numerous colleagues without whom Morpho would not exist in its current form, including M. Schildhauer, C. Jones, S. Andelman, E. Fegraus, and C. Bowles.

This material is based upon work supported by the National Science Foundation under Grants No. DBI99-04777 and No. DEB99-80154. Any opinions, findings and conclusions or recommendations expressed in this material are those of the author(s) and do not necessarily reflect the views of the National Science Foundation (NSF).

This work is hosted by the National Center for Ecological Analysis and Synthesis, a Center funded by NSF (Grant #DEB-0072909), the University of California, and UC Santa Barbara.

7. References

[1] The Knowledge Network for Biocomplexity, 2001, http://knb.ecoinformatics.org/

[2] Chad Berkley, Matthew Jones, Jivka Bojilova, Daniel Higgins, "Metacat: a Schema-Independent XML Database System", Thirteenth International Conference on Scientific and Statistical Database Management, ISBN 0-7695-1220-6, IEEE Computer Society Order Number PR01218.

[3] Jones, Matthew B., C. Berkley, J. Bojilova, M. Schildhauer, 2001. "Managing Scientific Metadata", *IEEE Internet Computing* 5 (5): 59-68.

[4] W. K. Michener et al.,"Non-Geospatial Metadata for the Ecological Sciences", *Ecological Applications*, vol. 7, 1997, pp 330-342.

[5] Ecological Metadata Language (EML), http://knb.ecoinformatics.org/software/eml/

[6] E. Miller, R. Swick, D. Brickley, Resource Description Framework (RDF), http://www.w3.org/RDF

[7] J. Clark, S. DeRose, XML Path Language (XPath), http://www.w3.org/TR/xpath.

[8] Apache XML Xalan-Java, http://xml.apache.org/xalan-j

[9] M. Marchioro, XML Query, http://www.w3.org/XML/Query

[10] Integrated Taxonomic Information System (ITIS), http://www.itis.usda.gov

[11] National Biological Information Infrastructure, Metadata Tools, http://www.nbii.gov/datainfo/metadata/tools/index.html

[12] Federal Geographic Data Committee:Metadata Tools, http://www.fgdc.gov/metadata/metatool.html

[13] Xanthoria: A Distributed Query System for XML Encoded Data, http://ces.asu.edu/bdi/subjects/xanthoria/

[14] Xylographa : An XML Schema-based Wizard for Editing Ecological Metadata, http://ces.asu.edu/bdi/subjects/xylographa/

Multiple Overlapping Classifications: Issues and Solutions

Cédric Raguenaud, Jessie Kennedy
School of Computing
Napier University
10 Colinton Road, Edinburgh, EH10 5DT
{c.raguenaud, j.kennedy}@napier.ac.uk

Abstract

This paper discusses issues and solutions for supporting multiple overlapping classifications in database systems. These classifications are commonly found in science, although they are often ignored in computing applications for scientific data, and inappropriate solutions adopted as their replacement. Known database models and classification techniques offer some degree of support for multiple overlapping classifications, but do not fully support the basic features we have identified as necessary: trees/graphs, traceability, semantics of classifications, independence of classification and data, and identity of classifications.

The approach to the problem adopted by the Prometheus project, based on an extended object-oriented database model and the independence of classification schemes from classified data, is presented and discussed.

1. Introduction

Classification is a widespread concept that helps categorise, and therefore simplify data or objects in order to facilitate their understanding and manipulation. Through representing the relationships between classified things, classifications may provide new insights into the things being classified, e.g. discovering that two groups thought to be independent are in fact related in some way or deducing from relationships between groups that they have similar properties. They also allow automatic reasoning, e.g. propagation of attributes in computing models [32] [15] or support user interactions, e.g. simplification of searches.

Examples of familiar classifications include library catalogues where books are placed into categories (e.g. genre) in order to ease access and simplify their management. Medical classification mechanisms, such as the International Classification of Diseases (ICD), which catalogues and relates diseases in order to make prevention, diagnosis, and cure possible. Classification of living organisms as found in for example plant and animal taxonomy and virology.

In order to model and support classifications, we need to understand how they work. Two aspects of classifications can be distinguished: the way objects are grouped into classes of equivalence, and the way classes of equivalence relate to each other and form hierarchies.

Classes of equivalence

In classifications, collections of objects are gathered and classes of equivalence are created within the collection (Figure 1 a-d) via an equivalence function. The equivalence function results in objects being *instances of their class(es)* and not of others.

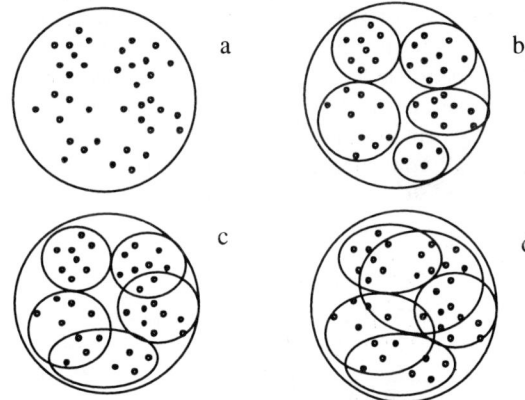

Figure 1: Creation of classes of equivalence

The equivalence function that partitions an application domain dictates how the set of objects to be classified will be clustered. The application of that function may result in the creation of distinct groupings without overlap. This is the case when the equivalence function offers exclusive partitioning. For example, eye colour offers an exclusively partitioning function: people normally have only one eye colour. Applying this equivalence function to a group of objects will result in several distinct sets as shown in Figure 1-b where each set represents a class of equivalence containing people with a particular eye colour.

In other cases, the partitioning function doesn't provide a clean clustering and objects may belong to several classes of equivalence. For example, the attribute genre of a library classification may have values "fiction", "crime", "drama", and "historical ". This equivalence function may lead to non-exclusive partitioning of the domain because its values are not exclusive. For example, a novel may be of genre "crime" and "historical" at the same time as shown in Figure 1-c. Therefore a book may belong to more than one category simultaneously. It may be argued that this shows a bad choice of equivalence function, but the fact is that these classifications do exist, are seen to be useful and therefore cannot be ignored.

The complexity of the equivalence function may also lead to non-exclusive partitioning of the domain. When the equivalence function is composite, i.e. it is made of several equivalence functions applied simultaneously (the classification is said to be polythetic, it is monothetic when only one function is used), objects to which the function is applied may fulfil the requirements of several classes simultaneously. For example, a book partitioning that takes into account the genre, the period, and the place of origin of the author of books will lead to a partitioning where a book may appear in, say, XXth century writing, crime, and Scottish writing (Figure 1-d).

Class hierarchies

In addition to gathering objects, classes of equivalence can be related to each other via the classification scheme. The mechanism underlying all classification schemes is membership. Indeed, all classification hierarchies imply that lower level classes are members of higher classes. They also imply that objects that are instances of lower classes of equivalence are also, through transitivity of the classification function, instances of higher classes of equivalence (Figure 2).

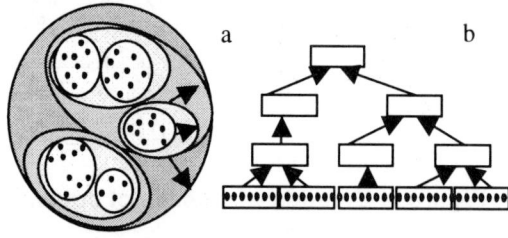

Figure 2: Two views of a classification hierarchy

We can identify at least three main kinds of classification semantics (the classification function): subsumption classifications (is-a), decomposition classifications (part-of), and similarity classifications.

Is-a classifications are based on the concept of specialisation/generalisation. Classes of equivalence are organised into a directed acyclic graph (DAG) where their

semantics are specialised in a top down fashion. An example of an is-a classification is the concept of object class hierarchy in computing, where objects are grouped into classes according to their structure and classes arranged into a specialisation/generalisation hierarchy based on e.g. attribute and structure. Another example is library information management, where the classification scheme may be based on genre. Books are placed into categories as shown in Figure 3. The generalisation/specialisation hierarchy places most specific classes at the bottom and more general ones at the top.

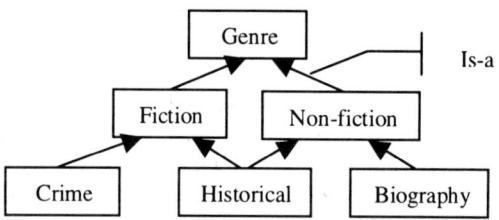

Figure 3: Is-a classification

Part-of classifications are based on the concept of decomposition of objects into smaller objects. Each object is related to its parent (whole) via a part-of relationship. These classifications form an abstract to concrete classification where higher concepts are transitively made of lower concepts. The semantics of part-of relationships have been extensively studied and include functional relationships, topological relationships, and homeomeric relationships [29]. An example of part-of classification is the International Classification of Diseases (ICD), where one classification is based on topology, i.e. it decomposes the human body into sub-parts in order to describe the illnesses that may affect each of them (Figure 4).

Similarity classifications, as other classifications, are based on the concept of membership. The classification function clusters classes of equivalence by similarity. They are related to is-a classifications, that also group classes by similarity (attribute structure) but do not imply the subsumption rules is-a classifications exhibit. An example of similarity classification is plant taxonomy where taxonomists create classifications based on similarity of specimens or *taxa*[1] (the classes of equivalence) according to phenotypic descriptions. Figure 5 shows an extract from a taxonomic classification built on similarity. The classification shows that "Caucalideae" and "Coriandreae" are similar in some respect and therefore belong to the same higher group, "Multiiudatae".

---

[1] Group of specimens or other *taxa*

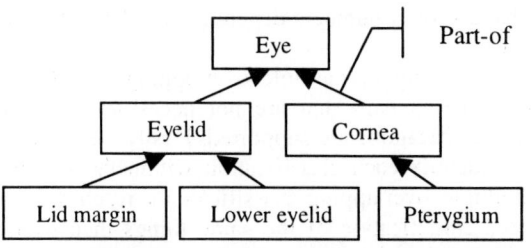

Figure 4: part-of classification

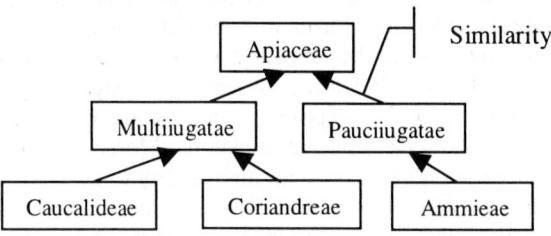

Figure 5: similarity classification

Other classification semantics are possible, as long as the classification function is transitive and a partitioning function for the domain is available.

Plant taxonomy

Plant classification is called a taxonomy because classes appear only once and in one place in each individual classification. Figure 3 shows a non-taxonomic classification because the class "historical" appears both in classes "Fiction" and "Non-fiction".

Plant taxonomy classifications exhibit additional peculiar properties. Plant taxonomy can be called a population-based classification mechanism. Theoretically [18], taxonomic classifications consist of one-level classifications where specimens are put into *piles* (classes) according to their characteristics. The classes (*taxa*) are in addition objects in their own right that are published when a name is assigned to them by application of nomenclature rules [18].

However, for practical reasons, i.e. it is hard to handle thousands or more specimens at once, these one-level classifications are merged into n-level classifications where elements of each level are *instances of* elements of the next higher level, i.e. *taxa* are made members of higher *taxa*. Specimens are members of all higher *taxa* by transitivity of the inclusion relationship.

This leads to classes that are both classes of equivalence that gather objects that fulfil certain requirements (e.g. they look similar), and act as surrogate objects for the objects they recursively contain, therefore are in turn classified.

Multiple classifications

In some application domains, revision of classifications is regular or common. For example, the ICD is revised by the World Health Organization every 10 years approximately and updated every year for minor changes. Each revision creates a new classification of diseases that replaces older ones, and each new version shows the history of classes (e.g. two new classes may correspond to a single class in an older version of the ICD).

In plant taxonomy, revisions are common as they are made when new data, new techniques, or new opinions appear and may lead to a different understanding of the world (e.g. DNA sequencing in the last few years). Unlike for the ICD, new revisions do not replace older ones therefore all classifications ever published are valid. These classifications also form the basis of new classifications through revision. This leads to a large number of classifications (hierarchies of *taxa*) of the same specimens or *taxa* as overlapping groups of specimens (i.e. groups sharing specimens). Figure 6 shows an example of multiple classification in plant taxonomy. The first classification (top), Berchtold & Presl 1820, is highlighted and all the classes that appear in that classification are highlighted in two subsequent classifications, Koch 1824 and De Candolle 1830. It is apparent that groups are moved around and are classified differently over time.

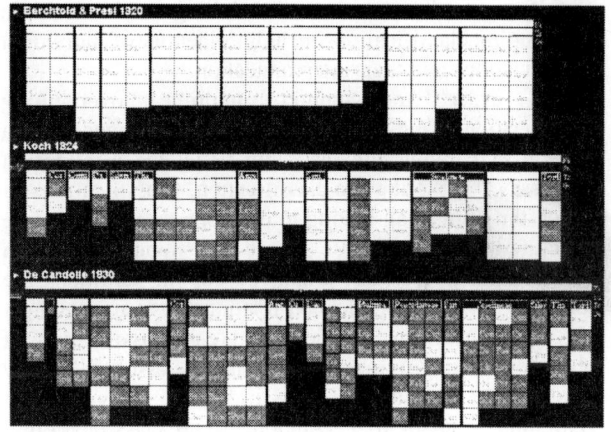

Figure 6: Multiple classifications

As we have seen, classifications are not always the straightforward unambiguous tree structures shown in Figure 2. We have seen that the objects that are classified may appear in many classes of equivalence when the partition function allows it. We have also seen that classes of equivalence can be related in several ways (one of which is is-a) and may appear in several higher classes (non-taxonomic classifications). We have also seen that some application domains generate several classifications overlapping in terms of classified objects and classes (e.g. plant taxonomy, ICD).

Because of the complexity of dealing with these multiple overlapping classifications, they are generally ignored in both biology (e.g. in plant taxonomy where consensus classifications are forced upon taxonomists

[36]) and computing (e.g. where classifications are declared suitable for biological classifications [28] when they only handle single classifications).

This paper first discusses the issues associated with supporting multiple overlapping classifications such as those found in plant taxonomy [36]. Then the ability of the major existing families of database models to handle multiple overlapping classifications is discussed. Our approach (Prometheus) to dealing with the issues is presented in section 4, and we conclude in section 5. A full survey and specification of Prometheus can be found in [37].

2. Issues

There are several issues associated with the handling of multiple overlapping classifications, which were identified during work on the Prometheus project [35]. Although raised in the context of plant taxonomy they apply to classification schemes in general. The first issue is the handling of single classifications, which includes the semantic of the classification mechanism, traceability of decisions and independence of things from their classification. The second is dealing with multiple overlapping classifications including identity of individual classifications and their interconnection.

In order to support multiple overlapping classifications, single classifications must be handled in a fashion that allows the recording of all the information necessary to describe the classification. Classifications in which the things can belong to only one category are tree structured, whereas those where the things appear in several categories are graph based. In addition the semantics of the classification relationships vary amongst classifications: is-a classifications, (e.g. classification of object-oriented programming language classes or library catalogues); similarity classifications (e.g. plant taxonomy); part-of (e.g. component-part classifications); other types of classifications, where the relationships between classes can be anything, e.g. a path that described costs of medical procedures [7], or ontologies. Therefore the selection of the appropriate classification representation for a given domain is important and the implementation of the chosen representation is not trivial, as will be seen in section 3. As a consequence of classification, traceability becomes fundamental. Traceability allows the explanation, in the data, of the motivation for a particular classification. For example, a plant taxonomist should be able to explain why a particular *taxon* has been placed in another. The ICD assigns unique numbers to diseases and operations based on their path in the classification (each branch of the classification carries a number). If classes appear in several placed in classifications, it is not sensible to make that unique number part of the class definition (one class

may have several numbers depending on the path used to reach it).

The handling of multiple overlapping classifications requires mechanisms that are not necessary when only single classifications are supported. Firstly we need to be able to identify each classification within the system. If the multiple overlapping classifications result from the repeated classification of the same things then overlaps occur in the classifications and it is necessary to be able to identify those overlaps.

An important feature of classification is that things should be independent from their classification. In the real world anything can be classified, not only things deemed classifiable. It would make no sense to design things so that they can be classified and need to maintain information about their classifications. Moreover, mixing the description of things with their ability to be classified would increase their complexity and reduce reuse and maintainability. This situation is exacerbated when things are multiply classified. It is therefore important that the objects that are classified do not participate directly in the classification process. This makes the management of basic data and the activity of classifying independent processes.

In summary, the requirements of a computer system to support multiple overlapping classifications are:
- support for trees/graphs
- support for semantic relationships
- support for traceability
- identification of distinct overlapping classifications
- orthogonality of classification and data

3. Supporting classifications with existing technology

There are many ways classifications could be handled in the main families of database models, however from the previous section it is clear that there are several requirements of the database to handle the range of classifications described above. In summary, for the simplest scenario of a single classification with no specific semantics, no overlap in categories nor traceability, they need to be able to represent and manipulate basic trees. For the more complex scenario of multiple overlapping classifications with specific semantics where traceability is required, they will require to represent and manipulate graphs with differing types of relationship with attributes to record their raison d'être.

This section examines the ability of relational, object-oriented, graph-based, and extended object-oriented models to support these requirements and discusses some specific techniques[2].

---

[2] Full details can be found in [37].

3.1. Representing single classifications

3.1.1. Trees and Graphs

The way classifications (trees or graphs) can be handled depends greatly on the structure of the database model. Relational models in general do not represent classifications easily, as they were originally designed for the manipulation of simple, flat data [10]. Extensions to the original relational model (extended model, in Third Manifesto [13]; object-relational models, e.g. Postgres [44], Oracle [30]) offer additional features such as extensible types or nesting that can be of use to describe more complex information. These models could handle graphs as relations, however these relations would have to play both the roles of nodes and edges in graphs, therefore their manipulation would need to be handled by user applications. Figure 7 shows a relation "Person" that could be related to its parents via the relation "Parents" playing the role of a relationship in a genealogy classification.

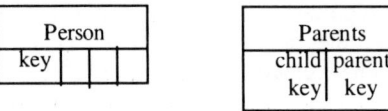

Figure 7: relation as relationship

Nested models could represent graphs via nesting, but this representation would still be simplistic as more complex graphs, e.g. weighted graphs, could not be captured. Finally, as all these options are too simple to accurately represent trees/graphs, they lead to complex processes/manipulations and possibly integrity constraint problems. These complex processes would have a negative impact on the efficiency of the overall system.

Object-oriented databases can support the definition of directed graphs (cyclic or not) using objects and references. However, only the simplest graphs (e.g. not weighted graphs) can be represented, as edges show only the existence of a link between two nodes (as a reference), without any additional information, which would be necessary for e.g. weighted graphs. An alternative approach would be the representation of graph edges by normal objects that user applications would recognise as edges. Figure 8 shows a class diagram where a class is used as relationship (with weight) to relate books to their class (category).

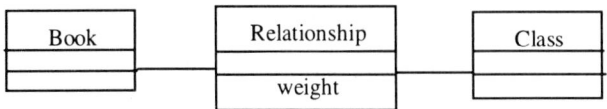

Figure 8: object as relationship

This has the advantage of providing a means to capture complex graphs such as weighted graphs. However, this would be limited in the sense that edge objects would not be recognised as relationships or references by the database system. Therefore, the insertion of an additional level of indirection would make user applications more complex and may lead to integrity problems (updating an object edge is more complicated than updating a reference). In addition, writing queries would be made harder by the additional level of indirection and the necessity to select these edge objects explicitly if they are required in the query result. Another approach could be to use the classification offered by some object-oriented models: classification of types by distinguishing types from classes [27] [22] [20]. However, changes to the classes may lead to important class reorganisation problems such as schema evolution (e.g. [4] [12]). They would also lead to very large schemas that could become unmanageable (a schema for a plant taxonomy flora would contain hundreds of thousands of classes).

In graph-based models, everything is represented by sets of nodes and edges that represent interaction between nodes. These models inherently support the description of trees/graphs, with various degrees of information: node-based models (e.g. TSIMMIS [9], Lore [25], GOOD [17]) have a weak representation of edges, and edge-based models (BDS95 [8]) limit the possibility to distinguish objects and interactions between objects. Models that support both kinds of features extensively (PROGRES [43], Telos [26], ConceptBase [21], [47], [46], Gram [3]), and in particular those that support nesting (Hyperlog [33], [49]), provide more freedom to choose the representation of information. By explicitly modelling nodes and edges of a graph, they allow the representation of any kind of graph, including weighted graphs or cyclic graphs.

Extended object-oriented models can represent graph structures as they are based on object-oriented models with first-class relationships. Their degree of support for graph structures varies: some support the definition of explicit but simple graphs (e.g. SORAC [16]); others support the definition of more complex graphs such as weighted graphs (e.g. GraphDB [19], ADAM [15], Albano [2]). In all cases, these graphs explicitly represent both nodes (objects for GraphDB and SORAC, or unary collections for OMS and Albano) and edges (specific relationship objects for GraphDB and SORAC, or binary/n-ary collections for OMS and Albano).

3.1.2. Semantics

Semantics of classifications is also an issue that is not handled well by most models. Relational models propose relations as the basic entity. These relations do not represent is-a, similarity, or part-of relationships in a system understandable manner, therefore relational models would be unable to capture these kinds of classifications naturally. All classifications in a relational database would be generic classifications, which can lead to problems interpreting their semantics (once again captured by user applications).

Object-oriented models could only model two kinds of classifications: is-a (inheritance) and another generic kind of classification (reference). Is-a and generic classifications may not be appropriate to all classification schemes (e.g. part-of classifications as in the ICD), and no other specific kind of classification (e.g. part-of) may be defined.

Graph-based models generally only support one kind of relationship between nodes. Only a few support relationship classes and is-a classifications (e.g. Hyperlog [33]). If relationships are sub-classed, then it is possible to create classifications that are of the required type. The limitation of graph-based models is that they do not interpret the semantics of relationships. Therefore it is possible to describe generic classifications, but the system would not be able to interpret their meaning. For example is-a edges may be created in models that support the extension of edge types, but they would only be called "is-a" edges and inheritance rules would not be enforced.

By defining different kinds of relationships and using them to capture classifications, extended object-oriented models make it possible to define classifications that are not is-a or part-of classifications. This can be done by creating new kinds of relationships, with their semantics (e.g. as constraints or rules), and linking objects together. However, for the models that do not support semantic relationships, similarity and part-of classifications are impossible.

3.1.3. Traceability

Traceability offers the ability to record the motivation behind the building of classifications. Depending on the approach chosen for representing classifcations, traceability may be supported by relational models. For example, if relations are used to represent edges (instead of nodes), then these edges can contain information that can capture classification motivation, therefore traceability. If nesting is used, then no traceability information can be recorded.

Traceability is an unresolved issue in object-oriented and graph-based models. Indeed, if it is decided that traceability should be part of edges/relationships, only models that allow the definition of edges/relationships with weights may provide a solution. However, as the previous section explains, this approach is not practical with object-oriented models (where references cannot contain values, and normal objects used as relationships introduce problems) and graph-based models offer relationships of a too simple kind to handle attributes.

Some of the extended object-oriented models support attributes on relationships (Albano, GraphDB, ADAM), others allow the combinations of relationships as relationships (attributes) of relationships (e.g. OMS), therefore these classification relationships could also record the motivations for classifications as attributes of relationships forming a graph.

3.2. Representing multiple overlapping classifications

Identity is the main issue associated with multiple overlapping classifications. Indeed, the fact that several classifications may share elements means that it is important to be able to make a distinction between all the classifications involved. One feature offered by most relational systems, views [11], may be of interest as they would allow the filtering of relations according to specific criteria in order to present a partial view of the information to the user, e.g. a single or several classifications at a time. The idea is seductive but practical problems arise: the definition of views may be very complex, as a single classification contains a high number of different concepts (e.g. composite entities). The selection of all these concepts would require an inordinate number of queries (at least one for each table of interest), which would not only be hard to express (new relations may need to be created), but would be extremely expensive to compute.

In object-oriented databases, view mechanisms allow the definition of different appearances for objects and classes [42] [38] [6] [41]. The view mechanism can maintain a global schema [38] [5] [23] that contains all classifications, and extracts individual classifications to present them to the user. Figure 9 shows a mechanism by which views (top plane) are extracted from a global schema (bottom plane). Views are more flexible than schema-based approaches, as they can be created with the query language [14] or a view definition language [40], and some view mechanisms allow reorganisation and possibly automatic class integration in existing schemas and views [39]. However, the cost of creating and modifying views, even if they are materialised [24], may be too high to allow dynamic classifications. For example conflicts must be detected and resolved, and mistakes (especially in taxonomic work) might not allow this.

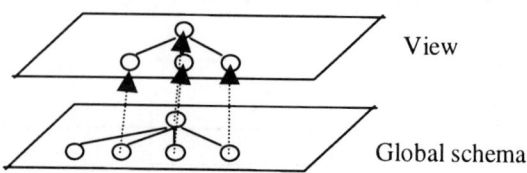

Figure 9: view as filtering mechanism

Views may also offer a way to express alternative classifications over graph-based models. View mechanisms have been proposed in the context of graph-based databases [49] [48] and semistructured databases [45] [1]. Graph views are essentially filtering mechanisms where sets of nodes and edges are selected. These proposals are all limited to specific aspects of graph databases: [49] only extracts views as sets of objects; [48] does not deal with updates and deletions; [45] only works with join-free queries and insert statements. Only [1] proposes a generic view mechanism that takes into account the specificity of graphs, and supports all operations. This view mechanism may allow the extraction of sub graphs from a general graph as classifications extracted from an overlapping larger classification.

Extended object-oriented models do not intrinsically support interconnected classifications. They only support the definition of unspecialised graphs. Additional mechanisms, at the model and/or at the query language level, would need to be developed in order to support multiple overlapping classifications. Unlike object-oriented and many graph-based models, no view mechanism is available for these approaches. In many cases, they are built from scratch in order to support uncommon features (e.g. GraphDB, SORAC, Albano) or are built on top of existing database systems that do not support views (e.g. ADAM). As a consequence, the representation of multiple overlapping classifications as views of a single larger classification is impossible.

4. Classification in context

The previous section has shown why existing technology fails to support multiple overlapping classifications and satisfy all the issues associated with their proper handling. Even mechanisms described in the literature for biological classifications (e.g. the *Materialization* relationship [31] and *power types* [28]) are too limited to support multiple overlapping classifications: they either work at class level (materialisation), which generate important classification reorganisation problems due to schema evolution, or do not support multiple overlapping classifications (materialisation, power types).

A new mechanism has been devised in the Prometheus project, which is described in this section. First, the technique for representing classifications is explained. This mechanism allows the representation of all types of classifications from single to multiple overlapping classifications. Then it is shown how the representation of multiple overlapping classifications does not impair the ability of the system to retain single points of views.

4.1. *Relationships as classifiers*

As the previous section explained, the model that offers the best support for classification representation is the extended object-oriented model. It combines the high level approach of object-oriented models with the decomposition, low level approach offered by graph-based models. This allows extended object-oriented models to capture some forms of classification. However, the previous section has also shown that this model fails to capture multiple overlapping classifications properly. The approach that has been taken for the Prometheus project is the use of such a model combined with additional mechanisms. We use relationships with the equivalent of weights (as in weighted graphs) to describe classifications.

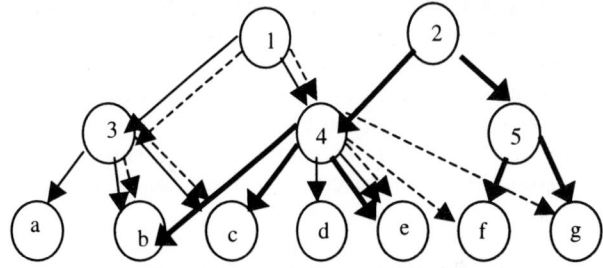

Figure 10: Multiple overlapping classification example

Relationships effectively act as classifiers (or the classifying mechanism). The action of creating such a relationship between two objects implies that these objects are classified. Furthermore, these relationships are the only objects in the system that are aware of the classifications and they contain all the necessary information to distinguish them from each other. They therefore support the independence of classifications and data. Figure 10 shows how the use of relationships as classifiers allows the description of multiple overlapping classifications. Each of these classifications represents a specific opinion, i.e. the context in which the specimens are classified. Figure 10 shows three distinct classifications: a dashed line classification, a thin line classification, and a thick line classification. In taxonomy, these distinct classifications would have been published by distinct authors and the publication information would replace the type of arrow in this example. The leaf nodes in these classifications could be for example books or plant specimens. The other nodes can be book subjects or taxa that are used to classify the leaf nodes.

4.2. *Traceability*

Traceability is handled by the relationships that act as classifiers. Traceability information is part of the classification information, therefore this is the right place to hold it. As relationships can have weights (captured by attributes), possibly also as part of the context (e.g. if following paths that only contain certain values is of interest), some of their attributes may be used to record decisions. For example, the edge between nodes 4 and b Figure 10 may have been created because 4 and b exhibit a specific property. This decision can be captured by an attribute of that relationship (with something as simple as free text if necessary, or more complex object structures).

4.3. Semantics

Semantics are provided by the fact that all relationships can participate in the classification description and that the model offers extension of relationships by sub-classing and description of behaviour and constraints. If part-of classifications are to be described, then aggregation relationships can be used and their semantics interpreted by the system. If other kinds of classifications are to be represented (e.g. similarity), new relationships, with specific semantics, can be created and used.

4.4. Multiple overlapping classifications

It can be seen in Figure 10 that the different classifications have elements in common: node 3 appears in the thin line classification and in the dashed line classification; node 4 appears in all three classifications. On the contrary, node 5 only appears in the thick line classification. Likewise, the leaf nodes can appear in one classification (node a), in two classifications (node b), or in all three (node e).

As Figure 10 shows, identity of distinct classifications can be handled through the type of the relationships that are used to describe them. Sub-classing of relationships (as first-class objects) allows for example the creation of specific relationships for each classification to be represented. It is also possible to manage this identity using attributes of relationships: specific values may represent each classification. In any case, distinct classifications are clearly identified and this distinction does not impair querying, as querying attributes and using types is inherent in object-oriented query languages.

4.5. Classifying in context

This new approach allows the generic classification of entities by context. By "context", one can understand "anything that uniquely identifies a view". In plant taxonomy, this can be a taxonomist, a publication, or a combination of both. For example, one taxonomist's view on the world is a context and in that context a set of specimens is classified in a certain way. Concurrently, another taxonomist's view of the world represents another context where the same specimens (or any other set of specimens) are classified differently. The overall graph

that is stored in the database represents a view of taxonomy out of context, or within all contexts concurrently. This view, although it is the most complete because it contains all existing information, does not suit some classification work (e.g. taxonomy work), as users tend to work in one particular context or in relation to a limited set of contexts for comparison purposes. By representing classification information on the hierarchies that constitute that graph, Prometheus captures single contexts that can be extracted as necessary.

Because the distinct hierarchies created in different contexts overlap (in terms of categories and classified concepts), the representation of all contexts in a single graph makes possible the comparison of classifications defined in different contexts and provides the ability to switch between contexts in order to gain knowledge. Indeed, by following relationships with specific values (e.g. publication information), it is possible to follow a path of a specific graph. But by switching between these values, it is possible to compare and navigate within and amongst classifications. For example in Figure 10, it is possible to compare nodes 3 and 4 and thereby to realise that they have some leaf nodes in common. This can give new insight into the data (e.g. in plant taxonomy when two groups partially contain the same specimens, they are partial synonyms). It is also possible to contrast the different meanings of node 4 according to the different classifications: it contains nodes d and e in the thin line classification, nodes e, f, and g in the dashed line classification, and nodes b, c, and e in the thick line classification.

5. Conclusion

This paper has presented the concept of classification as ranging from single taxonomy to multiple overlapping classifications, appearing in many areas of science. The latter case represents the multiple overlapping classification of objects or classes in separate but overlapping classifications. The features identified as necessary for handling these classifications include trees/graphs, traceability, semantics of classifications, independence of classification and data, and identity of classifications.

Common database models have been investigated for their support of multiple overlapping classification regarding the requirements expressed, and we have concluded that none offers full supports for the features outlined, but many provide a part of a satisfying solution. A new method of capturing multiple overlapping classifications has therefore been devised where context, i.e. what identifies one classification from another, plays a central role. The approach uses an extended object-oriented model to capture classification information and links, so that classes and classified objects can be related by classification information but stay independent from classifications.

The technique presented here has been implemented and tested in a plant taxonomy database system, and has been shown to be effective in handling multiple classifications and their associated processes. However this approach is applicable to all domains where contexts or multifaceted objects exist. For example, context is important for ontologies, as pointed out by Priss [34], and is often ignored. An approach such as that proposed here could be applied to ontology systems in order to introduce the concept of context.

6. References

[1]   S. Abiteboul, J. M. Hugh, M. Rys, V. Vassalos, and J. Wiener, "Incremental maintenance for materialized views over semistructured data," presented at VLDB 98, Proceedings of 24rd International Conference on Very Large Data Bases, New York City, New York, USA, 1998.

[2]   A. Albano, G. Ghelli, and R. Orsini, "A Relationship Mechanism for a Strongly Typed Object-Oriented Database Programming Language," presented at Proceedings of the seventeenth international conference on very large data bases, Barcelona, Spain, 1991.

[3]   B. Amann and M. Scholl, "Gram: A Graph Data Model and Query Language," INRIA, Le Chesnay, France Verso report number 046 (ECHT), 1992.

[4]   J. Banerjee, H. Chou, J. Garza, W. Kim, D. Woelk, and N. Ballou, "Data model issues for object-oriented applications," *ACM Transactions on Office Information Systems*, vol. 5, pp. pp 3-26, 1987.

[5]   Z. Bellahsene, "View Mechanism for Schema Evolution in Object-Oriented DBMS," presented at 14th British National Conferenc on Databases, BNCOD 14, Edinburgh, Scotland, 1996.

[6]   Z. Bellahsene, "Updating Virtual Complex Objects," presented at OOIS 97, 1997 International Conference on Object Oriented Information Systems, Brisbane, Australia, 1997.

[7]   G. C. Bowker and S. L. Star, *Sorting things out, classification and its consequences*: Massachusetts Institute of Technology, 1999.

[8]   P. Buneman, S. Davidson, and D. Suciu, "Programming Constructs for Unstructured Data," presented at DBPL-5 Proceedings of the Workshop on Database Programming Languages, Gubbio, Umbria, Italy, 1995.

[9]   S. Chawathe, H. Garcia-Molina, J. Hammer, K. Ireland, Y. Papakonstantinou, J. Ullman, and J. Widom, "The TSIMMIS Project: Integration of Heterogeneous Information Sources," presented at Proceedings of IPSJ Conference, Tokyo, Japan, 1994.

[10]   E. F. Codd, "A Relational Model of Data for Large Shared Data Banks," *Communications of the ACM (CACM)*, vol. 13, pp. 377-387, 1970.

[11]   S. J. Connan and G. A. M. Otten, *SQL - The Standard Handbook*: McGraw-Hill, 1992.

[12]   V. Crestana-Taube and E. A. Rundensteiner, "Schema Removal Issues for Transparent Schema Evolution," presented at Sixth International Workshop on Research Issues on Data Engineering, Interoperability of Non traditional Database Systems, RIDE 96, IEEE, New Orleans, Louisiana, 1996.

[13]   H. Darwen and C. J. Date, "The Third Manifesto," *SIGMOD Record 24*, vol. 24, pp. 39-49, 1995.

[14]   S. Deßloch, T. Härder, F.-J. Leick, N. M. Mattos, C. Laasch, C. Rich, M. Scholl, and H.-J. Schek, "COCOON and FRISYS - a comparison -," in *Objektbanken für Experten, Informatik Aktuell*, T. H. R. Bayer, P.C. Lockemann, Ed.: Springer, 1992, pp. pp 179-196.

[15]   O. Díaz and P. M. D. Gray, "Semantic-rich User-defined Relationship as a Main Constructor in Object Oriented Database," presented at Object-Oriented Databases: Analysis, Design & Construction (DS-4), Proceedings of the IFIP TC2/WG 2.6 Working Conference on Object-Oriented Databases: Analysis, Design & Construction, Windermere, UK, 1990.

[16]   M. Doherty, J. Peckham, and V. F. Wolfe, "Implementing Relationships and Constraints in an Object-Oriented Database Using a Monitor Construct," in *Rules in Database Systems*, M. H. W. Norman Paton, Ed. Edinburgh: Springer-Verlag, 1993, pp. 347-363.

[17]   M. Gemis, J. Paredaens, I. Thyssens, and J. V. d. Bussche, "GOOD, A graph-Oriented Database System," *Proceedings of SIGMOD, SIGMOD Record*, vol. 22, pp. 505–510, 1993.

[18]   W. Greuter, F. R. Barrie, H. M. Burdet, W. G. Chaloner, V. Demoulin, D. L. Hawksworth, P. M. Jørgensen, D. H. Nicolson, P. C. Silva, P. Trehane, and J. McNe, *International code of botanical nomenclature (Tokyo Code)*, vol. 131: Koeltz Scientific Books, 1994.

[19]   R. H. Güting, "GraphDB: Modeling and Querying Graphs in Databases," presented at Proc 20th Int. Conf. on Very Large Databases, Santiago, Chile, 1994.

[20]   N. Hori, M. Yoshikawa, and S. Uemura, "ASKA: An Object-Oriented Data Model with Multiple Hierarchies and Multiple Object-Perspectives," presented at 6th Int. Conf. and Workshop on Database and Expert Systems Applications (DEXA 95) - Workshop Proceedings, London, UK, 1995.

[21]   M. Jarke, R. Gallersdörfer, M. A. Jeusfeld, M. Staudt, and S. Eherer, "ConceptBase - a deductive object base for meta data management.," *Journal of Intelligent Information Systems. Special Issue on Advances in Deductive Object-Oriented Databases*, vol. 4, pp. 167-192, 1995.

[22]   J. Joseph, S. Thatte, C. Thompson, and D. Wells, "Object-Oriented Databases: Design and Implementation," *Proceedings of the IEEE*, vol. 79, pp. 42-64, 1991.

[23]   W. Kim and W. Kelley, "On View Support in Object-Oriented Database Systems," in *Modern database systems: The Object Model, Interoperability, and Beyond*, W. Kim, Ed. New York: Addison-Wesley Publishing Company, 1995, pp. 108-129.

[24]   H. A. Kuno and E. A. Rundensteiner, "Materialized Object-Oriented Views in MultiView," presented at

Fifth International Workshop on Research Issues on Data Engineering: Distributed Object Management (RIDE-DOM'95), IEEE, 1995, Taipei, Taiwan, Taipei, Taiwan, 1995.

[25] J. McHugh, S. Abiteboul, R. Goldman, D. Quass, and J. Widom, "Lore: A Database Management System for Semistructured Data," *SIGMOD Record*, vol. 26, pp. 54-66, 1997.

[26] J. Mylopoulos, A. Borgida, M. Jarke, and M. Koubarakis, "Telos: Representing Knowledge About Information Systems," *ACM Transactions on Information Systems*, vol. 8, pp. 325-362, 1990.

[27] M. C. Norrie, "Distinguishing Typing and Classification in Object Data Models," in *Information Modelling and Knowledge Bases VI*, vol. VI, Chapter 25, H. Kangassalo, H. Jaakola, S. Oshuga, and B. Wangler, Eds.: IOS, 1995.

[28] J. Odell, "Power types," *Journal of Object-Oriented Programming*, vol. 7, pp. 8-12, 1994.

[29] J. Odell, "Six different kinds of composition," *Journal of Object-Oriented Programming*, vol. 6, pp. 10-15, 1994.

[30] Oracle, "Oracle, http:///www.oracle.com," 2001.

[31] A. Pirotte, E. Zimányi, D. Massart, and T. Yakusheva, "Materialization: a powerfull and ubiquitous abstraction pattern," presented at Very Lage Data Bases (VLDB'94), Santiago, Chile, 1994.

[32] M. K. a. A. Pirotte, "An aggregation model and its C++ implementation," presented at 4th Int. Conf. on Object-Oriented Information Systems, OOIS'97, Brisbane, Australia, 1997.

[33] A. Poulovassilis and S. Hild, "Hyperlog: a graph-based system for database browsing, querying and update," *To appear in IEEE Knowledge and Data Engineering*, 1998.

[34] U. Priss, "Ontologies and Context," presented at 12th Midwest Artificial Intelligence and Cognitive Science Conference, Miami University, Oxford, OH, USA, 2001.

[35] Prometheus, "Prometheus project web page," 1998.

[36] M. R. Pullan, M. F. Watson, J. B. Kennedy, C. Raguenaud, and R. Hyam, "The Prometheus Taxonomic Model: a practical approach to representing multiple taxonomies," *Taxon*, vol. 49, pp. 55-75, 2000.

[37] C. Raguenaud, "Managing complex taxonomic data in an object-oriented database," in *School of Computing*. Edinburgh: Napier University, 2002.

[38] E. A. Rundensteiner, "MultiView: A Methodology for Supporting Multiple View in Object-Oriented Databases," presented at 18th International Conference on Very Large Data Bases, Vancouver, Canada, 1992.

[39] E. A. Rundensteiner, "A Classification Algorithm For Supporting Object-Oriented Views," presented at Proceedings of the Third International Conference on Information and Knowledge Management (CIKM'94), Gaithersburg, Maryland, 1994.

[40] E. A. Rundensteiner and L. Bic, "Automatic View Schema Generation in Object-Oriented Databases," Department of Information and Computer Science, University of California, Irvine 92-15, 01/92 1992.

[41] C. S. d. Santos, S. Abiteboul, and C. Delobel, "Virtual Schemas and Bases," presented at Advances in Database Technology - EDBT'94. 4th International Conference on Extending Database Technology, Cambridge, United Kingdom, 1994.

[42] M. E. S. a. H.-J. Schek, "Supporting Views in Object-Oriented Databases," *IEEE Database Engineering Bulletin, Special Issue on Foundations of Object-Oriented Database Systems*, vol. 14, pp. 43-47, 1991.

[43] A. Schürr, A. J. Winter, and A. Zündorf, "PROGRES: Language and Environment," in *Handbook on Graph Grammars: Applications*, vol. 2, G. Rozenberg, Ed., Singapur: World Scientific ed, 1998.

[44] M. Stonebraker, A. Jhingran, J. Goh, and S. Potamianos, "On rules, procedures, caching and views in database systems," presented at ACM SIGMOD International Conference on Management of Data, Atlantic City, NJ, USA, 1990.

[45] D. Suciu, "Query Decomposition and View Maintenance for Query Languages for Unstructured Data," presented at VLDB'96, Proceedings of 22th International Conference on Very Large Data Bases, Mumbai (Bombay), India, 1996.

[46] F. W. Tompa, "A data model for flexible hypertext database systems," *ACM Transactions on Information Systems*, vol. 7, pp. 85-100, 1989.

[47] C. Watters and M. A. Shepherd, "A Transient Hypergraph-Based Model for Data Access," *ACM Transactions on Information Systems*, vol. 8, pp. 77-102, 1990.

[48] P. T. Wood, "Graph Views and Recursive Query Languages," presented at BNCOD 8, University of York, UK, 1990.

[49] Y. Zhuge and H. Garcia-Molina, "Graph Structured Views and Their Incremental Maintenance," presented at Proceedings of the Fourteenth International Conference on Data Engineering, Orlando, Florida, USA, 1998.

Scientific Query Optimisation

ATreeGrep: Approximate Searching in Unordered Trees

Dennis Shasha[*]   Jason T. L. Wang[†]   Huiyuan Shan[‡]   Kaizhong Zhang[§]

Abstract

An unordered labeled tree is a tree in which each node has a string label and the parent-child relationship is significant, but the order among siblings is unimportant. This paper presents an approach to the nearest neighbor search problem for these trees. Given a database $\mathcal{D}$ of unordered labeled trees and a query tree Q, the goal is to find those trees in $\mathcal{D}$ that "approximately" contain Q. Our approach is based on storing the paths of the trees in a suffix array and then counting the number of mismatching paths between the query tree and a data tree. To speed up a search, we use a hash-based technique to filter out unqualified data trees at an early stage of the search. Experimental results obtained by running our techniques on phylogenetic trees and synthetic data demonstrate the good performance of the proposed approach. We also discuss the use of our work in XML and scientific database management.

1 Introduction

Unordered labeled trees represent data in many scientific and commercial disciplines.[1] For example, scientists model phylogenetic relations as unordered labeled trees and develop methods for constructing these trees [3, 5, 15]. A recent workshop report from Yale suggested

that more research be undertaken to improve heuristic search strategies using algorithms designed to meet the demands made by increasingly large tree datasets [11]. Recent efforts in Web computing model an XML document as a tree, offering further motivation for the need for efficient tree searching [4, 6]. Other potential applications include information retrieval in linguistic, taxonomic, and neuroanatomical databases, among others.

Many algorithms have been developed for tree searching and matching [1, 12, 18, 25]. Most of these algorithms focus on comparing two trees based on various distance metrics. Chawathe *et al.* [7, 8, 9] studied the tree matching problem in the context of change detection for structured and semistructured data. There are also efforts spent in the development of query languages [2, 14, 16, 20, 21] and query processing techniques [10] for trees, with applications to XML and object-oriented database management.

In this paper we propose a new approach for approximate search among unordered labeled trees. Our problem, denoted the *approximate nearest neighbor search* (ANN) problem for unordered labeled trees, is the following. Given an integer $DIFF$, a query tree Q and a database $\mathcal{D}$ of trees, the ANN problem is to find all the data trees D in $\mathcal{D}$ where D approximately contains Q within distance $DIFF$. That is, D contains a substructure D' and the distance from Q to D' is at most $DIFF$. We proved that this problem was NP complete [28] for editing distance. So we adopt a different method. We measure the distance from Q to D' by the total number of root-to-leaf paths in Q that do not appear in D'; the nodes in D' that do not appear in Q can be freely removed. To our knowledge, no previous work has addressed the ANN problem for unordered labeled trees.

To illustrate the distance measure, consider the query tree Q in Figure 1, which has two paths. The query tree will match data tree D_1 with distance of 0, and match data tree D_2 with distance of 1. This happens because every path in Q matches a path of the substructure D_1' in D_1. (D_1' is enclosed by the dashed line in D_1.) On

---

[*]Department of Computer Science, Courant Institute of Mathematical Sciences, New York University, 251 Mercer Street, New York, NY 10012, USA (shasha@cs.nyu.edu).

[†]Contact author: College of Computing Sciences, New Jersey Institute of Technology, University Heights, Newark, NJ 07102, USA (wangj@oak.njit.edu).

[‡]Department of Computer Science, New Jersey Institute of Technology, University Heights, Newark, NJ 07102, USA.

[§]Department of Computer Science, The University of Western Ontario, London, Ontario, Canada N6A 5B7.

[1]In this paper we will refer to unordered labeled trees simply as trees when the context is clear.

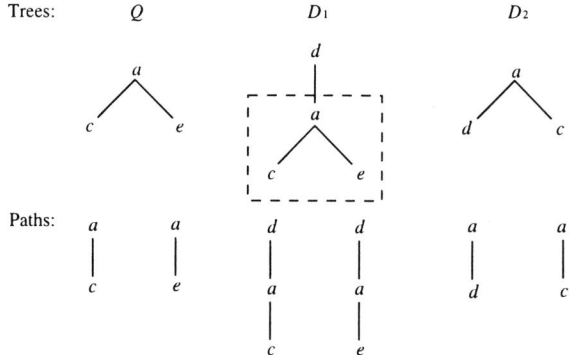

Figure 1. Example trees.

the other hand, there is one path '$a - e$' in Q that cannot be found in D_2. Counting mismatching paths is important in, for example, inferring evolutionary history, since that shows a deviation of ancestor-descendant relations. It is also a natural extension of path expressions in XML queries.

More specifically, we use this distance metric for two reasons, one semantic and one pragmatic:

1. In unordered trees, the parent-child relationship is the most significant one. This is reflected in paths. For example, each path in a phylogenetic tree stands for the evolutionary history of a taxon. When several siblings may have the same label, postprocessing must determine whether two paths of the form '$a - b - c$' and '$a' - b - c$' pertain to the same b or different bs.

2. The pragmatic reason is that there exist efficient algorithms for string searching. By decomposing trees to paths and by transforming tree searching to string searching, one can take advantage of the existing string searching algorithms and perform structural search efficiently.

In practice, it's likely that some portion of a query tree is unknown, uninteresting or unimportant. That portion is often represented by a don't care symbol. In general, there are two types of don't care symbols: variable length don't cares (VLDCs) [23, 27] and fixed length don't cares (FLDCs). In string matching, a VLDC, denoted "*", in the query string may substitute for zero or more characters in a data string. For example, if "com*er" is the query string, then the "*" would substitute for the substring "put" when matching with the data string "computer". On the other hand, a FLDC, denoted "?", in the query string substitutes for exactly a single character in

a data string. For example, if "com?uter" is the query string, then the "?" would substitute for the character "p" in the match with the data string "computer".

In this paper we discuss generalizations of don't cares to trees and present algorithms for processing them. In these cases, the labels on nodes can be "*" or "?".

The rest of the paper is organized as follows. Section 2 describes basic algorithms for processing query trees without don't cares. Section 3 presents algorithms for searching with don't cares. Section 4 describes a filtering technique for speeding up searching. Section 5 presents some experimental results. Section 6 discusses applications of our work. Section 7 concludes the paper.

2 Basic Algorithms

Our algorithm, called pathfix, consists of two phases. In the first phase, we build a suffix array database $\mathcal{SD}$ for all the trees in our database $\mathcal{D}$. $\mathcal{D}$ contains strings where each string corresponds to a root-to-leaf path in a data tree. We encode the paths (strings) into a suffix array [17]. In the second phase, which is the on-line search phase, we compare the root-to-leaf paths of the query tree Q with the paths in the suffix array database $\mathcal{SD}$ to locate those substructures approximately matching the query tree. In a later enhancement, we construct a filter to determine which data trees in $\mathcal{D}$ are possible matches.

2.1 Database Building Phase

The suffix array is a data structure designed for efficient searching in a large string [17]. This data structure is simply an array containing the pointers to all the string's suffixes sorted in lexicographical order. (A suffix is a substring starting at a certain position in the string and ending at the end of the string.) Searching for a query string can be performed by binary search using the suffix array.

In pathfix, we construct a suffix array for all the paths in a data tree and put it in a global set of suffix arrays for all the data trees. This global set is the database $\mathcal{SD}$. Figure 2 shows the algorithm.

As an example, consider again the data tree D_1 in Figure 1. D_1 has paths '$d - a - c$' and '$d - a - e$'. We can create a suffix array SA_1 for the two paths, separated by a delimiter #, as shown in Figure 3. In this figure, the parenthesized integer in front of each suffix indicates the position at which the suffix begins in the paths set. This integer, when stored in the suffix array, serves as a pointer

Procedure Build_Suffix_Array
Input: the database $\mathcal{D}$ of trees.
Output: the suffix array database $\mathcal{SD}$.

1. **for** every tree D in the database $\mathcal{D}$
2.   find all the root-to-leaf paths in D;
3.   concatenate those paths with a delimiter to form a long string S;
4.   form a suffix array for S and add it to the global set of suffix arrays;
5. **end for;**
6. return the global set of suffix arrays, which is the database $\mathcal{SD}$;

Figure 2. Procedure for building the suffix array database $\mathcal{SD}$.

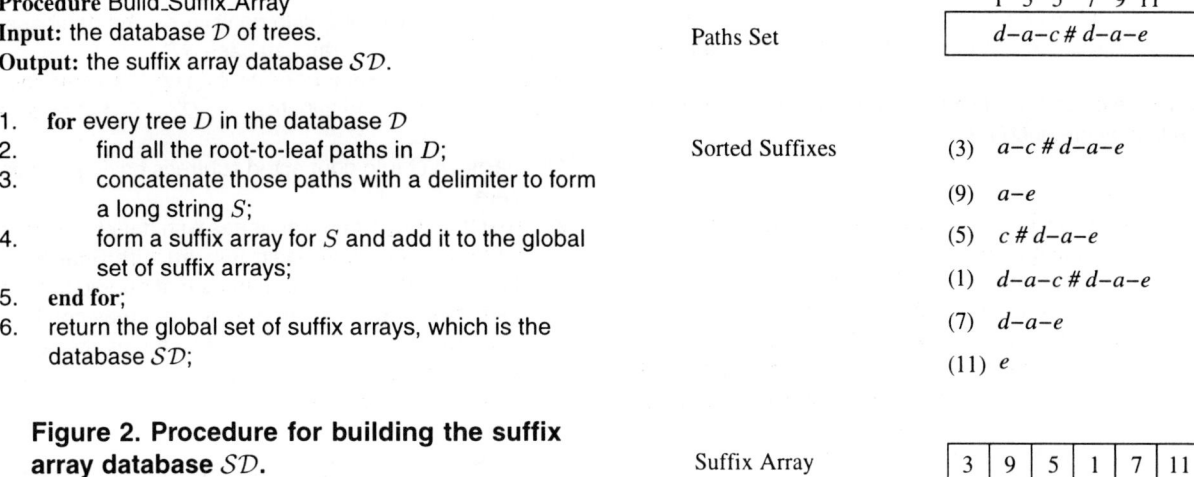

Figure 3. A suffix array.

to the corresponding suffix in the paths set. Likewise we can create a suffix array SA_2 for the paths of the data tree D_2 in Figure 1. The suffix array database $\mathcal{SD}$ contains SA_1 and SA_2.

Lemma 1. *Suppose each data tree has at most N nodes, and there are M trees in the database $\mathcal{D}$. The space complexity of the procedure* Build_Suffix_Array *in Figure 2 is $O(MN^2)$. The time complexity is also $O(MN^2)$.*

Proof. Each data tree has at most $O(N)$ paths if the tree is very bushy. No path can be longer than $O(N)$. We need at most N^2 pointers to the suffixes of the paths. Thus each data tree requires $O(N^2)$ space. There are totally M trees, so the space complexity is $O(MN^2)$. The expected time spent in constructing a suffix array for a string is linearly proportional to the string size [17], so the database $\mathcal{SD}$ can be built in $O(MN^2)$ time.

In practice, a tree with N nodes either has few paths or short depth, so the above upper bound is very pessimistic. In practice the complexity is linear, i.e. $O(MN)$. There are two alternative implementations of the suffix arrays. One suffix array can be constructed for all the trees or one suffix array can be constructed for each tree (as in our current implementation). Choosing one or the other depends on the effectiveness of filtering. If filtering yields very few candidate trees, then our current implementation works well. If there are many similar trees in the database, then a single suffix array for all trees is better.

2.2 On-Line Search Phase

In the on-line search phase, the query tree Q is compared to each data tree allowing a difference $DIFF$. When comparing Q with a data tree D, we take every root-to-leaf path p in Q and find roots of that path in D. (As a cutoff optimization, we stop searching D if more than $DIFF$ paths of Q are not found in D.) Suppose there are k root-to-leaf paths in Q. If a node n in D is the root of the k paths, then the subtree D' rooted at n matches Q with distance 0, provided there are no siblings having the same label. (If there are siblings having the same label, then post-processing can verify the match. Our technique will never miss a match.) If n is the root of $k - 1$ paths, then D' matches Q with distance 1 and D approximately contains Q with distance 1. Figure 4 shows the algorithm. Note that this algorithm can easily be modified to print out the subtree D' rooted at n that matches Q. We will refer to this modified algorithm as Basic_Substructure_Search.

Lemma 2. *Let q be the number of nodes in the query tree Q. Suppose that the size of a suffix array is S. The time complexity of the suffix array search portion of procedure* Basic_Search *in Figure 4 for such a suffix array is $O(q^2 \log S)$.*

Proof. We note that q is an upper bound on the number of paths in Q; q is also an upper bound on the lengths of the paths in Q. Searching for a path of length q takes time $O(q \log S)$, and hence the result follows.

Procedure Basic_Search

Input: the allowed distance threshold $DIFF$, the query tree Q, the database $\mathcal{D}$ of trees and the suffix array database $\mathcal{SD}$.

Output: the set $\mathcal{R}$ of data trees that approximately contain Q within distance $DIFF$.

1. $\mathcal{R} := \emptyset$;
2.   compute all the root-to-leaf paths of the query tree Q;
3.   let k be the number of paths of Q;
4.   **for** each data tree D in $\mathcal{D}$ (after filtering);
         /* Suffix array search portion */
5.   **for** each path p in Q from longest to shortest
6.   find the root set N_p in D such that for each n in N_p there is a node n' and the path from n' to n (ascending in D) is p;
7.   exit the for loop if the root sets for more than $DIFF$ paths are empty;
8.   **end for**;
         /* Intersection portion */
9.   **for** each n in N_p
10.   count the total number of occurrences $T(n)$ of n in all root sets N_p's for all paths p in Q;
11.   **if** $T(n) \geq k - DIFF$ **then**
12. $\mathcal{R} := \mathcal{R} \cup \{D\}$;
13.   **end for**;
14. **end for**;

Figure 4. Procedure for finding data trees approximately containing the query tree Q.

The time spent to count the number of occurrences of the entire tree depends on the number of matches of the paths. In the worst case, this can be nearly every node, but in practice is much smaller.

3   Query Trees with Don't Cares

When generalizing don't cares to trees, the semantics of the don't cares is as follows. The VLDC "*" in the query tree may substitute for a path of length zero or more in a data tree. The FLDC "?" in the query tree may substitute for a single node in the data tree. Figure 5 shows the algorithm for finding data trees exactly containing the query tree Q where the root of Q is not a don't care.

In general, for a query tree Q with don't cares, a node x in a data tree D is the root of a subtree that matches Q if all of the following hold:

Procedure Advanced_Search

Input: the query tree Q with don't cares, the database $\mathcal{D}$ of trees and the suffix array database $\mathcal{SD}$.

Output: the set of data trees containing Q and for each such data tree D, the substructure D' in D that matches Q.

1.   partition Q into connected subtrees having no don't cares;
2.   match each of those subtrees with data trees in $\mathcal{D}$ by invoking procedure Basic_Substructure_Search;
3.   For the matched substructures that belong to the same data tree, say D, determine whether they combine, forming D', to match Q based on the matching semantics of the don't cares explained in the text, and if so, return D and D';

Figure 5. Procedure for finding data trees containing the query tree Q with don't cares.

1. The partition of Q containing the root r_{all} of Q (call that the root partition of Q) matches D at x.

2. Consider the path p from the root r_{sub} of a subtree in Q to r_{all}. Suppose that r_{sub} matches D at possibly many nodes $x_1, x_2, \ldots$. The path from at least one such node in D, say x_j, has the property that the ascending path from x_j to x matches (with "*" and "?") the path from r_{sub} to r_{all}.

To avoid testing the roots of subtrees unnecessarily, the matching makes use of facts like the following: if Q is to match the data tree D at x, then the only relevant matches of a subtree of Q rooted at r_{sub} are nodes that are descendants of x.

Don't cares add to the time because each partition is much more likely to match than the whole tree, so there are many possible combinations to test. The basic time used for checking the suffix arrays, however, is less, since each tree is broken up into smaller trees.

Figure 6 illustrates how to match a query tree Q having don't cares with a data tree D. We first partition Q into three subtrees at its don't cares "*" and "?". The subtrees of Q match three subtrees in D, which are then glued. In the figure, nodes in D that are not touched by a dashed line are to be removed at no cost. The don't care "?" is instantiated into node o in D and the don't care "*" is instantiated into nodes h, j in D at no cost. The distance between Q and D is 0.

When a distance $DIFF$ is allowed in matching a

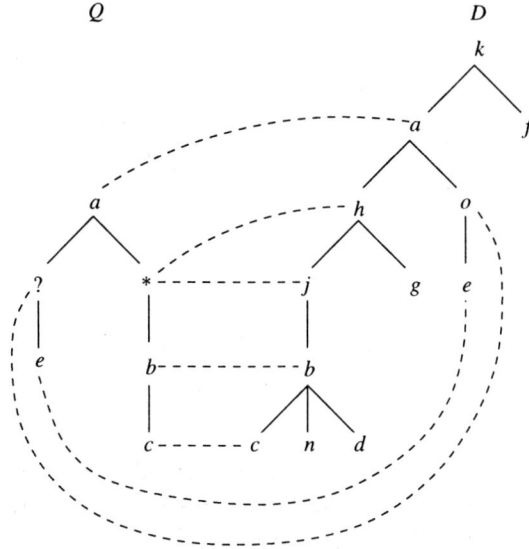

Figure 6. Illustration of matching a query tree having don't cares with a data tree.

query tree Q with a data tree, for each don't care free subtree Q' of Q, we find, by invoking procedure Basic_Substructure_Search, all subtrees of data trees that are within distance $DIFF$ of Q'. The gluing process involves testing whether the glued tree as a whole is indeed within distance $DIFF$ of the entire query tree Q.

4 Filtering

The above search process can be heuristically improved by using a hashing technique that works as follows on the don't care free portion of data and query trees. Compute and store all individual node labels and all parent-child label pairs in each data tree into a hash table, associating each parent-child pair with the set of data trees that contain the parent-child pair. Now suppose a query tree Q is given with a certain distance allowed in searching, $DIFF$. Take the multiset of labels from Q and see which data trees have a super-multiset of those possibly with $DIFF$ missing labels. Take the multiset of parent-child pairs from Q and see which data trees have a super-multiset of those parent-child pairs, again possibly with $DIFF$ missing pairs. This heuristic, referred to as pathfilter, eliminates irrelevant trees from consideration in the beginning of a search and yields a set of candidate trees to look for.

5 Experiments and Results

We have conducted a series of experiments to evaluate the performance of our algorithms pathfix and pathfilter, collectively referred to as ATreeGrep (reminiscent of AGrep [26] for approximate string searching and SGrep [13] for structure grep). The programs were written in K (http://www.kx.com) and run under Solaris on a SUN Ultra 10 workstation.

One thousand trees were randomly generated, each tree having 100 nodes. The number of children of non-leaf nodes ranged from 1 to 12. The length of paths in the trees ranged from 2 to 11. The number of paths in the trees ranged from 23 to 62. The string labels of nodes were randomly chosen from a dictionary. In each run, a tree was selected and modified into the query tree and the other trees were used as data trees. Ten runs were tested and the average was plotted. Figures 7 and 8 show the times spent in running ATreeGrep on the synthetic trees where the dictionary size is 50 and 1000, respectively. The curves in the figures correspond to query trees containing zero, one and two VLDCs, respectively.

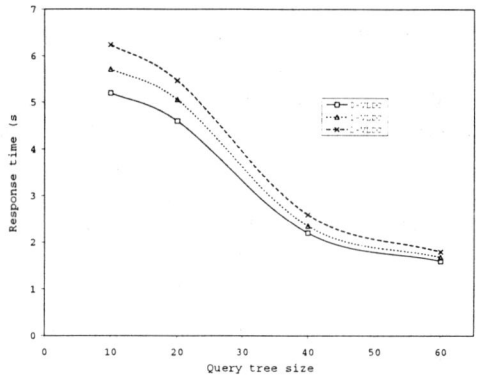

Figure 7. Running times of ATreeGrep **on the 1000 synthetic trees with a label dictionary size of 50.**

It can be seen from Figures 7 and 8 that the dictionary size has a significant impact on the running times of ATreeGrep. When both the dictionary size and query tree are small, pathfilter finds many candidate trees, requiring much time for checking. When the query tree is large, however, the parent-child pairs produce a selective filter and hence there are few trees to look for. Consequently, the total running time decreases.

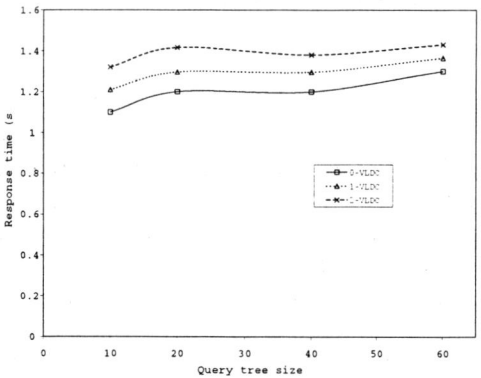

Figure 8. Running times of ATreeGrep **on the 1000 synthetic trees with a label dictionary size of 1000.**

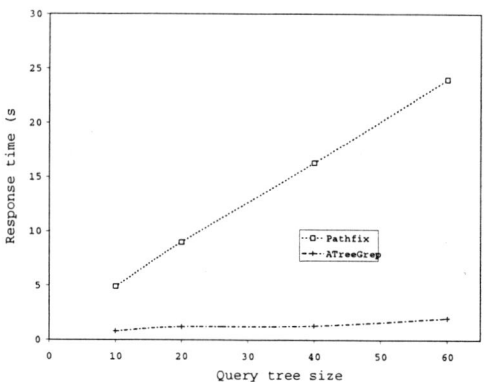

Figure 9. Running times of ATreeGrep **and** Pathfix **on the 1000 synthetic trees with a label dictionary size of 1000.**

On the other hand, when the dictionary size is large, very few parent-child pairs are the same regardless of the query tree size. As a consequence, many trees are eliminated by pathfilter.

Figure 9 compares ATreeGrep and pathfix for varying query tree sizes, where the dictionary size was fixed at 1000 and the query trees did not contain don't cares. The figure shows that pathfilter speeds up ATreeGrep considerably. It can be seen that the running time of pathfix is proportional to the size of a query tree (actually the total number of paths of the query tree). It was also observed that the running time of pathfix is proportional to the number of matches in the database.

Finally we ran the algorithms on the phylogenetic trees obtained from TreeBASE maintained at Harvard University Herbaria (`http://www.herbaria.ha rvard.edu/treebase`). Phylogenetic trees are structures used in biology to study the evolution of various life forms as well as the relationship of a particular life form with other life forms. We considered 1548 phylogenetic trees in TreeBASE. The number of nodes of the trees ranges from 50 to 200, and the dictionary size of node labels is 18870. The number of children of non-leaf nodes ranges from 2 to 25. The length of paths in these trees ranges from 2 to 26. The number of paths in these trees ranges from 8 to 153. The number of children of many non-leaf nodes is exactly 2. All the leaf-nodes and some non-leaf nodes have labels. For each non-leaf node without a label, we use "U" as its label. Figure 10 shows the results, where query trees do not contain don't cares.

The results are promising and consistent with those for the synthetic data.

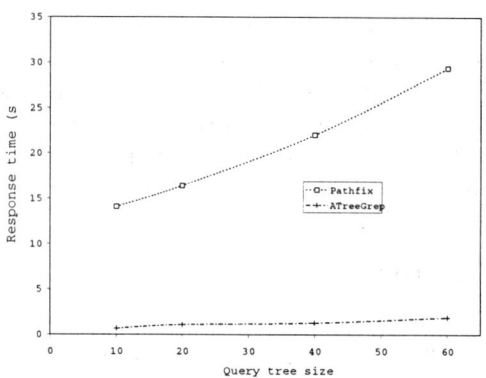

Figure 10. Running times of ATreeGrep **and** Pathfix **on the 1548 phylogenetic trees obtained from TreeBASE.**

6 Applications

We have incorporated ATreeGrep into two Web-based systems. The first one is a structure-based search engine [19], accessible at `http://aria.njit.edu/~biotool/`, and now incorporated into TreeBASE's keyword-based search engine, accessible at `http://www.treebase.org/t`

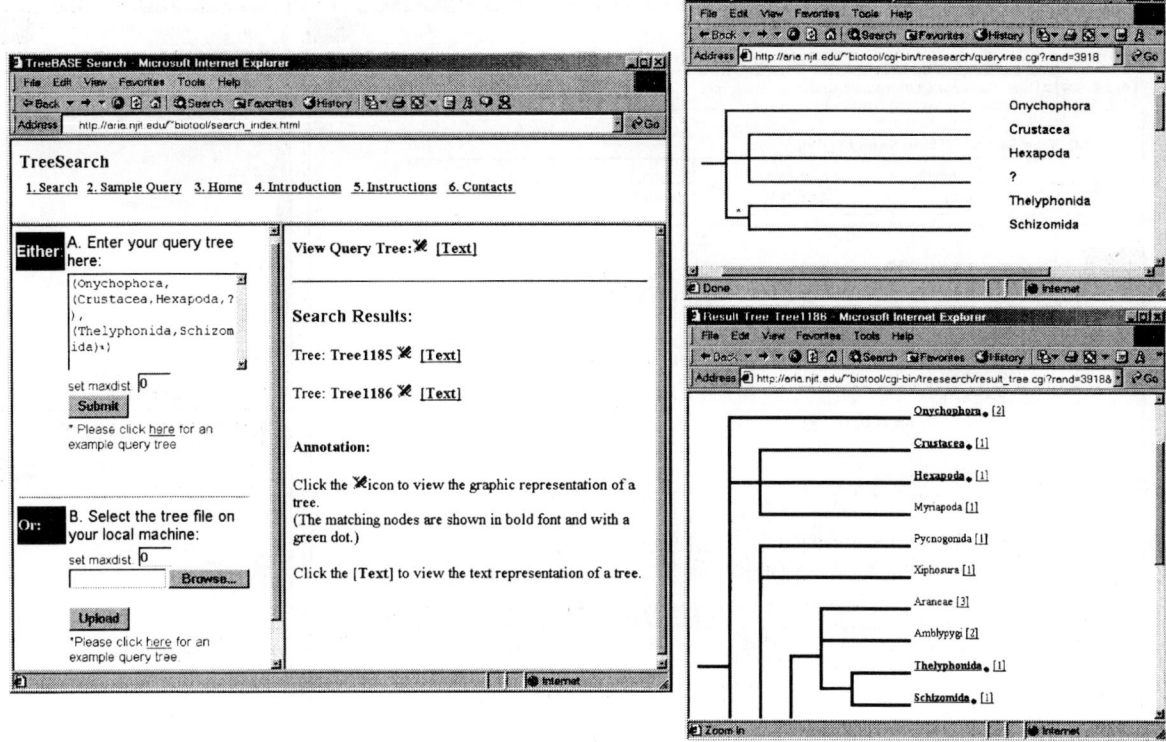

Figure 11. An example query and search results in the structure-based search engine for Tree-BASE.

reebase/console.html. This structure-based search engine is visited a few hundred times a month. Figure 11 shows its querying interface (in the left window), a query tree (in the right, top window) and a data tree (in the right, bottom window). In the figure, the query tree is entered using a parenthesized string format (see the leftmost frame in the left window). With this format, sibling labels are separated by commas and enclosed in parentheses, which is followed by the parent label, if it exists, of the siblings. The right, top window displays this same query tree in a dendrogram format. The query tree matches the data tree in the right, bottom window with distance 0, "?" matches "Myriapoda" and "∗" matches a path of the data tree. This query finds all the phylogenetic trees in TreeBASE containing the query tree. This structural search allows one to specify the relationship between taxa. Allowing don't care symbols further enhances the power of the query language, and offers more flexibility to the structural search.

The second system we have implemented, called XML Query by Example (or XML QBE) and accessible at http://aria.njit.edu/~mediadb/, allows the user to input an example XML fragment (query tree) and then finds those XMLs in an XML database that approximately contain the query tree. For example, the query in Figure 12 is to find all the XML documents describing movies in which Robert Redford is the director, Brad Pitt is an actor, and the movies are made in California, U.S.A. Shown in the figure are (counterclockwise, starting from upper left) the main menu, the querying window, the example XML (query) displayed via a Microsoft IE browser, a matching XML containing the query displayed via the IE browser, the query tree displayed via Java tree show applets, and the matching XML tree displayed via Java tree show applets. The matched portions in the matching XML tree are highlighted and marked with a bullet.

In general, when interacting with XML QBE, the user is able to type in his own query, load the query from a file, or use and modify a sample query provided by the

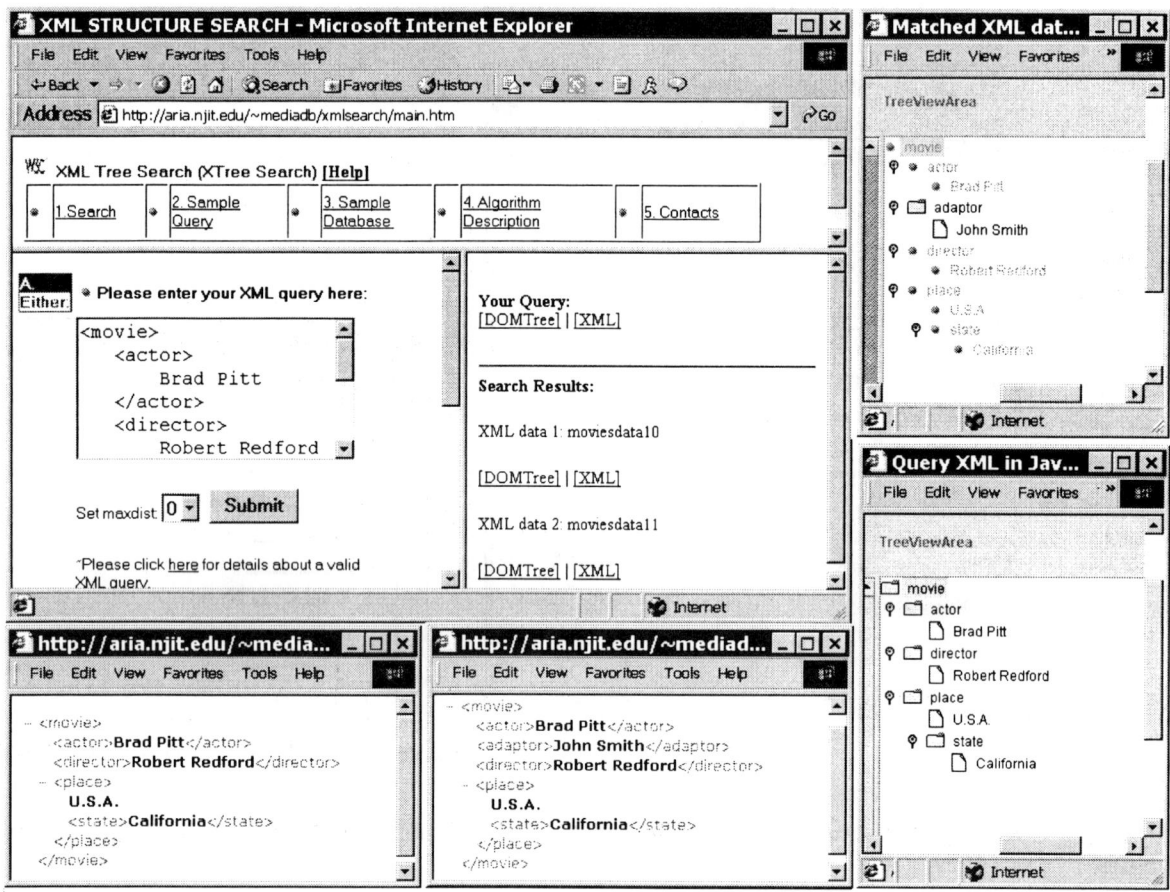

Figure 12. An example query and search results on a movie document database in XML QBE.

system. The user is also able to browse the underlying database and read the XML documents in the database.

databases, and for finding patterns in these databases [22, 24].

7 Conclusion

Unordered labeled trees find many applications in computer and natural sciences. In this paper we have presented a new approach, called **ATree-Grep**, for searching among these trees. Experimental results show that **ATreeGrep** is fast, particularly when the dictionary size of node labels is large. The algorithm and its code can be obtained at `http://cs.nyu.edu/cs/faculty/shasha/papers/treesearch.html`. We have implemented the algorithm into two Web-based search engines for phylogenetic databases and XML repositories. Future work includes extending our algorithm for ordered labeled trees, for searching other types of scientific

Acknowledgments

We would like to thank William Piel, the designer of TreeBASE, for his continuing collaboration. We also thank Ken Abe for performing experiments to determine the best filter parameters, and thank Greg Heil for a very compact K language implementation of suffix arrays. Finally we thank Sen Zhang, who implemented XML Query by Example and provided Figure 12. This work was supported in part by U.S. NSF grants IIS-9988345, IIS-9988636, and by the Natural Sciences and Engineering Research Council of Canada under Grant No. OGP0046373.

References

[1] E. N. Adams. Consensus techniques and the comparison of taxonomic trees. *Systematic Zoology*, 21:390–397, 1972.

[2] S. Amer-Yahia, S. Cho, L. V. S. Lakshmanan, and D. Srivastava. Minimization of tree pattern queries. In *Proceedings of the ACM SIGMOD International Conference on Management of Data*, pages 497–508, 2001.

[3] V. Berry and D. Bryant. Faster reliable phylogenetic analysis. In *Proceedings of the 3rd Annual International Conference on Computational Molecular Biology*, pages 59–68, 1999.

[4] P. Buneman, S. B. Davidson, M. F. Fernandez, and D. Suciu. Adding structure to unstructured data. In *Proceedings of the 6th International Conference on Database Theory*, pages 336–350, 1997.

[5] J. H. Camin and R. R. Sokal. A method for deducing branching sequences in phylogeny. *Evolution*, 19:311–326, 1965.

[6] G. Chang, M. J. Healey, J. A. M. McHugh, and J. T. L. Wang. *Mining the World Wide Web: An Information Search Approach*. Kluwer Academic Publishers, Norwell, Massachusetts, 2001.

[7] S. S. Chawathe, S. Abiteboul, and J. Widom. Representing and querying changes in semistructured data. In *Proceedings of the IEEE International Conference on Data Engineering*, pages 4–13, 1998.

[8] S. S. Chawathe and H. Garcia-Molina. Meaningful change detection in structured data. In *Proceedings of the ACM SIGMOD International Conference on Management of Data*, pages 26–37, 1997.

[9] S. S. Chawathe, A. Rajaraman, H. Garcia-Molina, and J. Widom. Change detection in hierarchically structured information. In *Proceedings of the ACM SIGMOD International Conference on Management of Data*, pages 493–504, 1996.

[10] Z. Chen, H. V. Jagadish, F. Korn, N. Koudas, S. Muthukrishnan, R. T. Ng, and D. Srivastava. Counting twig matches in a tree. In *Proceedings of the IEEE International Conference on Data Engineering*, pages 595–604, 2001.

[11] J. Cracraft and M. Donoghue. Assembling the tree of life: Research needs in phylogenetics and phyloinformatics. Report from NSF Workshop, Yale University, July 2000.

[12] A. Ferro, G. Gallo, R. Giugno, and A. Pulvirenti. Best-match retrieval for structured images. *IEEE Transactions on Pattern Analysis and Machine Intelligence*, 23(7):707–718, 2001.

[13] J. Jaakkola and P. Kilpelainen. Using sgrep for querying structured text files. University of Helsinki, Department of Computer Science, Report C-1996-83, November 1996; available at http://www.cs.helsinki.fi/u/jjaakkol/sgrep.html.

[14] H. V. Jagadish, L. V. S. Lakshmanan, D. Srivastava, and K. Thompson. TAX: A tree algebra for XML. In *Proceedings of the 8th Workshop on Data Bases and Programming Languages*, 2001.

[15] S. Kannan, E. Lawler, and T. Warnow. Determining the evolutionary tree. In *Proceedings of the 1st Annual ACM-SIAM Symposium on Discrete Algorithms*, pages 475–484, 1990.

[16] T. W. Leung, G. Mitchell, B. Subramanian, B. Vance, S. L. Vandenberg, and S. B. Zdonik. The AQUA data model and algebra. In *Proceedings of the 4th Workshop on Data Bases and Programming Languages*, pages 157–175, 1993.

[17] U. Manber and G. Myers. Suffix arrays: A new method for on-line string searches. In *Proceedings of the 1st Annual ACM-SIAM Symposium on Discrete Algorithms*, pages 319–327, 1990.

[18] A. S. Noetzel and S. M. Selkow. An analysis of the general tree-editing problem. In D. Sankoff and J. B. Kruskal, editors, *Time Warps, String Edits, and Macromolecules: The Theory and Practice of Sequence Comparison*, pages 237–252. Addison-Wesley, Reading, MA, 1983.

[19] H. Shan, K. G. Herbert, W. H. Piel, D. Shasha, and J. T. L. Wang. A structure-based search engine for phylogenetic databases. In *Proceedings of the 14th International Conference on Scientific and Statistical Database Management*, 2002.

[20] B. Subramanian, T. W. Leung, S. L. Vandenberg, and S. B. Zdonik. The AQUA approach to querying

lists and trees in object-oriented databases. In *Proceedings of the IEEE International Conference on Data Engineering*, pages 80–89, 1995.

[21] B. Subramanian, S. B. Zdonik, T. W. Leung, and S. L. Vandenberg. Ordered types in the AQUA data model. In *Proceedings of the 4th Workshop on Data Bases and Programming Languages*, pages 115–135, 1993.

[22] J. T. L. Wang, G.-W. Chirn, T. G. Marr, B. A. Shapiro, D. Shasha, and K. Zhang. Combinatorial pattern discovery for scientific data: Some preliminary results. In *Proceedings of the ACM SIGMOD International Conference on Management of Data*, pages 115–125, 1994.

[23] J. T. L. Wang, K. Jeong, K. Zhang, and D. Shasha. A system for approximate tree matching. *IEEE Transactions on Knowledge and Data Engineering*, 6(4):559–571, 1994.

[24] J. T. L. Wang, B. A. Shapiro, and D. Shasha (eds). *Pattern Discovery in Biomolecular Data: Tools, Techniques and Applications*. Oxford University Press, New York, 1999.

[25] J. T. L. Wang, B. A. Shapiro, D. Shasha, K. Zhang, and K. M. Currey. An algorithm for finding the largest approximately common substructures of two trees. *IEEE Transactions on Pattern Analysis and Machine Intelligence*, 20(8):889–895, 1998.

[26] S. Wu and U. Manber. Fast text searching allowing errors. *Communications of the ACM*, 35(10):83–91, 1992.

[27] K. Zhang, D. Shasha, and J. T. L. Wang. Approximate tree matching in the presence of variable length don't cares. *Journal of Algorithms*, 16(1):33–66, 1994.

[28] K. Zhang, R. Statman, and D. Shasha. On the editing distance between unordered labeled trees. *Information Processing Letters*, 42:133–139, 1992.

Compressing Bitmap Indexes for Faster Search Operations*

Kesheng Wu, Ekow J. Otoo and Arie Shoshani
Lawrence Berkeley National Laboratory
Berkeley, CA 94720, USA
Email: {kwu, ejotoo, ashoshani}@lbl.gov

Abstract

In this paper, we study the effects of compression on bitmap indexes. The main operations on the bitmaps during query processing are bitwise logical operations such as AND, OR, NOT, etc. Using the general purpose compression schemes, such as gzip, the logical operations on the compressed bitmaps are much slower than on the uncompressed bitmaps. Specialized compression schemes, like the byte-aligned bitmap code (BBC), are usually faster in performing logical operations than the general purpose schemes, but in many cases they are still orders of magnitude slower than the uncompressed scheme. To make the compressed bitmap indexes operate more efficiently, we designed a CPU-friendly scheme which we refer to as the word-aligned hybrid code (WAH). Tests on both synthetic and real application data show that the new scheme significantly outperforms well-known compression schemes at a modest increase in storage space. Compared to BBC, a scheme well-known for its operational efficiency, WAH performs logical operations about 12 times faster and uses only 60% more space. Compared to the uncompressed scheme, in most test cases WAH is faster while still using less space. We further verified with additional tests that the improvement in logical operation speed translates to similar improvement in query processing speed.

1 Introduction

This research was originally motivated by the need to manage the volume of data produce by a high-energy experiment called STAR[1] [23, 24]. In this experiment, in-

Figure 1. A sample bitmap index.

OID	X	bitmap index			
		=0	=1	=2	=3
1	0	1	0	0	0
2	1	0	1	0	0
3	3	0	0	0	1
4	2	0	0	1	0
5	3	0	0	0	1
6	3	0	0	0	1
7	1	0	1	0	0
8	3	0	0	0	1
		b_1	b_2	b_3	b_4

formation about each potentially interesting collision event is recorded. Tens of millions of such events are collected each year, amounting to multiple terabytes of raw data. All raw data go through a preliminary analysis where hundreds of summary attributes are generated for each event. Further analyses are typically only performed on some of the events. One important way of selecting the events is to search for events satisfying some condition on the summary attributes such as "Energy > 15 GeV and 7 <= NumParticles < 13" [4, 23]. This type of queries are known as *partial range queries*. One important data management task is to answer these partial range queries efficiently. Since these summary attributes are usually read, not modified, the indexing schemes used for commercial data warehouses should be useful for our task. Based on experiences in data warehouse applications, we know the bitmap index is efficient for partial range queries on relations with many attributes [5, 7, 19, 28]. Since our datasets have hundreds of attributes, bitmap index is even more appropriate [23].

Generally, a bitmap index consists of a set of bitmaps and queries can be answered using bitwise logical operations on the bitmaps. Figure 1 shows a set of such bitmaps for the attribute **X** of a tiny table (**T**) consisting of only eight tuples (rows). The attribute **X** can have one of four values, 0, 1,

---

*This work was supported by the Director, Office of Science, Office of Laboratory Policy and Infrastructure Management, of the U.S. Department of Energy under Contract No. DE-AC03-76SF00098. This research used resources of the National Energy Research Scientific Computing Center, which is supported by the Office of Science of the U.S. Department of Energy.

[1]Information about the project is also available at http://www.star.bnl.gov/STAR.

99

2 and 3. There are four bitmaps, each corresponding to one of the choices. For convenience, we have labeled the four bit sequences $b_1, \ldots, b_4$. To process the query "select * from T where X < 2," one performs the bitwise logical operation b_1 OR b_2. Since bitwise logical operations are well supported by computer hardware, bitmap indexes are very efficient to use [19]. In many data warehouse applications, bitmap indexes perform better than tree based schemes [5, 19, 28], such as the variants of B-tree [8] or R-tree [10]. According to the performance model proposed by Jürgens and Lenz [13], bitmap indexes are likely to be even more competitive in the future as disk technology improves. In addition to supporting queries on one single table as shown in this paper, researchers have also demonstrated that bitmap indexes can accelerate complex queries involving multiple tables [21]. Realizing the value of the bitmap indexes, most major DBMS vendors have implemented them.

The example shown in Figure 1 is the simplest bitmap index which we call the *basic bitmap index*. A bitmap index is typically generated for each attribute. The basic bitmap index produces one bitmap for each distinct attribute value and it may perform the logical OR operation on multiple bitmaps when answering a range query involving the attribute. For attributes with low cardinality, a bitmap index is small compared to tree based indexes and processes range conditions faster as well. To process the example query "Energy > 15 GeV and 7 <= NumParticles < 13," a bitmap index on attribute Energy and a bitmap index on NumParticles are used separately to generate two bitmaps representing objects satisfying the conditions on Energy and NumParticles. The final answer is the result of a bitwise logical AND operation on these two bitmaps. The whole process can be carried out efficiently if indexes for Energy and NumParticles involves only a small number of bitmaps. However, in real applications, especially scientific applications, there are many bitmaps in a bitmap index because the attribute cardinalities are high. In these cases, the bitmap indexes take a lot of space and processing range queries using these indexes take longer than without an index.

Compression is one way to reduce the size of the bitmap index and improve its effectiveness. To compress a bitmap, a simple option is to use one of the text compression algorithms, such as LZ77 (used in gzip) [16]. These algorithms are well-studied and effective in reducing file sizes. However, performing logical operations on the compressed bitmap are usually significantly slower than on the uncompressed bitmap. To address this performance issue, a number of special algorithms have been proposed. Johnson and colleagues have conducted extensive studies on their performances [12, 1]. From their studies, we know that the logical operations using these specialized schemes are usu-

ally faster than those using gzip. One such specialized algorithm, called the *Byte-aligned Bitmap Code* (BBC), is known to be very efficient. It is used in a commercial database system, ORACLE [2, 3]. However, even with BBC, in many cases logical operations on the compressed bitmap still can be orders of magnitudes slower than on the uncompressed bitmap.

When processing range queries using BBC compressed bitmap indexes, we observed that more than 90% of the time is spent on performing logical operations. The I/O time is only a small part of the total time. To reduce the total query processing time, we propose a "CPU-friendly" compression scheme. It improves the speed of logical operations by an order of magnitude over BBC at a cost of small increase in space. We call the method the Word-aligned Hybrid (WAH) compression scheme. This scheme not only supports faster logical operations but also enables the bitmap index to be applied to attributes with high cardinalities. Our tests show that by using WAH compression, we can achieve good performance on scientific datasets where most attributes have high cardinalities. From their performance studies, Johnson and colleagues came to the conclusion that one has to dynamically switch among different compression schemes in order to achieve the best performance [1]. We found that since WAH is significantly faster than earlier compression schemes, there is no need to switch compression schemes in a bitmap indexing software. The new compression scheme not only improves the performance of the bitmap indexes but also simplifies the indexing software.

Compression reduces the total size of a bitmap index by reducing the size of each bitmap, another strategy is to reduce the number of bitmaps used, for example, by using binning or more complex encoding schemes. With binning, multiple values are grouped into a single bin and only the bins are indexed [14, 23, 26]. This strategy reduces the number of bitmaps used but it produces precise answers only if range conditions fall on bin boundaries. In order to accurately answer an arbitrary query, one has to scan some of the attribute values after operating on the indexes. Many researchers have studied the strategy of using different encoding schemes [5, 6, 20, 25, 28]. One well-known scheme is the bit-sliced index, that encodes k distinct values using $log_2 k$ bits and creates a bitmap for each binary digit [20]. This is related to the binary encoding scheme discussed elsewhere [5, 25, 28]. A drawback of this scheme is that to answer each query, most of the bitmaps have to be accessed, and possibly multiple times. There are also a number of schemes that generate more bitmaps than the bit-sliced index but access less of them while processing a query, for examples, the attribute value decomposition [5], interval encoding [6] and the K-of-N encoding [25]. In all these schemes, an efficient compression scheme may further improve their effectiveness. Additionally, a number of other common indexing schemes such as the signature file [9, 11, 15] and the bit transposed files [25] may also benefit from efficient bitmap compression schemes.

The remainder of this paper is organized as follows. In Section 2 we review three commonly used compression schemes and identify their key features. These three were selected as representatives in our performance comparisons. Section 3 contains the description of the word-aligned hybrid code (WAH). We discuss the timing results of the bit-wise logical operations in Section 4, and the overall query processing performance in Section 5. A short summary is given in Section 6.

2   Review of byte based schemes

In this section, we briefly review three well known schemes for representing bitmaps and introduce the terminology needed to described our new scheme. These three schemes are selected as representatives from a number of schemes studied previously [12, 27].

A straightforward way of representing a bitmap is to use one bit of computer memory for each bit of the bitmap. We call this the *literal* (LIT) *bit vector*[2]. This is the uncompressed scheme and logical operations on uncompressed bitmaps are extremely fast.

The second type of scheme in our comparisons is the general purpose compression scheme such as gzip [16]. They are highly effective in compressing data files. We use gzip as the representative because it is usually faster than others in decompressing the data files.

As mentioned earlier, there are a number of compression schemes that offer good compression and also allow fast bit-wise logical operations. One of the best known schemes is the Byte-aligned Bitmap Code (BBC) [2, 3, 12]. The BBC scheme performs bitwise logical operations efficiently and it compresses almost as well as gzip. We use BBC as the representative for these types of schemes. Our implementation of the BBC scheme is a version of the two-sided BBC code [27, Section 3.2]. This version performs as well as the improved version by Johnson [12]. In both Johnson's tests [12] and ours, the time curves for BBC and gzip (marked at LZ in [12]) cross at about the same position.

Many of the specialized bitmap compression schemes, including BBC, are based on the basic idea of run-length encoding that represents consecutive identical bits (also called a *fill* or a *gap*) by their bit value and their length. The bit value of a fill is called the fill bit. If the fill bit is zero, we call the fill a *0-fill*, otherwise it is a *1-fill*. Compression schemes generally try to store repeating bit patterns in compact forms. The run-length encoding is among the simplest

of these schemes. This simplicity allows logical operations to be performed efficiently on the compressed bitmaps.

Different run-length encoding schemes commonly differ in their representations of the fill lengths and the short fills. A naive run-length code may use a word to represent all fill lengths. This is ineffective because it uses more space to represent short fills than in the literal scheme. One common improvement is to represent the short fills literally. The second improvement is to use as few bits as possible to represent the fill length. Given a bit sequence, the BBC scheme first divides it into bytes and then groups the bytes into *runs*. Each BBC run consists of a fill followed by a *tail* of literal bytes. Since a BBC fill always contains a number of whole bytes, it represents the fill length as the number of bytes rather than the number of bits. In addition, it uses a multi-byte scheme to represent the fill lengths [2, 12]. This strategy often uses more bits to represent a fill length than others such as ExpGol [18]. However it allows for faster operations [12].

Another property that is crucial to the efficiency of the BBC scheme is the byte alignment. This property limits a fill length to be an integer multiple of bytes. More importantly, it ensures that during any bitwise logical operation a tail byte is never broken into individual bits. Because working on individual bits is much less efficient than working on whole bytes on most CPUs, byte-alignment is crucial to the operational efficiency of BBC. Removing the alignment may lead to better compression. For example, the ExpGol scheme [18] can compress better than BBC partly because it does not obey the byte alignment. However, bitwise logical operations on ExpGol bit vectors are often much slower than on BBC bit vectors [12].

3   Word-aligned hybrid scheme

Most of the known compression schemes are byte based, that is, they access computer memory one byte at a time. On modern computers, accessing one byte usually takes as much time as accessing one word [22]. To take advantage of this and to minimize the logical operation time, we devised a compression scheme called the *word-aligned hybrid* (WAH) code. The main idea is to simplify the coding scheme so there are only two types of words in the compressed data and to design an alignment requirement so there is no need to extract individual bits or bytes during any logical operation. We have previously considered a number of word-based schemes and this is the most efficient one in our tests [27].

The word-aligned hybrid (WAH) code is similar to BBC in that it is a hybrid between the run-length encoding and the literal scheme. Unlike BBC, WAH is much simpler and it stores compressed data in words rather than in bytes. The two types of words in WAH are *literal* words and *fill* words,

128 bits	1,20*0,3*1,79*0,25*1				
31-bit groups	1,20*0,3*1,7*0	62*0		10*0,21*1	4*1
groups in hex	40000380	00000000	00000000	001FFFFF	0000000F
WAH (hex)	40000380	80000002		001FFFFF	0000000F

Figure 2. A WAH bit vector. Each WAH word (last row) represents a multiple of 31 bits from the bit sequence, except the last word that represents the four leftover bits.

A	40000380	80000002			001FFFFF	0000000F
B	C0000002			7C0001E0	3FE00000	00000003
C	40000380	80000003				00000003

Figure 3. A bitwise logical AND operation on WAH compressed bitmaps, C = A AND B.

where a literal word represents bitmap literally and a fill word represents a fill. Each word in WAH can be interpreted independently from others. In our implementation, we use the most significant bit of a word to distinguish between a literal word (0) and a fill word (1). This choice allows one to easily distinguish a literal word from a fill word without explicitly extracting the bit. The lower bits of a literal word contain the bit values from the bitmap. The second most significant bit of a fill word is the fill bit and the lower bits store the fill length. The word-alignment requirement is that fill lengths must be integer multiples of the number of bits in a literal word. If a computer word is 32-bit long, a literal word would represent 31 bits and the fill lengths must be multiple of 31 bits. If a computer word has 64-bit long, each literal word would store 63 bits from the bitmap and each fill would have a multiple of 63 bits.

Figure 2 shows a WAH bit vector representing 128 bits. In this example, we assume each computer word contains 32 bits. The second line in Figure 2 shows how the bitmap is divided into 31-bit groups and the third line shows the hexadecimal representation of the groups. The last line shows the values of the WAH words. The first three words are normal words, two literal words and one fill word. The fill word 80000002 indicates a 0-fill of two-word long (containing 62 consecutive zero bits). Note that the fill word stores the fill length as two rather than 62. In other word, we represent the fill length as multiples of the literal word size. The fourth word is the *active word* that stores the last few bits that can not be stored in a normal word, and another word (not shown) is needed to stores the number of useful bits in the active word.

The logical operation functions are easy to implement but are tedious to describe. To save space, we refer the interested reader to a technical report [27]. Here we only briefly describe one example, see Figure 3. In this example, the first operand of the logical operation is the one in Figure 2.

To perform a logical operation, we basically need to match each group of 31 bits from both operands and generate the groups for the result using the hardware support to perform the operations between groups of 31 bits. Each column of the table is reserved to represent one such group. A literal word occupies the location for the group and a fill word is given at the space reserved for the first group it represents. The first 31-bit group of the result C is the same as that of A because the corresponding group in B is part of a 1-fill. The next three groups of C contain only zero bits. The active words are always treated separated.

The logical operations can be directly performed on the compressed bitmaps and the time needed by one such operation on two operands is related to the sizes of the compressed bitmaps. Let the compression ratio be the ratio of size of a compressed bitmap and its uncompressed counterpart. When the average compression ratio of the two operands are less than 0.5, the logical operation time is expected to be proportional to the average compression ratio [27].

Compared against BBC, the logical operations on WAH compressed bitmaps should be more efficient mainly due to three reasons.

1. The encoding scheme of WAH is much simpler than BBC. WAH has only two kinds of words and one test is sufficient to determine the type of any given word. In contrast, our implementation of BBC has four different types of runs, other implementations have even more [12]. It may take up to three tests in order to decide the run type of a header byte. After deciding the run type, many clock cycles may still be needed to fully decode a run to determine the fill length or the tail value.

2. During the logical operations, WAH always accesses whole words, while BBC accesses bytes. On most bitmaps, BBC needs more time to load its data from the main memory to CPU registers than WAH.

3. BBC can encode shorter fills more compactly than WAH, however, this comes at a cost. Each time BBC encounters a short fill, say a fill with less than 8 bytes, it starts a new run. WAH typically represent such a short fill literally. It is much faster to operate on a WAH literal word than on a BBC run.

4 Performance of the logical operations

In this section, we discuss the performance of the logical operations. Ultimately we are interested in enhancing the speed of query processing. However, because logical operations are the main operations on the bitmaps and their performances are directly affected by the compression schemes, we discuss the performances of the logical operations first.

The WAH compression scheme are compared against the three schemes reviewed in Section 2. The tests are conducted on three sets of data, a set of random bitmaps, a set of bitmaps generated from a Markov process and a set of bitmap indexes on some real application data. Each synthetic bitmap has 100 million bits. The synthetic data are controlled through two parameters, the *bit density* and the *clustering factor*. In a bitmap, the bit density is the fraction of bits that are one and the clustering factor is the average length of the 1-fills. The random bitmaps are generated according to the bit density and the Markov process generates bitmaps with a specified bit density and clustering factor. The goal of this test is to examine the performance of the different compression schemes under various conditions. However to limit the number of test cases, we restrict all synthetic bitmaps to have bit density no more than 1/2. Since all compression schemes can compress 0-fills and 1-fills equally well, the performance on high bit density bitmaps should be the same as on their complements. When necessary to distinguish the two type of synthetic bitmaps, we refer to them as the random bitmaps and the Markov bitmaps according to how they are generated. The real application is a high-energy physics experiment called STAR [23, 24]. The data used in our tests can be viewed as one relational table consisting of about 2.2 million tuples and 500 attributes. The bitmaps used in this test are bitmap indexes on a set of 12 most frequently queried attributes.

We have conducted a number of tests on different machines and found that the relative performances among the different compression schemes are independent of the specific machine architecture. This characteristic was also observed in a different performance study [12]. The main reason for this is that most of the clock cycles are consumed by branching operations such as "if" tests and "loop condition" tests. These operations only depend on the clock speed. For this reason, we only report the timing results from a Sun En-

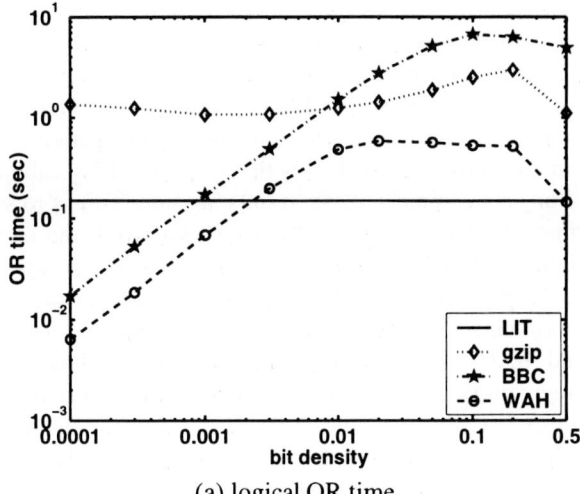

(a) logical OR time

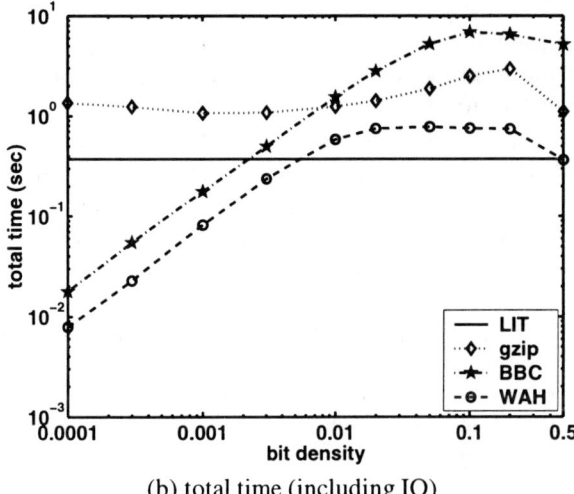

(b) total time (including IO)

Figure 4. CPU seconds needed to perform a bitwise OR operation on two random bitmaps.

terprise 450[3] that is based 400 MHz UltraSPARC II CPUs.

Because of space limitations, we only show performance of the logical OR operations in the following discussions. On the same machine, a logical AND operation typically takes slightly less time than a logical OR operation on the same bit vectors, and a logical XOR operation typically takes slightly more time. In general, if WAH is X times faster than BBC in performing a logical OR operation, the same would also be true for the two other logical operations.

The most likely scenario of using these bit vectors in a database system is to read a number of them from disks and then perform bitwise logical operations on them. In most

---
[3]Information about the E450 is available at http://www.sun.com/servers/workgroup/450.

cases, the bit vectors simply need to be read into memory and stored in the corresponding in-memory data structures. Only the gzip scheme needs a significant amount of CPU cycles to decompress the data files into the literal representation before actually performing the logical operations. In our tests involving gzip, only the operands of logical operations are compressed; the results are not. This is to save time. Had we compressed the result as well, the operations would take several times longer than those reported in this paper because the compression process is more time-consuming [27]. We use the *direct* method for both BBC and WAH. In other word, a logical operation directly operates on two compressed operands and produces a compressed result. It is one of the four strategies studied by Johnson [12]. We have chosen the direct method because it requires less memory and is often faster than the alternative methods.

Figure 4 shows the time it takes to perform the bitwise logical OR operations on the random bitmaps. Each data point shows the time to perform a logical operation on two bitmaps with similar bit densities. Figure 4(a) shows the logical operation time and Figure 4(b) shows the total time including the time to read the two bitmaps from files. In most cases, the IO time is a relatively small portion of the total time for BBC and WAH. Neglecting the IO time does not significantly change the relative performance between WAH and BBC. In an actual application, once the bitmaps are read into memory, they are likely to be used more than once. The average cost of a logical operation would be close to what is shown in Figure 4(a). From now on when showing the logical operation time, we will not include the IO time.

Among the schemes shown, it is clear that WAH uses much less time than either BBC or gzip. In all test cases, the gzip scheme uses at least three times more time than the literal scheme. In almost half of the test cases, BBC takes more than ten times longer than WAH.

When the bit density is about 1/2, the random bitmaps are not compressible by WAH. For convenience, we refer to the bit vectors only literal words as the decompressed bit vectors. Usually, each logical operation function takes two compressed bit vectors and generates a compressed result, but the functions that perform logical operations on decompressed bit vectors always generate decompressed results. It's easy to see that the logical operations on decompressed WAH bit vectors is nearly as fast as on the literal bit vectors. Unless one explicitly decompress a BBC bit vector, it is very unlikely to have a decompressed BBC bit vector. Even with bit density of 1/2, a BBC bit vector still contains a number of short fills. Even if we explicitly decompress the bit vectors, operations on decompressed BBC bit vectors are not as efficient as on literal bit vectors. In Figure 4, the line for WAH falls on top of the one for the literal scheme at bit

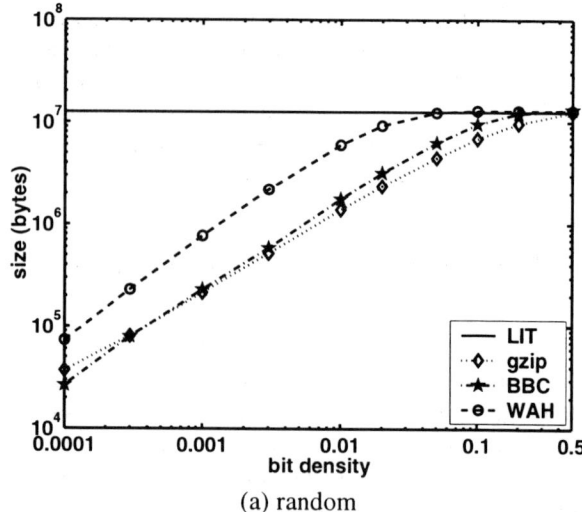

(a) random

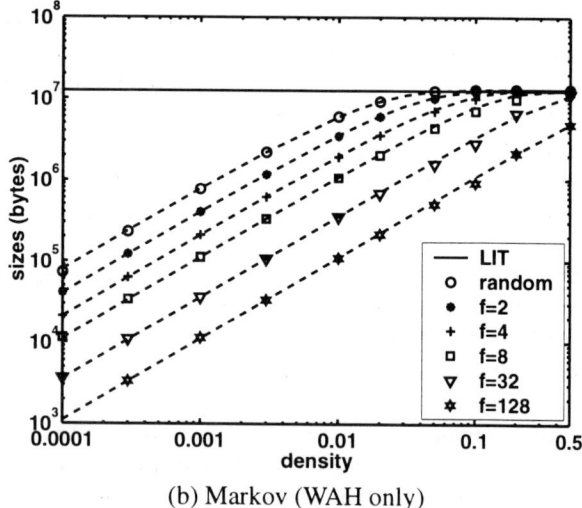

(b) Markov (WAH only)

Figure 5. The sizes of the compressed bit vectors. The symbols for the Markov bitmaps are marked with their clustering factors.

density of 1/2 but the line for BBC only shows a slight dip.

In Figure 4 we see that when bit density is above 0.01, WAH performs logical operations slower than the literal scheme. Since on the uncompressed bitmaps WAH can perform logical operations as well as the literal scheme, we might store those dense bitmaps without compression and expect the logical operations to be as fast as in the literal scheme. However, doing so significantly increases the space requirement and it does not even guarantee the speed of logical operation is always the fastest. This leads us to take a more careful look at the compression effectiveness and factors that determine the logical operation speed.

Figure 5 shows the sizes of the four types of bit vectors. Each data point in this figure represents the average size of a number of bitmaps with the same bit density and clustering factor. As the bit density increases from 0.0001 to 0.5, the bit sequences become less compressible and it takes more space to represent them. When the bit density is 0.0001, all four compression schemes use less than 1% of the disk space required by the literal scheme. At a bit density of 0.5, the test bitmaps become incompressible and the compression schemes all use slightly more space than the literal scheme. In most cases, WAH uses more space than the two byte based schemes, BBC and gzip. For bit density between 0.001 and 0.01, WAH uses about 2.5 ($\sim 8/3$) times the space as BBC bit vectors. In fact, in extreme cases, WAH may use four times as much space as BBC. Fortunately, these cases do not dominate the total space required by a bitmap index. In a typical bitmap index, the set of bitmaps contains some that are easy to compress and some that are hard to compress, and the total size is dominated by the hard to compress ones. Since most schemes use about the same amount of space to store these hard to compress ones, the differences in total sizes are usually much smaller than the extreme cases. For example, on the set of STAR data, the bitmap indexes compressed using WAH are about 60% bigger than those compressed using BBC, see Figure 7. This is a fairly modest increase in space compared to the increase in speed.

To verify that the logical operation time is proportional to the sizes of the operands, we plotted the timing results of the two sets of synthetic bitmaps together in Figure 6(a) and the results on the STAR bitmaps in Figure 6(b). In both cases, the compression ratio is used as the horizontal axes. Since in each plot, the bitmaps are of the same length, the sizes are directly proportional to the compression ratios. In each plot, a symbol represents the average time of logical operations on bitmaps with the same size. The dashed and dotted lines are produced from linear regressions. Most of the data points near the center of the graphs are close to the regression lines. Those logical operations involving bit vectors with high compression ratios are nearly constant. For very small bit vectors, where the logical operation time is measured to be a few microseconds, the logical operations time deviates from the linear relation because of the overheads such as the timing overhead, function call overhead and other lower order terms in the complexity expression. The regression lines for WAH and BBC are about a factor of ten apart in both plots.

If we sum up the execution time of all logical operations performed on the STAR bitmaps for each compression scheme, the total time for BBC is about 12 times that of WAH. Much of this difference can be attributed to reason 3 discussed in the previous section. There are a number of bitmaps that can not be compressed by WAH but can

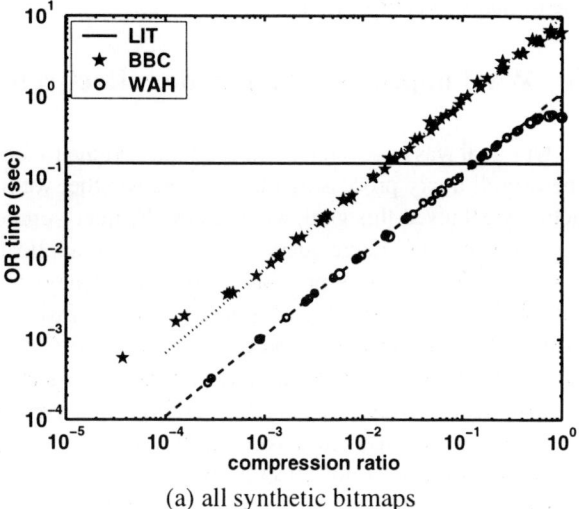

(a) all synthetic bitmaps

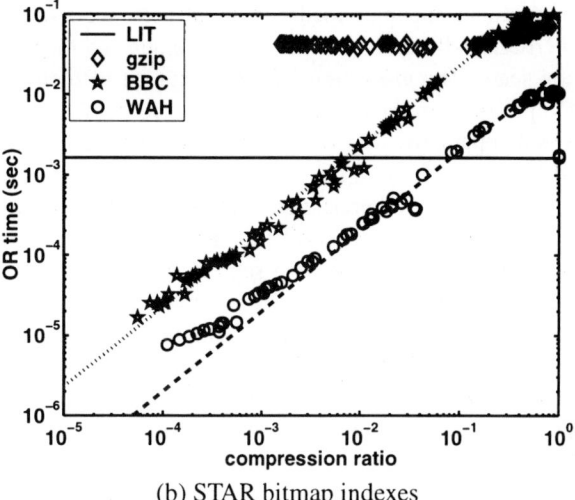

(b) STAR bitmap indexes

Figure 6. Logical operation time is almost proportional to compression ratio. The STAR bitmap indexes are on the 12 most queried attributes.

be compressed by BBC. When operating on these bitmaps, WAH is nearly 100 times faster than BBC. On very sparse bit vectors, WAH is about four to five times faster than BBC.

Compared to the literal scheme, BBC is faster in a fraction of the test cases, however, WAH is faster in more than 60% of the test cases. In the worst case, BBC can be nearly 100 times slower than the literal scheme, but WAH is only 6 times slower. It might be desirable to use the literal scheme in some cases. To reduce the complexity of the software, we suggest one to use WAH but only use the literal words. Regarding whether to store random bitmaps with bit density greater than 0.01 without compression, we recommend that the bitmaps be compressed.

5 WAH improves bitmap index effectiveness

Our goal was to develop a compression scheme to reduce the overall query processing time. To see whether we have actually achieved this goal, we measure the query processing time of partial range queries on a set of real application data from STAR. In the previous section, we demonstrated that WAH can perform logical operations much faster than BBC, but WAH also uses more space than BBC. Since query processing involves many operations other than bitwise logical operation, e.g., I/O operation, query parsing and locking, we need to make sure the decrease in logical operation time is not offset by the increase in I/O time caused by the increase index size. In addition to testing the performance of our own implementations of compressed and uncompressed bitmap indexes, we also tested ORACLE's BBC compressed bitmap index and the projection scan. The *projection scan* is a scheme of performing comparisons on the attribute values where each attribute is stored separately. It is also know as the projection index [20].

Our goal is to demonstrate that WAH compression can improve the performance of the bitmap indexing scheme. To do this, we perform two sets of tests. The first one is on some low cardinality attributes and the second is on some high cardinality attributes. The bitmap index is usually thought to be efficient for low cardinality attributes. In this case, we show that the WAH compressed indexes are not only smaller than the uncompressed ones but are also more efficient in answering range queries. When the cardinalities are high, it is impractical to generate the uncompressed indexes. In this case, we show that the WAH compressed indexes are still of reasonable sizes and can process range queries faster than the BBC compressed indexes and the projection index. The high cardinality case are of particular interests to us because the most frequently queried attributes of the STAR data have high cardinality.

In our tests, the low cardinality attributes are the 12 attributes with the lowest cardinalities from the STAR data, and the high cardinality attributes are the 12 attributes that are most likely to be queried by physicists. All low cardinality attributes are four-byte integers; the frequently queried attributes are mostly four-byte integers and floating-point values except one attribute is eight-byte floating-point value. The total size for the first set is about 104 MB and the second one is 113 MB.

Figure 7 shows the total sizes of the bitmap indexes. Four columns are displayed in each table. Columns marked 'ours' are our own implementation of the compressed bitmap indexes based on WAH and BBC compression schemes. The columns marked 'ORACLE' show the

	ours		ORACLE	
N	WAH	BBC	BBC	B-tree
12 low cardinality attributes				
312	7	4	7	370
12 most frequently queried attributes				
2,673,646	186	117	111	408

Figure 7. Sizes (MB) of the bitmap indexes. "N" indicates the number of bitmaps in the bitmap indexes.

sizes of the two kinds of indexes available, the bitmap index and the B-tree index. The bitmap index is marked with the compression scheme 'BBC'. Conceptually, ORACLE's BBC compressed index is equivalent to our BBC compressed bitmap index.

In the first data set, there are a total of 312 distinct values, i.e., there are 312 bitmaps in all bitmap indexes. Without compression, 312 bitmaps use about 84 MB. All three versions of the compressed bitmap indexes are less than 10% of this size and are less than 7% of the data size. In the second data set, there are nearly 2.7 million distinct values. Without compression, the bitmap index size would be more than 720 GB. Both BBC and WAH are very effective in reducing the sizes of the bitmap indexes because the majority of the bitmaps are very sparse. For both datasets, the compressed bitmap indexes are significantly smaller than the B-tree indexes.

Figure 8 shows the average query processing time of three compressed bitmap indexes. In this figure, the performance of ORACLE's BBC compressed indexes is marked as ORACLE. The partial range queries are generated by randomly selecting two attributes and constructing a query with the specified query box size. The *query box* is defined to be the ratio of the volume of the hypercube formed by the ranges and the total volume of the attributes [17]. For example, let the values of Energy be in the range of 0 to 30 GeV and NumParticles in the range of 1 to 15, the query box size of "Energy > 15 GeV and 7 <= NumParticles < 13" is $15/31 \times 6/15 = 0.19$. Given a query box size, the shape of the query box is allowed to vary. For simplicity, we only use conjunctive queries; that is the conditions on each attribute are joined together using the AND operator. Typically, as the query box size increases and the number of attributes increases, it takes more time to process the query.

Since the projection scan, "p scan" in Figure 8, only access the attributes involved in a query, it is quite faster [20]. For example, on our test machine, ORACLE takes about 6.5 seconds to scan a table with 12 attributes while the projection scan only need 0.56 ($\approx 6.5/12$) seconds. Had we ac-

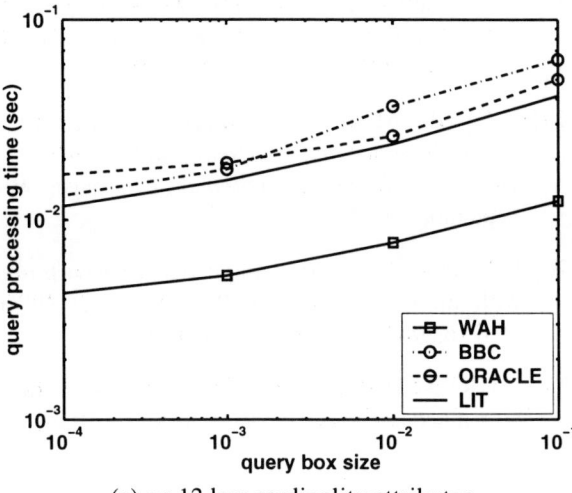

(a) on 12 low cardinality attributes

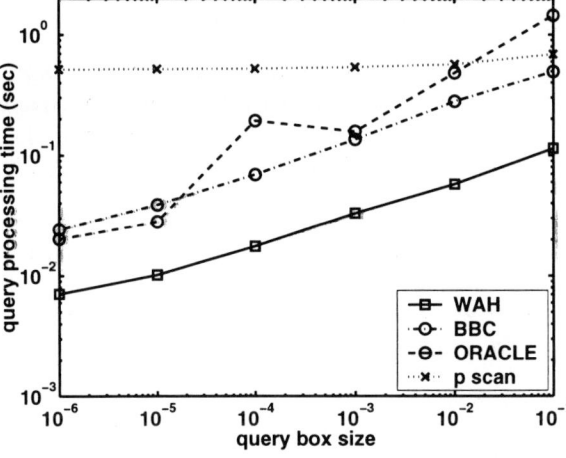

(b) on 12 most frequently queried attributes

Figure 8. The average query processing time of random range queries on the STAR data. Each query contains range conditions on two attributes.

tually stored all 500 attributes in the table, ORACLE would take nearly 5 minutes to perform its scan operation. We also take full advantage of the fast bitmap data structure to store the intermediate results. When evaluating conjunctive queries, the result of the left side can be used as the mask to limit the amount work needed to evaluate the right side. This is an efficient strategy since the projection scan time is always close to 0.56 seconds in Figure 8.

Our tests indicate that the bitmap indexes are efficient for both low cardinality attributes and high cardinality attributes. On the low cardinality attributes, WAH compressed indexes are not only smaller but are also faster. In contrast, BBC compressed indexes are slower than the

uncompressed ones. On high cardinality attributes, BBC compressed indexes require about as much time as the projection indexes, but WAH compressed indexes are always faster. The relative differences between WAH compressed indexes and BBC compressed indexes are larger on high cardinality attributes than on lower cardinality attributes. On high cardinality attributes, the average query processing time using the ORACLE bitmap indexes is nearly 10 times longer than using the WAH compressed bitmap indexes.

6 Summary

This research was motivated by the need to improve the query response time of a scientific data management project. Based on the characteristics of the dataset and queries, the bitmap indexing strategy is a good choice. However because most of the commonly queried attributes have a large number of distinct values, the basic bitmap index takes too much space and query response time is too long. This paper describes a compression scheme for addressing these performance issues. The best existing bitmap compression schemes are byte-aligned. In this paper, we presented a word-aligned scheme WAH, that is not only much simpler but is also very CPU-friendly. This ensures that the operations on compressed bitmaps can be performed efficiently. Tests on a set of real application data show that it is 12 times as fast as BBC while using only 60% more space.

Since the total query processing time includes both I/O time and logical operation time, we need to verify that the decrease in logical operation time is not offset by the increase in I/O time due to the 60% increase in index size. Our tests on a set of real application data show that WAH compressed indexes can indeed reduce the overall query processing time. Compared to uncompressed indexes, WAH compressed indexes are not only smaller but also take less time to answer partial range queries. Compared to the indexes compressed with BBC, the WAH compressed indexes can be 10 times faster.

References

[1] Sihem Amer-Yahia and Theodore Johnson. Optimizing queries on compressed bitmaps. In *Proceedings of VLDB 2000*, pages 329–338. Morgan Kaufmann, 2000.

[2] G. Antoshenkov. Byte-aligned bitmap compression. Technical report, Oracle Corp., 1994. U.S. Patent number 5,363,098.

[3] G. Antoshenkov and M. Ziauddin. Query processing and optimization in ORACLE RDB. *The VLDB Journal*, 5:229–237, 1996.

[4] Luis M. Bernardo, Arie Shoshani, Alex Sim, and Henrik Nordberg. Access coordination of tertiary storage for high energy physics applications. In *IEEE Symposium on Mass Storage Systems*, pages 105–118, 2000.

[5] C.-Y. Chan and Y. E. Ioannidis. Bitmap index design and evaluation. In *Proceedings of SIGMOD 1998*. ACM press, 1998.

[6] C. Y. Chan and Y. E. Ioannidis. An efficient bitmap encoding scheme for selection queries. In *Proceedings of SIGMOD 1999*. ACM Press, 1999.

[7] S. Chaudhuri and U. Dayal. An overview of data wharehousing and OLAP technology. *ACM SIGMOD Record*, 26(1):65–74, March 1997.

[8] Douglas Comer. The ubiquitous B-tree. *Computing Surveys*, 11(2):121–137, 1979.

[9] K. Furuse, K. Asada, and A. Iizawa. Implementation and performance evaluation of compressed bit-sliced signature files. In *Proceedings of CISMOD'95*, pages 164–177. Springer, 1995.

[10] V. Gaede and O. Günther. Multidimension access methods. *ACM Computing Surveys*, 30(2):170–231, 1998.

[11] Y. Ishikawa, H. Kitagawa, and N. Ohbo. Evalution of signature files as set access facilities in OODBs. In *Proceedings of SIGMOD 1993*, pages 247–256. ACM Press, 1993.

[12] T. Johnson. Performance measurements of compressed bitmap indices. In *Proceedings of VLDB'99*, pages 278–289. Morgan Kaufmann, 1999. A longer version appeared as AT&T report number AMERICA112.

[13] M. Jürgens and H.-J. Lenz. Tree based indexes vs. bitmap indexes - a performance study. In *Proceedings of DMDW'99*, 1999.

[14] Nick Koudas. Space efficient bitmap indexing. In *Proceedings of CIKM 2000*, pages 194–201. ACM, 2000.

[15] D. L. Lee, Y. M. Kim, and G. Patel. Efficient signature file methods for text retrieval. *IEEE Transactions on Knowledge and Data Engineering*, 7(3), 1995.

[16] Jean loup Gailly and Mark Adler. *zlib 1.1.3 manual*, July 1998. Source code available at http://www.info-zip.org/pub/infozip/zlib.

[17] V. Markl and R. Bayer. Processing relational OLAP queries with UB-trees and multidimensional hierarchical clustering. In *Proceedings of DMDW 2000*, 2000.

[18] A. Moffat and J. Zobel. Parameterised compression for sparse bitmaps. In *Proceedings of ACM-SIGIR 1992*, pages 274–285. ACM Press, 1992.

[19] P. O'Neil. Model 204 architecture and performance. In *2nd International Workshop in High Performance Transaction Systems, Asilomar, CA*, pages 40–59, September 1987.

[20] P. O'Neil and D. Quass. Improved query performance with variant indices. In *Proceedings of SIGMOD 1997*, pages 38–49. ACM Press, 1997.

[21] P. E. O'Neil and G. Graefe. Multi-table joins through bitmapped join indices. *SIGMOD Record*, 24(3):8–11, 1995.

[22] D. A. Patterson, J. L. Hennessy, and D. Goldberg. *Computer Architecture : A Quantitative Approach*. Morgan Kaufmann, 2nd edition, 1996.

[23] A. Shoshani, L. M. Bernardo, H. Nordberg, D. Rotem, and A. Sim. Multidimensional indexing and query coordination for tertiary storage management. In *Proceedings of SSDBM 1999*, pages 214–225. IEEE Computer Society, 1999.

[24] K. Stockinger, D. Duellmann, W. Hoschek, and E. Schikuta. Improving the performance of high-energy physics analysis through bitmap indices. In *Proceedings of DEXA 2000*, September 2000.

[25] H. K. T. Wong, H.-F. Liu, F. Olken, D. Rotem, and L. Wong. Bit transposed files. In *Proceedings of VLDB 85, Stockholm*, pages 448–457, 1985.

[26] K.-L. Wu and P. Yu. Range-based bitmap indexing for high cardinality attributes with skew. Technical Report RC 20449, IBM Watson Research Division, Yorktown Heights, New York, May 1996.

[27] Kesheng Wu, Ekow J. Otoo, Arie Shoshani, and Henrik Nordberg. Notes on design and implementation of compressed bit vectors. Technical Report LBNL/PUB-3161, Lawrence Berkeley National Laboratory, Berkeley, CA, 2001.

[28] M.-C. Wu and A. P. Buchmann. Encoded bitmap indexing for data warehouses. In *Proceedings of ICDE 1998*, pages 220–230. IEEE Computer Society, 1998.

Accelerating High-dimensional Nearest Neighbor Queries*

Christian A. Lang   Ambuj K. Singh

Department of Computer Science
University of California
Santa Barbara, CA 93106
{clang,ambuj}@cs.ucsb.edu

Abstract

The performance of nearest neighbor (NN) queries degrades noticeably with increasing dimensionality of the data due to reduced selectivity of high-dimensional data and an increased number of seek operations during NN-query execution.

If the NN-radii would be known in advance, the disk accesses could be reordered such that seek operations are minimized. We therefore propose a new way of estimating the NN-radius based on the fractal dimensionality and sampling. It is applicable to any page-based index structure. We show that the estimation error is considerably lower than for previous approaches.

In the second part of the paper, we present two applications of this technique. We show how the radius estimations can be used to transform k-NN queries into at most two range queries, and how it can be used to reduce the number of page reads during all-NN queries. In both cases, we observe significant speedups over traditional techniques for synthetic and real-world data.

1 Introduction

Nearest neighbor (NN) queries are an important query type for high-dimensional data, such as multimedia data, strings, and time sequences. One example for such a query is "Get the 10 closest images to a given query image" where similarity is typically defined on color distributions or texture features. Another example is DNA sequencing. Here, for each gene segment of one genome, all closest matching gene segments of another genome have to be found. Instead of asking for one best match, this query type asks for all best matches.

On large databases, such queries can be very time-consuming, especially when the number of feature dimensions is high. The main reason for the typically high response times lies in the fact that index structures supporting these queries are disk-based and therefore require a large amount of disk I/Os for answering queries.

Much research has therefore focused on reducing the number of page reads and seek operations during query execution. Techniques that try to reduce the number of seeks cannot guarantee a minimum number of page reads, and vice versa. One extreme example is the linear scan which minimizes the number of seeks but maximizes the number of page reads. An example for the other extreme is Hjaltason and Samet's k-NN query algorithm [11] which accesses only a minimal number of disk pages but induces a high amount of seek operations.

An algorithm that obtains the best of both worlds would be reading only the pages that contain the NNs with a read schedule that minimizes seeks. In other words, the NN query is transformed into a range query. The problem with this approach is that we need to know the required pages in advance. In order to determine these pages, it would be helpful to know the range of the query containing the NNs. Much work in the last years went into predicting such query ranges. However, current prediction techniques are not sufficient for this task since they predict an *average* query range for the entire dataspace rather than for a single query.

A major contribution of this paper is therefore an improved estimate of the query range based on the fractal dimensionality of data and sampling. We distinguish between two types of estimates: query-independent ones and query-dependent ones. The former type only assumes knowledge of the data distribution and estimates the average query range based on that. The latter type assumes that the data distribution and the query is known and a range estimate

---

*Work supported partially by NSF under grants EIA-0080134, EIA-9986057, IIS-9877142, ANI-0123985, and NSFIIS98-17432.

is desired. We show how the query-independent estimate can be used to obtain an (even tighter) query-dependent query range estimate through sampling. The basic idea is to get a first estimate on the sample and then to use the query-independent estimates to compensate for the effects due to sampling.

As a second contribution, we introduce a way of obtaining a close upper bound on the NN range for page-based index structures. This bound is obtained by inspecting all points stored in pages that are intersected by the query-dependent range estimate. The NNs found in these pages are then used to obtain an upper bound on the query range estimate. Our experiments show that this estimate is within 1% of the actual radius.

As a third contribution, we introduce two new NN query algorithms based on these estimates. The first algorithm is for single k-NN queries, and the second algorithm is for all-NN queries. Both new algorithms reduce the amount of disk seeks and page reads at the same time by using our new query-dependent range estimates to determine which pages will have to be fetched from disk in the future. For both query types, we observed a significant reduction in overall query time.

The paper is organized as follows. In Section 2, we show how the fractal dimensionality of data can be used to obtain a query-independent NN radius estimate. This is then used together with sampling to obtain a query-dependent way of calculating a query range estimate in Section 3. Section 4 analyzes the quality of the obtained estimates. Section 5 introduces a technique to compute a tight upper bound on the query-dependent query range. Sections 6 and 7 present two applications of these estimates. Related work is discussed in Section 8 and we conclude in Section 9.

2 Query-independent Estimation

In this section, we show how the notion of fractal dimensionality of data can be used to obtain an estimate for the average NN-query range.

Before going into details, let us first define some of the terms used in the following. Given a query point q, the k *NNs of* q are defined as a set $NN_k(q)$ of at least k points such that

$$\forall o \in NN_k(q), \forall o' \in DB - NN_k(q) : d(o, q) < d(o', q)$$

where DB denotes the full dataset and $d(o, q)$ denotes the distance between o and q.

The k-*NN radius* is then defined as

$$\max_{o \in NN_k(q)} d(o, q).$$

Table 1. Notation used in the paper

N	number of data points
σ	data sampling rate
E	embedding dimensionality
D_2	Correlation fractal dim. of full dataset
D_2'	Correlation fractal dim. of data sample
r_{sample}	sampling-based k-NN radius estimate
$r_{expected}$	query-dependent k-NN radius estimate
r_{upper}	upper bound on k-NN radius
$d_{nn}(k)$	radius estimate based on D_2 (full dataset)
$d_{sample-nn}(k)$	radius estimate based on D_2 (data sample)
k	number of NNs

Real datasets oftentimes exhibit self-similarity. This led to the use of the theory of fractals for characterizing datasets. More precisely, Pagel et al. [14] showed how two different fractal dimensionality measures can be used to describe certain properties of datasets. One of these measures is the so-called *correlation fractal dimensionality* D_2. It is defined as follows:[1]

Definition 1 (Correlation Fractal Dim.) *For a point set that has the self-similarity property in the range of scales* $r \in (r_1, r_2)$, *its Correlation fractal dimensionality* D_2 *for this range is measured as*

$$D_2 \equiv \frac{\partial \log \sum_i p_i^2}{\partial \log(r)} = \text{constant} \quad r \in (r_1, r_2)$$

where p_i *is the percentage of points which fall inside the* ith *cell when dividing the data space into hypercubic grid cells of side* r.

Korn et al. [14] show how this measure can be used to estimate the k-NN query radius for arbitrary datasets. They show that for the L_∞-norm, the average k-NN radius is given by

$$d_{nn}(k) = \frac{1}{2} \cdot \left(\frac{k}{N-1}\right)^{\frac{1}{D_2}} \quad (1)$$

where N is the number of data points.

The correlation dimension D_2 of point datasets is typically computed using a box-counting algorithm, as discussed in [1]. This algorithm subdivides the dataspace successively into smaller and smaller partitions and counts the number of occupied partitions in each step.

In our experiments, we use a slightly modified version of this algorithm which is more suitable for high-dimensional data. After the second subdivision of the original algorithm, most partitions tend to be empty due to the sparsity of high-dimensional data. We overcome this problem by using more intermediate partitioning steps. More details can be found in the full version of this paper [17].

---

[1]A summary of all symbols used in this paper can be found in Table 1.

3 Query-dependent Estimation

The box counting algorithm allows the estimation of the average k-NN radius. This calculation can be done statically for a given dataset. It is possible to achieve better results if we know the actual query location and have a data sample available. We will refer to this case as *query-dependent radius estimation.*

Assume we have a dataset of size N and we obtain an in-memory sample of size $N \cdot \sigma$; σ is called the *sampling rate*[2]. Furthermore, let q be a query point and k the number of its NNs we are looking for. We can obtain a first (rough) estimate of the query radius by computing the k NNs of q on the sample. Since the sample is stored entirely in memory, this can be done very efficiently.

Let us denote this estimate by r_{sample}. Since there might be a point that is a k-NN of q but is not in the sample, r_{sample} will usually be larger than the real k-NN-radius. If we knew how the radius changes with the sampling rate, we could compensate for this change. We can use Formula 1 for this purpose. Similar to $d_{nn}(k)$, we can calculate the k-NN radius for the sample dataset:

$$d_{sample-nn}(k) = \frac{1}{2} \cdot \left(\frac{k}{N \cdot \sigma - 1} \right)^{\frac{1}{D'_2}} \qquad (2)$$

where D'_2 denotes the Correlation fractal dimensionality of the sample dataset. Note that $d_{sample-nn}(k)$ can also be precomputed for a given dataset sample.

Once $d_{nn}(k)$ and $d_{sample-nn}(k)$ are known, we also know the expected rate of change in the NN-radius when moving from the sample to the full dataset, namely $\frac{d_{nn}(k)}{d_{sample-nn}(k)}$. Let us denote the query-dependent estimate of the k-NN radius by $r_{expected}$. It is given as

$$r_{expected} = \frac{d_{nn}(k)}{d_{sample-nn}(k)} \cdot r_{sample}. \qquad (3)$$

The quality of these query-dependent estimates is discussed in the next section.

4 Quality of Query-dependent Radius

We first examine analytically the deviation of the radius estimate from the correct NN-radius for uniform datasets (Section 4.1). We then show that similar numbers hold for real datasets (Section 4.2).

---

[2]We assume the sample is obtained by randomly selecting $N \cdot \sigma$ points of the dataset.

4.1 Expected Error for Uniform Data

In order to derive a formula for the expected error of $r_{expected}$, let us assume a normalized dataspace. According to Belussi and Faloutsos [1], the average number of neighbors k of a point within a region of regular shape and radius $\tilde{r}$ is then given by

$$k = (N-1) \cdot vol(\tilde{r})^{\frac{D_2}{E}},$$

where $vol(\tilde{r})$ denotes the volume of the region with radius $\tilde{r}$. One the other hand, we can use this formula to compute the expected query radius $\tilde{r}$ via some root finding method. In that case, the value $vol(\tilde{r})$ has to denote the volume of the query shape with radius $\tilde{r}$ after being cut off at the dataspace boundary. Details on the computation of this volume can be found in the full version of the paper [17].

Since $D_2 = E$ for uniform data, the above equation is equivalent to

$$k = (N-1) \cdot vol(\tilde{r}).$$

The same holds for the sample, resulting in the following equation for the expected sample radius $r_{\widetilde{sample}}$:

$$k = (N \cdot \sigma - 1) \cdot vol(r_{\widetilde{sample}}).$$

Thus, we know the expected correct query radius for the full dataset and the sample. However, what is the expected value of $r_{expected}$? Since we assume uniformity, $D_2 = D'_2$ and therefore

$$\frac{d_{nn}(k)}{d_{sample-nn}(k)} \approx \left(\frac{N \cdot \sigma}{N} \right)^{\frac{1}{D_2}}.$$

Therefore,

$$r_{expected} \approx r_{\widetilde{sample}} \cdot \sigma^{\frac{1}{D_2}}. \qquad (4)$$

The relative error of $r_{expected}$ is then given as

$$\frac{r_{expected} - \tilde{r}}{\tilde{r}}$$

which is plotted in Figure 1 for varying k ($N = 100,000$, $E = 60$). With increasing k, the error of the radius estimate increases. However, it stays below 22% even for a sampling rate of 1/1000. This shows that (at least for uniform data) our query-dependent radius estimate is very close to the correct NN-radius. We can also see that the relative error is always positive for uniform data, meaning that $r_{expected}$ overestimates the correct radius. Real data, however, can cause an underestimation as we will see in the next section.

Table 2. Datasets used in the experiments

UNIFORM	$100,000$ 8-dimensional points distributed uniformly.
LANDSAT	$275,465$ 60-dimensional points representing texture feature vectors of Landsat images (transformed using KLT).

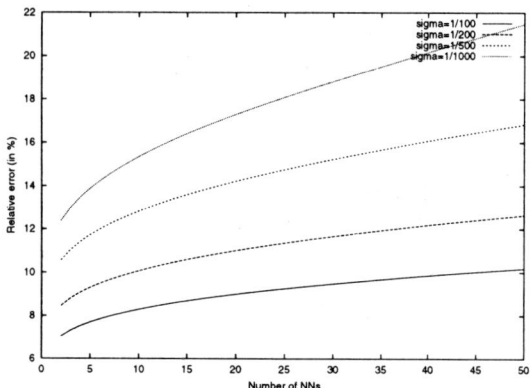

Figure 1. Rel. error of $r_{expected}$ (analytical)

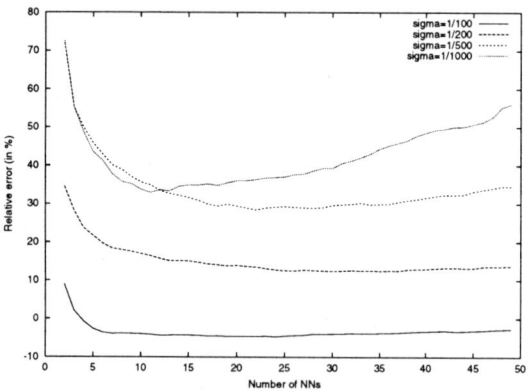

Figure 2. Rel. error of $r_{expected}$ (LANDSAT)

4.2 Measured Error for Real Data

This section shows how the relative error of $r_{expected}$ varies for a real dataset. Figure 2 shows the results. Our first dataset, LANDSAT[3], contains more than a quarter million 60-dimensional points and is highly clustered. Figure 2 shows that — similar to the uniform case — with increasing k, the relative error of $r_{expected}$ increases slightly. The same holds for the sampling rate. The smaller the sample, the higher the relative error. However, two differences can be noted compared to the uniform case of the last section. First, for a sample rate of $1/100$, the relative error drops below zero for $k > 3$. This means that $r_{expected}$ underestimates the real radius.

---

[3]Table 2 summarizes all datasets used in this paper.

The second difference that can be seen from the graphs is that the error is higher for very small k and then drops quickly until it reaches a minimum. This can be explained as follows. Since the data is clustered, the distances between data points vary a lot (compared to uniform data). Since our queries are density-biased, the distance of data points from the query point varies a lot as well. If k is large, the effect of sampling is alleviated by the large number of points in the query range. If k is small, sampling can cause $r_{expected}$ to be much larger than the real radius because the point distances have such a high variance. Even the compensation via the fractal dimensionality cannot counteract this effect because it describes the dataset globally, whereas this effect is local.

As a comparison, the statically computed radius $d_{nn}(k)$ bottomed out at about 400% relative error. This clearly shows the advantage of query-dependent radius estimation.

5 Bounding the NN-radius

In the previous sections, we saw how a query-dependent NN-radius estimate can be obtained via fractal dimensionalities and sampling. However, since this estimate may also underestimate the correct radius, its application can lead to false misses. If this is not acceptable, as is the case in our upcoming applications (cf. Sections 6 and 7), an upper bound on the radius has to be obtained. This section shows how a close upper bound on the NN-radius can be obtained if an approximate NN-radius is known and if the underlying index structure is page-based.

5.1 Upper Bound Computation

The basic idea behind the upper bound computation is shown in Figure 3. Assume, we are given the radius estimate as shown (labeled with "approx. NN-sphere"). Furthermore assume that the index structure distributed the dataset (indicated as dots) into four pages (labeled with A, B, C, and D). By reading all points from the pages that are intersected by the radius estimate (in this case, A, B, and C), we can determine the closest point in these pages (labeled with x). The distance to this point is an upper bound on the 1-NN radius (shown as "upper bound on NN-sphere"). Note that it is not necessarily the smallest upper bound (as point y shows).

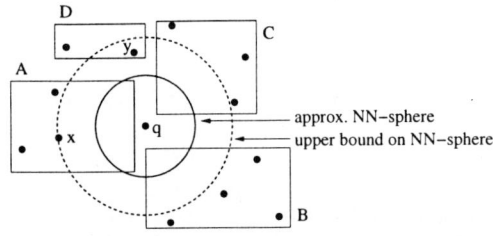

Figure 3. Upper Bound Computation

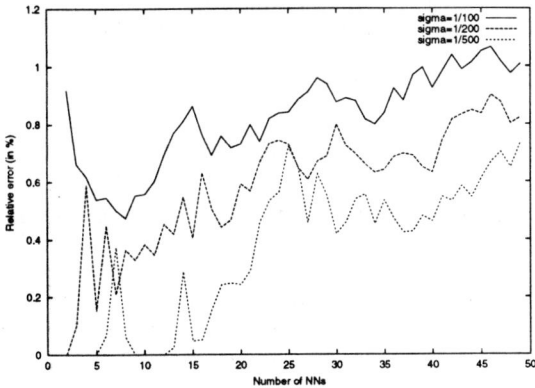

Figure 4. Rel. error of r_{upper} (LANDSAT)

Let us describe this estimation more formally now for arbitrary values of k. Let P be the set of index pages intersected by the approximate k-NN-sphere. Let C_{eff} be the effective capacity of each disk page. This is usually less than the page capacity due to the reduced page utilization in index structures. If $C_{eff} \cdot |P| \geq k$, we can simply pick q's k closest points from the pages in P and calculate the smallest radius that covers them. We will denote this second radius estimate by r_{upper}. Obviously, r_{upper} is larger than the real radius since the corresponding query sphere centered at q covers k points read from the pages, but not necessarily the k globally closest ones. If $C_{eff} \cdot |P| < k$, we pick the smallest distance between q and the points in the data sample (which is kept in memory) as r_{upper}.

5.2 Quality of the Upper Bound

In order to examine the tightness of r_{upper}, we performed some experiments on real data. Figure 4 shows the results for the LANDSAT dataset and an R-tree [10] index structure. The graph was obtained as follows. First, 500 random query points were picked. For each query point q, we computed r_{upper} as described above by choosing $r_{expected}$ to be the initial radius estimate. We then computed the relative error (the amount of overestimation) of r_{upper} and plotted

---

Input:	Query point q, number of NNs k
Output:	k NNs of q

(1) Let $queue := \{root\}$.
(2) Pick page or point p from $queue$ which has smallest MINDIST from q.
(3) If p is a page, read p into memory and place its children in $queue$.
(4) Otherwise, report p as a NN.
(5) If k NNs were reported, stop. Otherwise, goto (2).

Figure 5. Access optimal k-NN query

the averages for varying k-values. As can be seen, with increasing k, the error increases slightly but it stays overall below 1.2%. This shows that the points read from the pages touched by the initial radius estimate produce a very accurate upper bound.

6 Accelerated k-NN Query Algorithm

In this section, we present the first application of our query-dependent radius estimators. It aims at accelerating k-NN query processing by reducing the number of random disk accesses.

6.1 The Access Optimal Algorithm

Before going into the details, let us first revisit the optimal k-NN query algorithm for page-based tree index structures by Hjaltason and Samet [11]. Their algorithm works as follows (cf. Figure 5). First, a priority queue is initialized with the root page of the tree (Step 1). This queue is sorted by the MINDIST between q and the page. For a query q, an element p is removed from the queue whose MINDIST is closest to q (Step 2). If this element was a point, it is reported as a NN (Step 4). Otherwise, the element was a page. In this case, the page has to be fetched from disk and the elements stored in it are sorted into the waiting queue by their MINDIST. If the page is an inner page, the elements are typically pointers to child pages. If it is a leaf page, the elements are data points.

We assume in the following that all levels of an index tree is stored in memory. Or, in other words, only the leaf level pages have to be fetched from disk. This means that Step (3) causes only disk I/O if p was a leaf page. From a disk access point of view, the whole algorithm then reduces to the following: sort all leaf pages by their MINDIST from q and fetch them from disk in that order while keeping a list of the current k closest NNs found in the read pages. The algorithm terminates as soon as the most distant NN in the list is closer to q than the next leaf page to be read. Hjaltason and Samet prove that this algorithm is optimal in that it accesses a minimal number of pages, namely only the pages intersected by the k-NN query sphere. We will refer to this optimality as *access optimal*.

It is clear that especially for high dimensional data, the pages close to q can be scattered widely over the disk. This causes a large amount of seek operations during the query processing. If all pages to be read during a NN query would be known in advance, these expensive seeks could be avoided. We refer to an algorithm that minimizes the overall response time as *response time optimal* algorithm. Our new accelerated algorithm reduces the response time significantly by utilizing our new radius estimators. As we will see, its performance is very close to a (hypothetical) response time optimal k-NN query algorithm.

6.2 The New Algorithm

This basic idea of our new k-NN query algorithm is as follows. For every incoming NN-query, we compute $r_{expected}$ (cf. Section 3) as a first estimate of its NN-query radius. Using this estimate, we can perform a range query on the full dataset. Since all pages are known in advance for range queries, the amount of seek operations can be minimized by employing a specific page read scheduler. If the first estimate was too large, at least k points were in the range, and we can return the k closest ones as the result set. If the first estimate was too small, less than k points were in the range and we need to increase the search radius and perform another range query. Since we have already a first radius estimate, we can use r_{upper} (cf. Section 5) as this second estimate. Since r_{upper} is an upper bound on the real radius, we read at least k points during the second range query. Therefore, at most two range queries are necessary to answer the k-NN query, resulting in a worst case cost of two linear scans over a subset of the data pages (if an optimal page read scheduler is used). Figure 6 gives the algorithm in more detail.

Note that a simple modification yields an approximate k-NN query algorithm with less I/O cost. Instead of performing a second range query, we can stop the algorithm after step (3) and simply report the k closest points encountered during the first range query. As we saw in Section 4.2, for typical applications, $r_{expected}$ is already close to the real radius. Therefore, the quality of the approximate NNs can be expected to be high.

Before being able to run these algorithms for a given dataset, we need to precompute a sample of the dataset, the Correlation fractal dimensionality D_2 and D_2' of the full dataset and the sample dataset, respectively. This can be done in $O(N \log N)$ time with the box counting algorithm discussed in [1]. Note that this cost has to be paid only once and is therefore amortized over time.

Input:	Query point q, number of NNs k
Output:	k NNs of q

(1) Perform a k-NN query on sample in memory; this yields a k-NN-radius r_{sample}.

(2) Let $r_{expected} := r_{sample} \cdot \frac{d_{nn}(k)}{d_{sample-nn}(k)}$.

(3) Perform a range query with radius $r_{expected}$ on full dataset; this results in a set R of data points and a set P of index leaf pages accessed during the query.

(4) If $|R| \geq k$, return the k closest points from R, stop.

(5) Otherwise (more NNs need to be found), compute the k closest points stored in P and calculate the smallest radius r_{upper} enclosing them.

(6) Perform a modified range query around q with radius r_{upper} on full dataset; this results in a set R' of data points and a set P' of index leaf pages accessed during the query.

(7) Return the k closest points from R', stop.

Figure 6. Accelerated k-NN query algorithm

Disk Page Read Strategy Only steps (3) and (6) of our accelerated k-NN query algorithm induce disk I/O due to the two range queries. The range query in step (3) is a regular range query provided by the indexing system. The range query in step (6) is modified as follows. Since the k closest points from the pages P are already known, this range query needs to read only the pages from $P' - P$. Note that it is possible to perform step (6) with a regular range query without affecting the correctness of the algorithm but the modified version reduces the amount of unnecessary page reads and thereby leads to lower I/O cost. In our experiments, we make use of the heuristic suggested by Seeger et al. [20] to minimize the I/O cost for reading a set of disk pages during a range query execution.

6.3 Experimental Results

In order to evaluate our accelerated k-NN-query algorithm, we performed experiments for a large number of datasets. Here we present the results for a synthetic (UNIFORM) and a real dataset (LANDSAT). For each dataset we ran 100 k-NN queries where k varied between 10 and 50 and measured the amount of seeks and page transfers. All queries are density-biased, i.e. query points are picked with higher probability from a region with higher density. The sampling rate is always $1/100$. For the underlying harddisk we assume a 20 MB/s transfer rate and an average seek time of 10 ms. This, together with the measured seek and transfer numbers, is then used to compute the average query response time. All experiments were conducted on a prototype implementation.

Due to space restrictions, we report only on the results for the X-tree [3] index structure (a discussion on other index structures (such as R-tree [10] and VA-file [21]) can be found in the full version of the paper [17]). For comparison, we use the optimal NN-query algorithm proposed by Hjaltason and Samet [11]. The index page capacity is 8 KBytes and the upper part of the index tree is kept in memory. Therefore, only leaf page accesses cause disk I/O.

In the next paragraphs, we compare the performance of the access optimal NN-query algorithm with our accelerated version, a hypothetical algorithm with perfect radius estimator, and the linear scan. The hypothetical algorithm "ORACLE" always picks the correct query radius for the first range query and reads therefore always the minimal number of pages possible for our algorithm. It provides a lower bound of our algorithm's response time. The linear scan reads all pages in a linear fashion and therefore requires no seek operations. It is therefore a good point of reference for judging the impact of filtering out pages by our radius estimates.

Results for Synthetic Data Figure 7 shows the results for the X-tree and 8-dimensional uniform data. For each k-value, we show four I/O costs: for the ac-

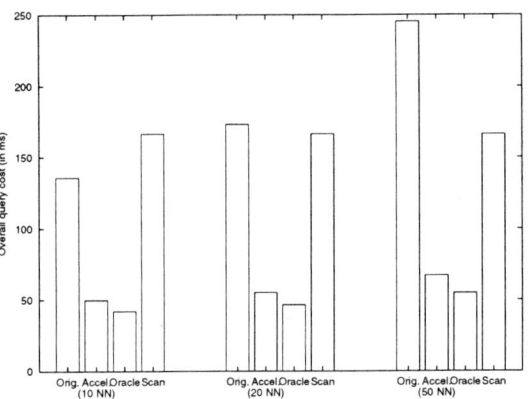

Figure 7. Overall query cost (UNIFORM)

cess optimal query algorithm (denoted by "Orig."), for our accelerated query algorithm (denoted by "Accel."), for ORACLE (denoted by "Oracle"), and for the linear scan (denoted by "Scan"). Each cost value is an average over 100 queries.

Since we use an elaborate bulkloading algorithm to build the tree, the page utilization is high and no page overlaps occur. This improves the query performance drastically. For 10-NN queries, the index is even faster than the linear scan. For larger k, the performance deteriorates again and becomes worse than scanning. When using our acceleration technique, the performance is improved even further, as can be seen in the second bars. Compared to the original X-tree, we

achieve speed-ups between 3 and 5. Additionally, the accelerated query algorithm is at least 3 times faster than the linear scan.

Results for Real Data The experimental results for the LANDSAT dataset can be found in Figure 8. Our accelerated query algorithm outperforms the well-

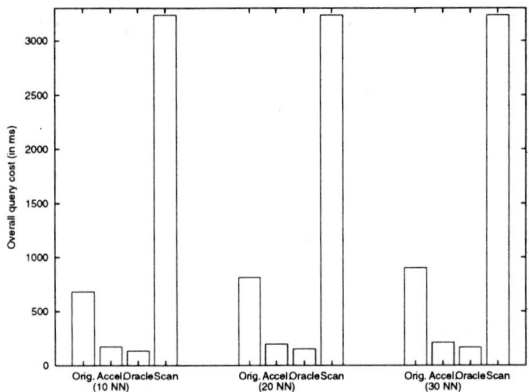

Figure 8. Overall query cost (LANDSAT)

tuned X-tree index structure by a factor of $3 - 5$. The reason is simple. Even though the accelerated query algorithm needs to read more pages, it knows them in advance and can read as many of them sequentially as necessary. In a way, it gets the best of both worlds, the X-tree and the linear scan: it reads nearly as few pages as the optimal NN-query algorithm of the X-tree, and it performs nearly as few seeks as the linear scan.

We want to emphasize that our technique is not to be understood as an improved X-tree or another index structure. It should rather be seen as a general technique for accelerating existing indexing schemes.

7   Accelerated All-NN Algorithm

This section discusses the second application of our improved radius estimators: an accelerated all-NN query algorithm.

7.1   The Block-nested-loops Join Algorithm

Assume, we are given two point datasets, DS_1 and DS_2. For each point in DS_1 we want to determine its NN among all points in DS_2. This problem can easily be generalized to all k-NN queries.

One straightforward way of answering such queries is by performing $|DS_1|$ regular NN queries on DS_2 (which may then be accelerated using our technique presented in the last section). However, this solution causes a

Input:	Dataset DS_1, dataset DS_2
Output:	NNs of all $q \in DS_1$ taken from DS_2

(1) Scan DS_1 in chunks of size $M/2$ (half of the buffer);
    for each chunk of DS_1-pages:
(2)   Read $M/2$ pages from DS_2 (fill other half);
        for each chunk of DS_2-pages:
(3)   Update NNs for each point in the DS_1-pages.
(4)   Output NNs for all points in the DS_1-pages.
(5) Stop.

Figure 9. Block-nested-loops join algorithm

high number of disk seeks for high-dimensional datasets since typical caches are not large enough to store common pages between queries. It would be beneficial to perform these operations in bulks.

One way of achieving this is by utilizing a block-nested-loops join algorithm (cf. for example [19]) since the all-NN problem can be viewed as the problem of joining DS_1 and DS_2 with the join condition being $\sigma(x,y) = $ "x has y as a NN" where $x \in DS_1$ and $y \in DS_2$. The pseudocode for the block-nested-loops join algorithm is given in Figure 9. The algorithm consists mainly of two nested loops, one reading pages from DS_1 and one reading pages from DS_2. These page reads are performed in a sequential fashion in order to reduce disk seeks. We assume that a buffer of size M (in number of pages) is provided. The algorithm fills the first half with DS_1-pages and the second half with DS_2-pages. For each point in a DS_1-page we store additionally its best NNs so far. This information is updated in Step (3).

7.2 The New Algorithm

In this section, we show how our new radius estimators can be used to accelerate the block-nested-loops join algorithm for answering all-NN queries. The basic idea is as follows: instead of estimating the NN-radius for a single point, we estimate it for all points in the DS_1-pages stored in the buffer. Once we have calculated the NN-radius-estimates for all points of a page of DS_1, we can predict which pages of DS_2 have to be accessed in order to compute the join. This is where the I/O cost is reduced significantly compared to a block-nested-loops join algorithm which accesses *every* page of DS_2 for a page of DS_1.

Let us discuss the new algorithm in more depth. It is given in Figure 10. In steps (1) and (2), a so-called *influence region* $R_{expected}(P)$ is computed for every page P in DS_1. This can be seen as a generalization of the computation of $r_{expected}$ for the accelerated k-NN query algorithm. The influence region can be regarded as the union of the expected NN-spheres of all points in P, as shown in Figure 11. In this example, page P

Input:	Dataset DS_1, dataset DS_2
Output:	NNs of all $q \in DS_1$ taken from DS_2

(1) Scan DS_1 in chunks of size M (whole buffer);
    for each chunk of DS_1-pages:
(2)   Compute expected influence region $R_{expected}(P)$
        for each DS_1-page P.
(3) Scan DS_1 in chunks of size $M/2$ (half of the buffer);
    for each chunk of DS_1-pages:
(4)   Read $M/2$ consecutive pages from DS_2
        that intersect at least one influence region
        of the DS_1-pages (fill other half of M);
        for each chunk of DS_2-pages:
(5)   Update NNs for each point in the DS_1-pages.
(6)   Update $R_{expected}(P)$ for each DS_1-page P.
(7)   Read all pages (at most $M/2$ at a time)
        of DS_2 that intersect at least one influence
        region of the DS_1-pages (fill other half of M);
        for each chunk of DS_2-pages:
(8)   Update NNs for each point in the DS_1-pages.
(9)   Output NNs for all points in the DS_1-pages.
(10) Stop.

Figure 10. Accelerated all-NN query algorithm

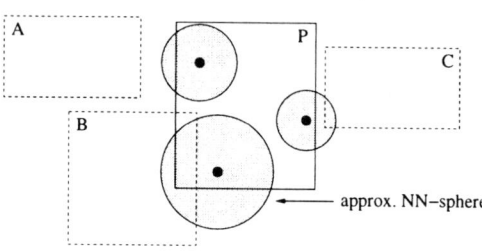

Figure 11. Influence region of page P

contains three points. The influence region is shown as the shaded area. Two of the DS_2-pages (shown as dashed boxes) are intersected by P's influence region and have to be retrieved in Step (4) of the algorithm. Page A would not be retrieved at this time.

In case DS_2-pages that have to be retrieved are less than $M/2$ pages apart on disk, they can be read together in Step (4). This helps reducing disk seeks. In our implementation, we access the $M/2$ pages around the intersected DS_2-page in disk-placement order. Since most index structures try to place pages nearby on the disk that are spatially close in the data space, pages stored before and after the intersected page have a high probability to contain points that are also close by. This leads to better updated estimates in Step (6) of the algorithm.

After $M/2$ DS_2-pages were read in Step (4), the NN-information is updated in Step (5) similar to the block-nested-loops join algorithm. Then the points found in the $M/2$ DS_2-pages are used to improve $R_{expected}(P)$ for each DS_1-page in the buffer. This can be seen as the equivalent to Step (5) of our accelerated k-NN query algorithm. Similar to there, the updated $R_{expected}(P)$ represents an upper bound on the correct influence region of P.

The updated $R_{expected}(P)$ is used in Step (7) to read the pages from DS_2 that are intersected by this upper bound influence region. In the remainder of the algorithm, the NN-info of all points in DS_1-pages are updated and output once all DS_2-pages are processed.

An important observation is that *the accelerated all-NN search algorithm can never perform more than twice as many page accesses as the block-nested-loops join algorithm*. This can be seen as follows: in the worst case, all pages of DS_2 are accessed in line (4) and again in line (7), since Step (4) reads additional (possibly unnecessary) pages that may be refetched in Step (7).

7.3 Experimental Results

For experimental evaluation, we ran two experiments: on two-dimensional synthetic uniform data, and on 60-dimensional real data. The uniform datasets consist of 10,000 points each and the real datasets consist of approximately 13,000 points each extracted randomly from the LANDSAT dataset.

In the case of joins, the overall running times are not necessarily I/O-bound. Due to the high number of distance computations required, the CPU cost becomes an important factor. Hence, we include both costs in our comparisons. For the uniform data, our accelerated join algorithm dropped the overall response time from about 4 minutes to less than 1 minute. In the LAND-SAT case, the response time dropped from 150 minutes to about 40 minutes. The buffer size M (which was varied between 3 and 25% of all pages) had no noticeable impact on this drop.

The largest percentage of the cost savings stems from the reduced number of distance computations performed by our new algorithm. For the lower-dimensional data, only 1/5 of all distance computations have to be performed. For the higher-dimensional data, only 1/4 of the computations are necessary. This shows that the pruning caused by our influence region estimation is very effective. The I/O cost is also reduced in both cases but since distance computations are very costly for high-dimensional data, the algorithm becomes CPU-bound in such settings. More details on the different costs can be found in the full version of the paper [17].

8 Previous Work

Fractal Dimensionality and NN-radius Estimation Faloutsos et al. [9] present the first cost model for R-trees based on the fractal dimensionality which is claimed to be the "inherent dimensionality" of a dataset. This first model is restricted to range queries but later work by Papadopoulos and Manolopoulos [18] extends it for 1-NN queries in R-trees. Korn et al. [14] present a version for k-NN queries.

All of these approaches provide only a query-independent radius estimation which, as we saw in the experiments, leads to high estimation errors. The use of data sampling for estimation purposes in high-dimensional spaces to reduce this error, was first presented by Lang et al. [16].

Accelerated k-NN Query Algorithms The work that comes most closely to our technique was presented by Berchtold et al. [2]. They propose a new index structure, called IQ-tree, and give a probability-based method to optimize page reads during NN queries. The authors estimate the probability that a page will be read during a query and use this probability to decide which pages should be read together with the next page in order to avoid expensive seeks. The probability that a page will be accessed is defined as the percentage of the page volume covered by the NN sphere.

In contrast to our technique, this model assumes uniform data distribution within the pages. This can lead to high estimation errors in the access probability. Since the page reads are controlled by this probability estimate, many unnecessary pages are accessed. Furthermore, at least one page needs to be accessed to get an initial radius estimate.

Chaudhuri and Gravano [7] propose a new technique for translating top-k queries into range queries for relational databases. Donjerkovic and Ramakrishnan [8] present a way of reducing the query space by augmenting the query with some selection similar to our initial range query. In contrast to our work, both techniques focus on low-dimensional queries with 3–4 attributes. Moreover, they do not establish an upper bound on the query radius in order to guarantee at most two range queries.

Accelerated All-NN Query Algorithms Several papers (e.g. [6, 13]) focus on efficient processing of spatial joins on R-trees. However, they do not take seek and transfer costs into account. Instead, they try to minimize the number of page misses during a join. Since these algorithms perform the join by navigating through the index tree rather than through the disk pages, they cause many random page accesses.

Other papers (e.g. [15, 4]) investigate new indexing and query processing techniques in order to reduce the number of page accesses during spatial joins. In contrast to our approach, these techniques do not make use of existing index structures and are restricted to join conditions of the type $\sigma(x, y) =$ "the distance between x and y is less than d".

Hjaltason and Samet [12] propose an incremental join algorithm that returns the closest join pair first. This work does not try to accelerate the search for the full join result as we do.

A different approach by Braunmüller et al. [5] reduces the I/O and CPU-cost of multiple NN-queries. They process one query after the other but take results of older queries into account in order to accelerate newer ones. This process is certainly faster than single NN-queries but it can be expected to perform worse when searching for all NNs since the page accesses are still random. Our approach, on the other hand, tries to maximize the number of sequential page accesses.

9 Conclusions

We showed in this paper how query-dependent query range estimates can be used to accelerate two important query algorithms. As a major contribution, we introduced the notion of query-dependent query range estimates and gave an algorithm for their computation based on sampling. Our analysis indicates that the estimation error for the expected radius is below 14% and that errors in the fractal dimensionality estimation have only minor impact on the accuracy. For the upper bound estimate, the observed error was even lower, namely below 1%.

In the experimental section, we introduced two new algorithms that we accelerated using the estimates discussed in the first part of the paper: one for answering k-NN queries, and one for answering all-NN queries. Our new k-NN query algorithm accelerates a bulk-loaded X-tree index structure by a factor of nearly 5. Even the VA-file, which does not cluster points in pages, benefits from our acceleration. The all-NN-query algorithm we presented outperforms a block-nested-loops join algorithm by a factor of $3-4$. Surprisingly, our technique was able to accelerate significantly query algorithms proposed as part of index structures.

For the future, we plan on applying our technique to index structures storing only a subset of the data dimensions (e.g. reduced via KLT).

References

[1] Alberto Belussi and Christos Faloutsos. Estimating the selectivity of spatial queries using the 'correlation' fractal dimension. In *Proceedings of the Int. Conf. on Very Large Data Bases*, pages 299–310, 1995.

[2] S. Berchtold, C. Böhm, H. V. Jagadish, H.-P. Kriegel, and J. Sander. Independent quantization: An index compression technique for high-dimensional data spaces. In *Proc. Int. Conf. on Data Engineering*, 2000.

[3] Stefan Berchtold, Daniel A. Keim, and Hans-Peter Kriegel. The X-tree : An index structure for high-dimensional data. In *Proceedings of the Int. Conf. on Very Large Data Bases*, pages 28–39, 1996.

[4] Christian Böhm, Bernhard Braunmüller, Florian Krebs, and Hans-Peter Kriegel. Epsilon grid order: An algorithm for the similarity join on massive high-dimensional data. In *Proc. ACM SIGMOD Int. Conf. on Management of Data*, 2001.

[5] B. Braunmüller, M. Ester, H.-P. Kriegel, and J. Sander. Efficiently supporting multiple similarity queries for mining in metric databases. In *Proc. Int. Conf. on Data Engineering*, pages 256–267, 2000.

[6] T. Brinkhoff, H. Kriegel, and B. Seeger. Efficient processing of spatial joins using R-trees. In *Proc. ACM SIGMOD Int. Conf. on Management of Data*, 1993.

[7] Surajit Chaudhuri and Luis Gravano. Evaluating top-k selection queries. In *VLDB*, pages 397–410, 1999.

[8] Donko Donjerkovic and Raghu Ramakrishnan. Probabilistic optimization of top n queries. In *VLDB*, pages 411–422, 1999.

[9] Christos Faloutsos and Ibrahim Kamel. Beyond uniformity and independence: Analysis of R-trees using the concept of fractal dimension. In *Proc. ACM Symp. on Principles of Database Systems*, pages 4–13, 1994.

[10] Antonin Guttman. R-trees: A dynamic index structure for spatial searching. In *Proc. ACM SIGMOD Int. Conf. on Management of Data*, pages 47–57, 1984.

[11] Gisli R. Hjaltason and Hanan Samet. Ranking in spatial databases. In *Advances in Spatial Databases — Fourth International Symposium (SSD)*, pages 83–95, 1995.

[12] Gisli R. Hjaltason and Hanan Samet. Incremental distance join algorithms for spatial databases. In *Proc. ACM SIGMOD Int. Conf. on Management of Data*, pages 237–248, 1998.

[13] Yun-Wu Huang, Ning Jing, and Elke A. Rundensteiner. Spatial joins using r-trees: Breadth-first traversal with global optimizations. In *Proceedings of the Int. Conf. on Very Large Data Bases*, pages 396–405, 1997.

[14] Flip Korn, Bernd-Uwe Pagel, and Christos Faloutsos. Deflating the dimensionality curse using multiple fractal dimensions. In *Proc. Int. Conf. on Data Engineering*, 2000.

[15] Nick Koudas and Kenneth C. Sevcik. High dimensional similarity joins: Algorithms and performance evaluation. In *Proc. Int. Conf. on Data Engineering*, pages 466–475, 1998.

[16] Christian A. Lang and Ambuj K. Singh. Modeling high-dimensional index structures using sampling. In *Proc. ACM SIGMOD Int. Conf. on Management of Data*, 2001.

[17] Christian A. Lang and Ambuj K. Singh. Accelerating high-dimensional nearest neighbor queries. Technical Report CS-TR-0204, Univ. of California at Santa Barbara, January 2002.

[18] Apostolos Papadopoulos and Yannis Manolopoulos. Performance of nearest neighbor queries in R-trees. In *Proc. Int. Conf. Database Theory*, volume 1186 of *Lecture Notes in Computer Science*, pages 394–408, 1997.

[19] R. Ramakrishnan and J. Gehrke. *Database Management Systems (2nd Edition)*. McGraw-Hill, 2000.

[20] Bernhard Seeger, Per-Åke Larson, and Ron McFayden. Reading a set of disk pages. In *Proceedings of the Int. Conf. on Very Large Data Bases*, pages 592–603, 1993.

[21] R. Weber and S. Blott. An approximation based data structure for similarity search. Technical Report 24, ESPRIT project HERMES (no. 9141), October 1997.

Spatio-Temporal Data

Efficient k Nearest Neighbor Queries on Remote Spatial Databases Using Range Estimation *

Danzhou Liu   Ee-Peng Lim   Wee-Keong Ng

Centre for Advanced Information Systems

School of Computer Engineering

Nanyang Technological University, Singapore 639798

{P149571472, aseplim, awkng}@ntu.edu.sg

Abstract

K-Nearest Neighbor (k-NN) queries are used in GIS and CAD/CAM applications to find the k spatial objects closest to some given query points. Most previous k-NN research has assumed that the spatial databases to be queried are local, and that the query processing algorithms have direct access to their spatial indices; e.g., R-trees. Clearly, this assumption does not hold when k-NN queries are directed at remote spatial databases that operate autonomously. While it is possible to replicate some or all the spatial objects from the remote databases in a local database and build a separate index structure for them, such an alternative is infeasible when the database is huge, or there are large number of spatial databases to be queried. In this paper, we propose a k-NN query processing algorithm that uses one or more window queries to retrieve the nearest neighbors of a given query point. We also propose two different methods to estimate the ranges to be used by the window queries. Each range estimation method requires different statistical knowledge about the spatial databases. Our experiments on the TIGER data allow us to study the behavior of the proposed algorithm using different range estimation methods. Apart from not requiring direct access to the spatial indices, the window queries used in the proposed algorithm can be easily supported by non-spatial database systems containing spatial objects.

1   Introduction

1.1   Motivation

The Nearest Neighbor (NN) queries in spatial databases refer to finding the spatial objects nearest to some given query points. NN queries are used in a wide range of applications, such as Geographic Information Systems (GIS), Computer Aided Design (CAD), computational biology, decision support, and pattern recognition [27]. NN queries in spatial databases can be classified into five major categories: simple k-NN queries [2, 6, 8, 9, 17, 21, 22, 26], approximate k-NN queries [3, 10, 14], reverse NN queries [20, 28], constrained k-NN queries [13], and k-NN join queries [16]. In this paper, we focus on simple k-NN queries. Given a set of spatial objects denoted by $\mathbb{S}$, and a distance function d, a simple k-NN query for a query point q is to find the k objects in $\mathbb{S}$ with smallest $d(q, o)$, where $o \in \mathbb{S}$. The query result can be represented as

$$NN(q, k, \mathbb{S}) = \{o_1, \ldots, o_k\},$$

where $d(q, o_i) \leqslant d(q, o') \ \forall i, 1 \leqslant i \leqslant k, o_i \in \mathbb{S}$ and $\forall o' \in \mathbb{S} - \{o_1, \ldots, o_k\}$.

With the rapid growth of the World Wide Web (WWW), large volume of spatial data are now available for access on the Web. For example, the Clearinghouse sponsored by the Federal Geographic Data Committee (FGDC) is a collection of over 250 servers that provides geospatial data on the Web [12]. While some spatial data on the Web are stored in databases managed by spatial database systems, a large proportion of these data may still be stored in data files, or SQL databases. The storage methods used affect the way the spatial data can be queried. For example, for spatial data stored in SQL databases, it is clearly not

---

*This work is funded by the SingAREN Project M48020004

121

possible to adopt a k-NN query algorithm that requires the use of a spatial index such as R-tree. Moreover, to query spatial data on the Web, it is often necessary to use some Web-based query interfaces such as HTML forms. The Web-based query interfaces, unlike spatial query languages, can only support very simple spatial queries but not the complex ones including the k-NN queries. Hence, in our research on evaluating k-NN queries against remote spatial databases, we only assume that the Web-based query interfaces support *window queries*. A window query retrieves spatial objects within a given bounding rectangle also known as the *window*. Window queries are relatively simple, and can be easily supported by most Web-based query interfaces.

In a literature survey, we found that almost all existing k-NN query algorithms require direct access to the spatial indices. Therefore, they cannot be directly applied to remote spatial databases that do not support remote index accesses. One may consider creating a new local spatial index for all the data downloaded from a remote spatial database and directly applying the existing algorithms. Such a strategy, however, does not scale well for large numbers of remote spatial databases and for remote spatial databases containing large amount of data. It also violates the local autonomy of these databases. In other words, new strategies for k-NN query evaluation are required.

1.2 Objectives and Contributions

The main objective of this research is to develop an algorithm to evaluate k-NN queries on remote spatial databases efficiently using window queries. When the window used is just right, one expects exactly k nearest spatial objects to be returned by the window query. When the window used is *loose*, more than the required k nearest neighbors will be returned. On the other hand, when the window is *tight*, fewer than k nearest neighbors will be returned.

We would like to propose the use of statistical knowledge about the remote databases to derive the window queries. The windows used in the window queries can be obtained by different range estimation methods. We would also like to evaluate the performance of our proposed k-NN algorithm using different range estimation methods.

The main contributions of this paper can be summarized as follows. First, it presents a generic k-NN query processing algorithm that can accommodate different range estimation methods. Second, we have developed two range estimation methods; namely, the density-based method and the bucket-based method. Each

method requires a different type of summary knowledge about the remote spatial database and it allows us to derive the window queries. Lastly, the paper describes a series of experiments conducted to evaluate the performance of our proposed k-NN algorithm using the two range estimation methods. To compare them, we have adopted several performance metrics, such as the *iteration*, *efficiency* and *accuracy*.

Apart from assuming that the remote spatial databases support window queries, we have also assumed that the statistical knowledge about each spatial database can be made available. This assumption requires cooperation from the spatial database owners. While the assumption may not always hold, we still adopt it for this preliminary work on k-NN queries on remote spatial databases. We also believe that our proposed methods can be further extended to handle cases where such statistical knowledge cannot be provided by the database owners. For example, we can adopt sampling to collect remote database information. Nevertheless, these extensions of our range estimation methods are beyond the scope of this paper. For simplicity, we have also assumed that the spatial objects are points in 2-D space. With some modifications, our methods should be able to handle more complex spatial objects and spatial objects in a higher dimensional space.

1.3 Outline of the Paper

The remaining sections of this paper are organized as follows. In Section 2, related work is presented. In Section 3, we present a generic k-NN query processing algorithm that can accommodate different range estimation methods. In Section 4, two methods for range estimation are proposed. Section 5 describes our performance experiments and presents the results. Finally, we conclude the paper and highlight our future research directions in Section 6.

2 Related Work

In this section, we survey existing research on k-NN query algorithms. Since our work involves simple k-NN queries, we have only examined work related to simple k-NN queries [2, 6, 8, 9, 17, 21, 22, 26]. Algorithms for simple k-NN queries may be divided into three major groups: *partition-based* algorithms, *graph-based* algorithms, and *range-based* algorithms.

Partition-based algorithms partition the space containing spatial objects recursively to create spatial indices such as quadtrees, K-D trees, and R-trees. The algorithms retrieve the k nearest neighbors from the spatial indices by pruning away nodes that cannot lead to the k nearest neighbors. In addition, cost models for these algorithms are presented in [5, 25]. For example, Roussopoulos *et al.* [26] proposed an algorithm using the R*-tree [4] for simple 1-NN queries and the algorithm can be generalized to handle k nearest neighbors. The algorithm was further extended to reduce more unnecessary disk accesses by Hjaltason and Samet [17]. On the whole, the above partition-based algorithms can be very efficient but they cannot be adopted for querying remote spatial databases on the Web. As integral components of spatial database systems, spatial indices are usually not available to non-local applications. The partition-based algorithms are also not applicable to spatial data managed by non-spatial database systems.

Graph-based algorithms pre-calculate the nearest neighbors of spatial objects and create new index structures for the pre-calculated nearest neighbor information for efficient search [5]. Examples of such algorithms include the RNG* algorithm [2] and the algorithms using Voronoi diagrams [7, 11]. For example, one can first derive the Voronoi diagram for a given set of spatial points followed by indexing the cells in the Voronoi diagram. Given a query point q, finding the nearest neighbor can be simplified to finding the Voronoi cell that contains q. Although graph-based algorithms are very efficient and can be extended to support k-NN queries, they again requires some spatial index to be maintained for the collection of cells. The index may also occupy much storage space. Hence, these algorithms are not feasible in the Web environment.

Range-based algorithms refer to using range queries to retrieve k nearest neighbors. Lang *et al.* [21] proposed an algorithm for transforming a k-NN query into at most two range queries. The first range query is obtained by estimating the required range using sampled spatial data and fractal dimensionality [23] of spatial objects. If the k nearest neighbors are found, the goal is achieved. Otherwise, the second range estimation is needed. The second range is defined as the distance between the query point and the kth nearest neighbor within the index leaf nodes accessed during the first range query. This method, however, still need to access the underlying spatial index. Papadopoulos *et al.* [24] presented alternatives in answering k-NN queries by using range queries in a parallel environment (i.e., spatial objects are stored in a spatial index and the whole index is distributed over a number of servers). Their work, motivated by [26] and to exploit parallelization as much as possible, also assumes that the spatial index is accessible. Ciaccia *et al.* [9] employed the relative distance distributions of several "witnesses" se-

lected in some way among spatial objects to estimate the range. Yu *et al.* [22] transformed k-NN queries to one-dimensional range queries by partitioning spatial objects, selecting a reference point for each partition, and using B$^+$-tree to index distance between spatial objects and their corresponding reference point. Both Ciaccia's and Yu's methods approximate the range by employing some sampled spatial objects. However, determining the sample size and selecting samples of spatial objects properly are still a challenge, and improper sampling will result in too much inaccuracy in estimating the data distribution, and/or storage overhead.

Compared with the above three types of k-NN algorithms, our proposed solution does not rely on access to the spatial index of the databases. Our density-based method derives some statistical knowledge about the distribution density of a set of spatial objects. The storage requirement of the statistical knowledge required is much smaller than that of the spatial indices used in the above k-NN algorithms. In [1, 18, 19, 29], various two-dimensional histograms have been proposed to estimate the selectivity of range queries. Our bucket-based range estimation method is very much inspired by their work.

3 k-NN Query Algorithm based on Range Estimation

In this section, we will outline our proposed algorithm for k-NN queries. Unlike the other algorithms, our algorithm transforms a k-NN query into one or more **window queries** without using any spatial index. The **window** used in a window query can be represented by $[(x_l, y_l), (x_u, y_u)]$, where (x_l, y_l) and (x_u, y_u) refer to its lower-left and upper-right corners respectively. Given a set of spatial objects denoted by $\mathbb{S}$ and a window w, a window query about w refers to finding all spatial objects in $\mathbb{S}$ located in w [15]. The window query result can be represented as

$$\{o \in \mathbb{S} \mid (x_l \le o.x \le x_u) \land (y_l \le o.y \le y_u)\}.$$

Ideally, we would like to retrieve exactly k nearest neighbors for a given query point by retrieving all spatial objects within a circle with the query point as the center and the distance between the query point and the kth nearest neighbors as the radius. Since our spatial databases are assumed to support window queries but not circle queries, we approximate a circle query by defining a window query with a window that inscribes the circle as illustrated in Figure 1.

Our proposed k-NN query algorithm is shown in Figure 2. This algorithm is designed to be generic enough

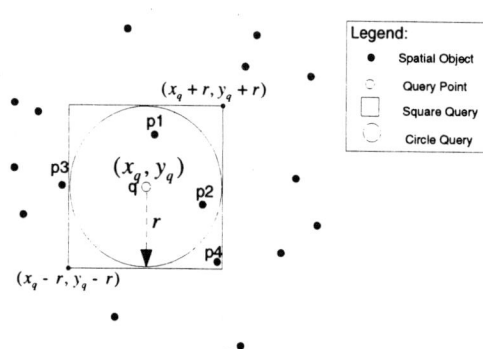

Figure 1. Example of Window Query

Input:
   required number of nearest neighbors k
   query point q
Output:
   k nearest neighbors $nnqueue$

```
01    range ← ESTIRANGE1(k, q)
02    window ← [(x_q − range, y_q − range), (x_q +
      range, y_q + range)]
03    rlist ← WINDOWQUERY(window)
04    nnqueue ← NEWPRIORITYQUEUE(k)
05    count ← 0
06    for each object in rlist do
07        distance ← DIST(object, q)
08        if distance ≤ range then
09            ENQUEUE(nnqueue, (object, distance))
10            count ← count + 1
11        endif
12    enddo
13    while count < k do
14        range ← ESTIRANGE2(k, q, window, count)
15        window ← [(x_q − range, y_q − range), (x_q +
          range, y_q + range)]
16        rlist ← WINDOWQUERY(window)
17        count ← 0
18        for each object in rlist do
19            distance ← DIST(object, q)
20            if distance ≤ range then
21                ENQUEUE(nnqueue, (object, distance))
22                count ← count + 1
23            endif
24        enddo
25    enddo
26    return nnqueue
```

Figure 2. k-NN Query Algorithm

4   Range Estimation Methods

In this section, we will present two different range estimation methods; namely, *density-based* and *bucket-based* methods. The first method yields tight ranges, and the second one yields loose ranges.

4.1   Density-Based Method

Density-based range estimation method is based on uniform distribution assumption. This method requires only a very simple piece of statistical knowledge about a spatial database; i.e., the *density* of the database. Given a database of spatial objects, we define the minimum bounding box (MBB) of the database as the minimum rectangle containing all the spatial objects. The **density** of the database, denoted by $D(MBB)$, is defined as $\frac{N}{Area(MBB)}$ where N is the total number of spatial objects. Given k and a query point q, the density-based range estimation method computes the first range estimate by

to accommodate different range estimation methods. At line 1, the EstiRange1 function first estimates the range (or radius) of the circle query to retrieve the k nearest neighbors of a given query point. The algorithm then derives the window query from the estimated range and evaluates the window query. The window query result is inserted into *rlist* at line 3. At line 4, we create an empty priority queue *nnqueue* of size k to maintain the k nearest neighbors. Only those spatial objects with a distance from q that is not larger than *range* are inserted into *nnqueue* from lines 6 to 12. For example, in Figure 1, $p4$ will not be inserted into *nnqueue* because it is not within the circle with the query point as the center and r as the range. At line 13, *count* represents the number of nearest neighbors retrieved so far. If *count* is larger than or equal to k, all the k nearest neighbors would have been found and stored in *nnqueue*. Otherwise, more nearest neighbors have to be obtained by expanding the range. The expanded range can be computed from the current window and *count* using the EstiRange2 function at line 14. With a revised range, a new window query is evaluated against the spatial database. The whole process ends when all the k nearest neighbors are found.

Depending on the number of window queries required to retrieve all k nearest neighbors, this generic k-NN algorithm may involve one or more iteration. We say that a range estimation method is **loose** when the range given by the EstiRange1 function is large enough to cover all the k nearest neighbors, and EstiRange2 is not required at all. On the other hand, we say that a range estimation method is **tight** when the EstiRange1 function may return less than k nearest neighbors. Hence, the EstiRange2 function may be invoked to derive the revised range(s) to be used in the second or further window queries.

ESTIRANGE1(k, q)

Input:
   required number of nearest neighbors k
   query point q
Output:
   estimated range r

01    $r \leftarrow \sqrt{\frac{k}{\pi D(MBB)}}$
02   **return** r

ESTIRANGE2(k, q, $window$, $count$)

Input:
   required number of nearest neighbors k
   query point q
   previous window $window$
   the number of nearest neighbors retrieved so far $count$
Output:
   estimated range r

01   **if** $count == 0$ **then**
02      $r \leftarrow 2 \times \frac{(window.x_u - window.x_l)}{2}$
03   **else**
04      $D(window) \leftarrow \frac{count}{(window.x_u - window.x_l) \times (window.y_u - window.y_l)}$
05      $r \leftarrow \sqrt{\frac{k}{\pi D(window)}}$
06   **endif**
07   **return** r

Figure 3. Density-based Range Estimation Method

$r = \sqrt{\frac{k Area(MBB)}{\pi N}} = \sqrt{\frac{k}{\pi D(MBB)}}$. The corresponding ESTIRANGE1 function is shown in Figure 3. Given the range, the window used in the first window query is $[(x_q - r, y_q - r), (x_q + r, y_q + r)]$.

The above range estimate is not loose, and the corresponding window query may return less than k nearest neighbors. When that happens, further refinement on the range estimate is required. If the first window query does not return any spatial objects, we simply double the original range r. Otherwise, we compute the density of spatial objects within the window and derive the next range estimate by the new density information. Let $D(window)$ be the density of the $window$ used in the last window query, the new range estimate is derived by $\sqrt{\frac{k}{\pi D(window)}}$. The corresponding range estimation function ESTIRANGE2 is shown in Figure 3.

The above density-based range estimation method demonstrates some important features. Firstly, the space required to store the density information takes only several bytes. Thus, the space complexity is $\mathcal{O}(1)$. Secondly, every time a new range estimate is required; it is derived from the density of the window used in the previous window query. This range estimation method further guarantees that the estimated range increases monotonically. The upper bound on the number of times the range estimation functions are called (or number of window queries) for a k-NN query can be determined using the following theorem.

Theorem 1 *Let the MBB of a spatial database that contains N spatial objects be $[(x_l, y_l), (x_u, y_u)]$. Assume that the query point q is inside the MBB. In order to retrieve k nearest neighbors, the upper bound on the number of times the range estimation functions are called (denoted by I_{max}) for the density-based method is*

$$I_{max} = \begin{cases} \left\lceil \frac{\ln r_{max} - \ln r_0}{\ln 2} \right\rceil + 1 & if \quad k = 1; \\[2em] \left\lceil \frac{\ln r_{max} - \ln r_0}{\ln \sqrt{\frac{4k}{\pi(k-1)}}} \right\rceil + 1 & if \quad k > 1. \end{cases}$$

where

$$r_0 = \sqrt{\frac{k(x_u - x_l)(y_u - y_l)}{\pi N}}$$

and

$$r_{max} = \sqrt{(x_u - x_l)^2 + (y_u - y_l)^2}.$$

Proof:
According to the given conditions, the first range, denoted by r_0, estimated by ESTIRANGE1 is

$$r_0 = \sqrt{\frac{k}{\pi D(MBB)}} = \sqrt{\frac{k(x_u - x_l)(y_u - y_l)}{\pi N}}.$$

The function ESTIRANGE2 always returns a range value (denoted by r_i) that is larger than the previous range (denoted by r_{i-1}). That is,

$$r_i = \begin{cases} 2r_{i-1} & if \quad count = 0; \\[1.5em] \sqrt{\frac{4k}{\pi \times count}} r_{i-1} & if \quad 0 < count \le k - 1. \end{cases}$$

The maximum value for r_i, denoted by r_{max}, is the length of the MBB's diagonal, $\sqrt{(x_u - x_l)^2 + (y_u - y_l)^2}$. For any query point inside the MBB, the window query with the range r_{max} will retrieve all the spatial objects including the k nearest neighbors.

There are two cases to be considered for calculating I_{max}. Consider the first case when $k = 1$. The maximum number of calls to the range estimation functions is achieved when the last window query retrieves the nearest neighbor with a range not less than r_{max} and the previous window queries retrieve none; i.e., $count = 0$. In other words, ESTIRANGE1 is called once, while EstiRange2 is called $I_{max} - 1$ times with $count = 0$. That is, I_{max} should be the smallest integer such that

$$r_{max} \leq 2^{I_{max}-1} r_0$$

Hence,

$$I_{max} = \left\lceil \frac{\ln r_{max} - \ln r_0}{\ln 2} \right\rceil + 1, \text{ if } k = 1.$$

Consider the second case when $k > 1$. To have maximum number of calls to EstiRange2, r_i should increase with the smallest rate. This is achieved when $\sqrt{\frac{4k}{\pi \times count}}$ is as small as possible. In other words, the value of $count$ need to be $k - 1$. The range r_i will therefore increase at the smallest rate of $\sqrt{\frac{4k}{\pi \times (k-1)}}$.

Then we can derive I_{max} as the smallest integer such that:

$$r_{max} \leq \left(\sqrt{\frac{4k}{\pi(k-1)}} \right)^{I_{max}-1} r_0$$

Hence,

$$I_{max} = \left\lceil \frac{\ln r_{max} - \ln r_0}{\ln \sqrt{\frac{4k}{\pi(k-1)}}} \right\rceil + 1, \text{ if } k > 1.$$

4.2 Bucket-Based Method

In the bucket-based range estimation method, we use summary information about partitions or buckets of spatial objects to estimate ranges. Buckets are created by dividing the entire space into different groups [1]. For each bucket, we maintain its minimum bounding box (MBB) and the total number of spatial objects inside the bucket.

In [1], Acharya *et al.* proposed four strategies to create buckets. They are the *Equi-Count*, *Equi-Area*, *Min-Skew* and *Min-Overlap* partitioning strategies. The *Equi-Count* partitioning strategy creates buckets containing roughly the same number of spatial objects. The *Equi-Area* partitioning strategy creates buckets with MBBs having the same area. The *Min-Skew* partitioning strategy divides spatial objects into buckets such that each bucket contains uniformly distributed spatial objects. The *Min-Overlap* partitioning strategy is derived from R*-tree [4] and it creates buckets that have minimal overlaps among them.

Our bucket-based range estimation method may adopt any of the above partitioning strategies. In our experiments, we have implemented the *Equi-Area* partitioning strategy. As shown in Figure 5, the range estimation method first calculates the maximum distance

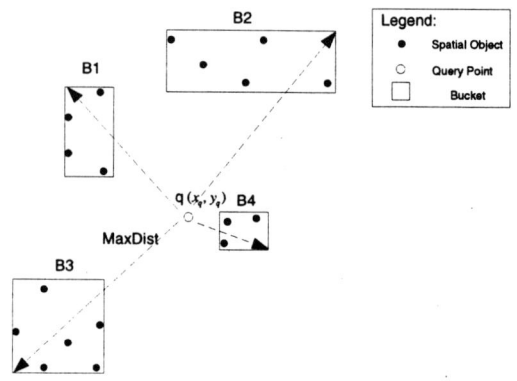

Figure 4. Example of Bucket-based Range Estimation Method

between the query point and each bucket from lines 1 to 3. The **maximum distance** here is defined as the farthest distance between the query point and the MBB of the bucket. Afterwards, we sort the buckets by their maximum distances in ascending order at line 4. From lines 7 to 13, the method finds the first few buckets in the ordered list that can return k or more spatial objects. For example, as shown in Figure 4, the buckets are sorted by their maximum distance in the following order: B4, B1, B2, B3. Suppose k is 4. B4 and B1 together return 7 spatial objects while B4 only returns 3. Hence, the estimated range is assigned the maximum distance between B1 and query point. It is quite obvious the bucket-based range estimation method is *loose* since it always provides an estimated range that covers all the k nearest neighbors. The EstiRange2 function is therefore not required.

The performance of the bucket-based range estimation method is affected by the distribution of spatial objects in the buckets and the number of buckets. If the number of buckets is too small, it will likely overestimate the range leading to the retrieval of many unwanted spatial objects. While more buckets will yield better performance, it will require more storage overhead. Suppose we need 16 bytes to store both the upper-right corner and lower-left corner of the MBB of a bucket, and 4 bytes to store the number of spatial objects within the bucket (i.e., *bucket.count*). The storage space required to store all the bucket information is $20 N_B$, where N_B is the total number of buckets. Hence, the space complexity is $\mathcal{O}(N_B)$.

ESTIRANGE1(k, q)

Input:
   required number of nearest neighbors k
   query point q
Output:
   the estimated range r

```
01    for each bucket in the BucketList do
02        bucket.maxdist ← MAXDIST(bucket, q)
03    enddo
04    BucketList.sort()      // sort buckets by the
maximum distance to q
05    count ← 0
06    maxdist ← 0
07    for each bucket in the BucketList do
08        count ← count + bucket.count
09        maxdist ← MAX(maxdist, bucket.maxdist)
10        if(count >= k) then
11            break
12        endif
13    enddo
14    return maxdist
```

Figure 5. Bucket-based Range Estimation Method

5   Experiments

In our experiment, we used the *NJ Road* dataset from TIGER [30]. This dataset contains the road data for the state of New Jersey in the line segment format. In our experiments, we calculated the centroid of each line segment, and randomly selected almost 5,000 points among all centroids. The selected centroids are shown in Figure 6. The latitude and longitude of the original centroids were shifted for the computational convenience.

To measure the performances of our methods, we adopted three measures; namely, *iteration*, *average accuracy*, and *average efficiency* [31]. *Iteration* refers to the number of window queries needed to obtain all k nearest neighbors. Average accuracy refers to the average ratio of the number of actual nearest neighbors retrieved to k in each window query. Mathematically, $accuracy_{avg} = \frac{\sum_{i=1}^{iteration} accuracy_i}{iteration}$, where $accuracy_i = \frac{nn_i}{k}$ (nn_i denotes the number of actual nearest neighbors retrieved in the ith window query). Average efficiency is the average ratio of the number of actual nearest neighbors retrieved to the number of spatial objects retrieved in each window query. Formally, $efficiency_{avg} = \frac{\sum_{i=1}^{iteration} efficiency_i}{iteration}$, where $efficiency_i = \frac{nn_i}{o_i}$ (o_i denotes the number of spatial objects retrieved in the ith window query). For some window queries that do not return any spatial objects, their efficiency is undefined and hence excluded from the com-

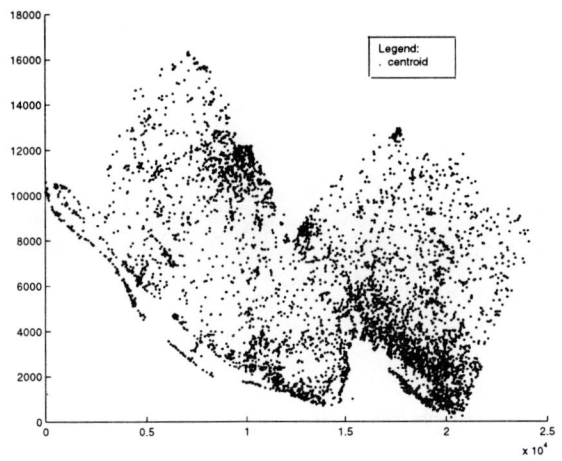

Figure 6. NJ Road Dataset

Iteration	k						
	1	5	10	15	20	25	50
minimum	1	1	1	1	1	1	1
maximum	8	7	7	6	6	6	5
upper bound	9	20	25	26	27	27	26

Table 1. Minimum, Maximum and Upper Bounds on the number of Iterations for Density-based Method

putation of the average efficiency measure. Ideally, all the measures are 1. If a method requires fewer iterations to retrieve k nearest neighbors and the average accuracy and efficiency are high, this method is considered good.

In our experiments, we used different k values, $k \in \{1, 5, 10, 15, 20, 25, 50\}$. We randomly selected 100 points in the space as query points. For each k, the average of the three measures (i.e., iteration, average accuracy and average efficiency) were taken for the 100 query points. In addition, for the bucket-based method, we only calculated average efficiency. Other measures are omitted since the method uses loose range estimation (i.e., *iteration* = 1, and $accuracy_{avg} = 1$).

Figures 7 to 9 depict the performance results of the k-NN query algorithm using density-based and bucket-based range estimation methods. For the bucket-based method, we experimented with different bucket numbers; i.e., 64, 100, and 256.

As shown in Figure 7, the k-NN query algorithm based on density-based range estimation method requires fewer number of iterations (window queries) for each k-NN query as k increases. When $k = 1$, an average of about 4.25 iterations (window queries) are re-

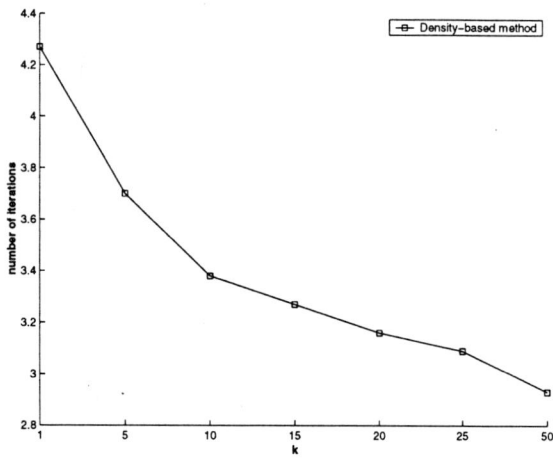

Figure 7. Number of Iterations for Density-based Range Estimation Method

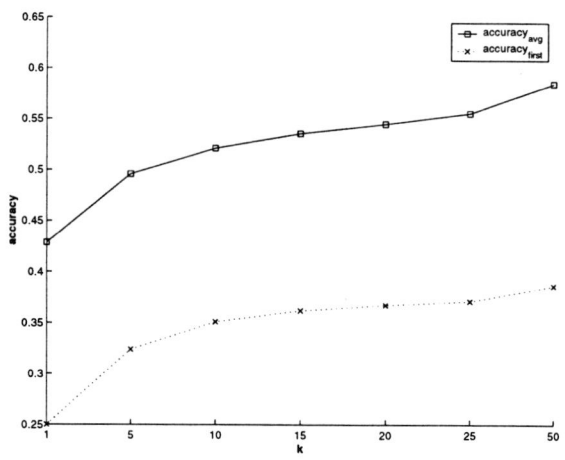

Figure 8. Accuracy for Density-based Range Estimation Method

quired. The number of window queries drops to about 3 when $k = 50$. This suggests that the density-based method works better for larger k's. In Table 1, the minimum and maximum numbers of iterations for each k value are given. The maximum number of iterations among all the k-NN queries ranges from 5 to 8. Using Theorem 1, we derived the upper bounds on the number of iterations and show them in Table 1. It was encouraging to see the maximum numbers of iterations occurred during our experiments stayed well within the upper-bounds.

Figure 8 shows the accuracy of k-NN query algorithms based on density-based method. The average accuracy ranges from 0.43 (when $k = 1$) to 0.58 (when $k = 50$). Again, the density-based method delivers better performance for larger k. To give an idea how the density-based method performs in the first window query, we also showed the average accuracy for the first window query ($accuracy_{first}$) in the figure. The figure shows that the accuracy of the first window query is about 0.2 lower than the average of all window queries. On average, about 25% of the k-NN queries got the nearest neighbors for $k = 1$. As k approaches 50, the first window query returns on average more than 35% of the nearest neighbors.

In Figure 9, we observe that both the density-based and bucket-based methods improve their efficiency as k increases. The figure also depicts that the density-based method has a better efficiency. Efficiency is a measurement of the proportion of nearest neighbors in the spatial objects returned by the window query. The figure suggests that the bucket-based method always over-estimates the ranges required to find nearest

neighbors as it uses only one window query for each k-NN query. On the other hand, the density-based method usually gives much smaller ranges for the window queries. The figure, in some way, surprised us as the efficiency of the density-based method was lower than our original expectation. Due to its conservative approach of increasing the window size, we expected perfect efficiency (i.e., 1) for the density-based method for its first few window queries. After a more detailed investigation, we realized that the efficiency of some initial window queries of the density based method could be zero as some window queries retrieved spatial objects (some may qualify as nearest neighbors) which are not within their corresponding circular ranges. Since the k-NN algorithm can only consider objects within these circular ranges as nearest neighbors, such window queries will have zero efficiencies. Since density-based range estimation method only returns all the k nearest neighbors in the last iteration, we also computed the efficiency of the last iteration (denoted by $efficiency_{last}$) for the density-based method as shown in Figure 9. The efficiency of the last iteration (window query) for the density-based method is clearly better than that of bucket-based method although the gap between them becomes closer as k increases.

6 Conclusions

In this paper, we describe a window query approach to evaluate k-NN queries on remote spatial databases. Our research has been motivated by the large amount of spatial information on the Web and their limited query interface. While most existing k-NN research as-

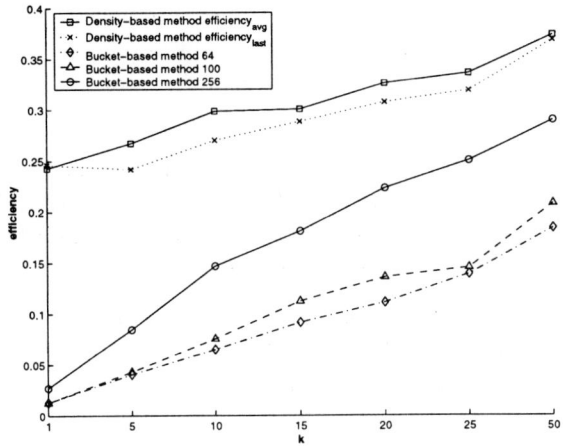

Figure 9. Efficiency for Density-based and Bucket-based Range Estimation Methods

sumes direct access to the indices of spatial databases, we have adopted a less intrusive approach to evaluate k-NN queries.

We have described a k-NN query algorithm that incorporates different range estimation methods for determining the window queries to be used for the remote databases. We have also proposed the density-based and bucket-based range estimation methods, and conducted experiments to evaluate their performance. Our experiments have shown that the k-NN query algorithm based on both range estimation methods improves as k increases. While the density-based method has better efficiency, it requires an average of 3 to 4 window queries to find all the nearest neighbors.

In the following, we outline two interesting topics for future research. These include:

- Extending our range estimation methods with sampling techniques. At present, our range estimation methods depend on statistical knowledge provided by the database owners. Although this is less intrusive compared with the k-NN algorithms using indices, it is still not good enough in the realistic environment. We therefore plan to investigate how the statistical knowledge can be automatically constructed using sampling techniques.

- Developing strategies to select the appropriate range estimation methods for evaluating k-NN queries. The density-based and bucket-based methods are only the first two range estimation methods proposed so far. We anticipate more range estimation methods could be developed in the future. It is therefore important to study how

the different methods should be chosen for a given remote spatial database.

References

[1] S. Acharya, V. Poosala, and S. Ramaswamy. Selectivity estimation in spatial databases. In *Proc. of the ACM SIGMOD Int. Conf. on Management of Data*, pages 13–24, Philadelphia, 1999.

[2] S. Arya. *Nearest Neighbor Searching and Applications*. PhD thesis, Department of Computer Science, University of Maryland, College Park, MD, USA, 1995.

[3] S. Arya, D. M. Mount, N. S. Netanyahu, R. Silverman, and A. Y. Wu. An optimal algorithm for approximate nearest neighbor searching fixed dimensions. *Journal of the ACM*, 45(6):891–923, 1998.

[4] N. Beckmann, H. P. Kriegel, R. Schneider, and B. Seeger. The R*-tree: an efficient and robust access method for points and rectangles. In *Proc. of the ACM SIGMOD Int. Conf. on Management of Data*, pages 322–331, Atlantic City, NJ, USA, 1990.

[5] S. Berchtold, C. Bohm, D. A. Keim, and H. P. Kriegel. A cost model for nearest neighbor search in high-dimensional data space. In *Proc. of the ACM Symposium on Principles of Database Systems*, pages 78–86, Tucson, AZ, USA, 1997.

[6] S. Berchtold, B. Ertl, D. A. Keim, H. P. Kriegel, and T. Seidl. Fast nearest neighbor search in high-dimensional spaces. In *Proc. of the 14th Int. Conf. on Data Engineering*, pages 23–27, Orlando, FL, USA, 1998.

[7] S. Berchtold, D. A. Keim, H. P. Kriegel, and T. Seidl. Indexing the solution space: A new technique for nearest neighbor search in high-dimensional space. *IEEE Trans. on Knowledge and Data Engineering*, 12(1), 2000.

[8] K. L. Cheung and A. W. C. Fu. Enhanced nearest neighbor search on the R-tree. *SIGMOD Record*, 27(3):16–21, 1998.

[9] P. Ciaccia, A. Nanni, and M. Patella. A query-sensitive cost model for similarity queries with M-tree. In *Proc. of the 10th Australasian Database Conf. (ADC'99)*, pages 65–76, Auckland, New Zealand, 1999.

[10] P. Ciaccia and M. Patella. PAC nearest neighbor queries: Approximate and controlled search in high-dimensional and metric spaces. In *Proc. of the 16th Int. Conf. on Data Engineering*, pages 244–255, San Diego, CA, USA, 2000.

[11] K. Clarkson. A randomized algorithm for closest-point queries. *SIAM Journal of Computing*, 17:830–847, 1988.

[12] The Federal Geographic Data Committee. *The Clearinghouse.* URL:http://www.fgdc.gov/clearinghouse.

[13] H. Ferhatosmanoglu, I. Stanoi, D. Agrawal, and A. E. Abbadi. Constrained nearest neighbor queries. In *Proc. of the 7th Int. Symposium on Spatial and Temporal Databases (SSTD)*, Los Angeles, CA, USA, 2001.

[14] H. Ferhatosmanoglu, E. Tuncel, D. Agrawal, and A. E. Abbadi. Approximate nearest neighbor searching in multimedia databases. In *Proc. of the 17th Int. Conf. on Data Engineering*, Heidelberg, Germany, 2001.

[15] V. Gaede and O. Günther. Multidimensional access methods. *ACM Computing Surveys*, 30(2):170–231, 1998.

[16] G. R. Hjaltason and H. Samet. Incremental distance join algorithms for spatial databases. In *Proc. of the ACM SIGMOD Int. Conf. on Management of Data*, pages 237–248, Seattle, WA, USA, 1998.

[17] G. R. Hjaltason and H. Samet. Distance browsing in spatial databases. *ACM Trans. on Database Systems*, 24(2):265–318, 1999.

[18] J. Jin, N. An, and A. Sivasubramaniam. Analyzing range queries on spatial data. In *Proc. of the 16th Int. Conf. on Data Engineering*, pages 589–598, San Diego, CA, USA, 2000.

[19] F. Korn, T. Johnson, and H. V. Jagadish. Range selectivity estimation for continuous attributes. In *Proc. of Int. Conf. on Scientific and Statistical Database Management*, pages 244–253, Cleaveland, OH, USA, 1999.

[20] F. Korn and S. Muthukrishnan. Influence sets based on reverse nearest neighbor queries. In *Proc. of the ACM SIGMOD Int. Conf. on Management of Data*, Dallas, TX, USA, 2000.

[21] C. A. Lang and A. K. Singh. A framework for accelerating high-dimensional NN-queries. Technical Report TRCS01-04, University of California, Santa Barbara, 2001.

[22] B. C. Ooi, C. Yu, K. L. Tan, and H. V. Jagadish. Indexing the distance: an efficient method to KNN processing. In *Proc. of the 27th Int. Conf. on Very Large Data Bases*, Roma, Italy, 2001.

[23] B. U. Pagel, F. Korn, and C. Faloutsos. Deflating the dimensionality curse using multiple fractal dimensions. In *Proc. of the 16th Int. Conf. on Data Engineering*, pages 589–598, San Diego, CA, USA, 2000.

[24] A. Papadopoulos and Y. Manolopoulos. Nearest neighbor queries in shared-nothing environments. *GeoInformatica*, pages 369–392, 1997.

[25] A. Papadopoulos and Y. Manolopoulos. Performance of nearest neighbor queries in R-trees. In *Proc. of the 6th Int. Conf. on Database Theory*, pages 394–408, Delphi, Greece, 1997.

[26] N. Roussopoulos, S. Kelley, and F. Vincent. Nearest neighbor queries. In *Proc. of the ACM SIGMOD Int. Conf. on Management of Data*, pages 71–79, San Jose, CA, USA, 1995.

[27] S. Shekhar, S. Chawla, S. Ravada, A. Fetterer, X. Liu, and C. T. Lu. Spatial databases - accomplishments and research needs. *IEEE Trans. on Knowledge and Data Engineering*, 11(1):45–55, 1999.

[28] I. Stanoi, D. Agrawal, and A. E. Abbadi. Reverse nearest neighbor queries for dynamic databases. In *ACM SIGMOD Workshop on Research Issues in Data Mining and Knowledge Discovery*, pages 44–53, Dallas, TX, USA, 2000.

[29] Y. Theodoridis and T. Sellis. A model for the prediction of R-tree performance. In *Proc. of the 14th ACM Symposium on Principles of Database Systems*, pages 161–171, Montreal, Canada, 1996.

[30] U.S. Bureau of the Census. *Tiger/line files.* URL:http://www.census.gov.

[31] C. Yu, P. Sharma, W. Y. Meng, and Y. Qin. Database selection for processing k nearest neighbors queries in distributed environments. In *Proc. of the ACM/IEEE Joint Conf. on Digital Libraries*, pages 215–222, Roanoke, VA, USA, 2001.

A Cost Model for Interval Intersection Queries on RI-Trees

Hans-Peter Kriegel[*], Martin Pfeifle[*], Marco Pötke[**], Thomas Seidl[*]

[*]University of Munich, Institute for Computer Science
{kriegel, pfeifle, seidl}@dbs.informatik.uni-muenchen.de
[**]sd&m AG software design & management, marco.poetke@sdm.de

Abstract

The efficient management of interval data represents a core requirement for many temporal and spatial database applications. With the Relational Interval Tree (RI-tree[1]), an efficient access method has been proposed to process interval intersection queries on top of existing object-relational database systems. This paper complements that approach by effective and efficient models to estimate the selectivity and the I/O cost of interval intersection queries in order to guide the cost-based optimizer whether and how to include the RI-tree into the execution plan. By design, the models immediately fit to common extensible indexing/ optimization frameworks, and their implementations exploit the built-in statistics facilities of the database server. According to our experimental evaluation on an Oracle database, the average relative error of the estimated cost to the actual cost of index scans ranges from 0% to 23%, depending on the resolution of the persistent statistics and the size of the query objects.

1. Introduction

There is a growing demand for database applications to handle intervals which, for instance, occur as transaction time and valid time ranges in temporal databases [31] [26] [4] or as line segments on a space-filling curve in spatial applications [9] [3]. The SQL:1999 standard provides the datatype PERIOD with the predicates *precedes, succeedes, meets, equals, overlaps (= intersects), contains,* and *during* [30]. With the Relational Interval Tree[1] (RI-tree), a relational access method has been proposed which supports all of the PERIOD predicates [18].

Highly accurate but still efficient selectivity estimation and cost prediction are the fundamentals of effective query optimization. As pointed out in [28], standard selectivity estimation does not estimate well the result cardinalities of se-

lections having temporal predicates, and standard built-in methods are not directly suitable for interval intersection queries, in particular. For complex query objects and query predicates, the recent object-relational database servers provide extensible optimization frameworks that come along with the extensible indexing frameworks, in order to complete the seamless integration of user-defined index structures into the declarative DML. As an example for such an extension, we propose a cost model for the RI-tree that fits well to the extensible frameworks by design. Though the RI-tree immediately maps intervals to built-in B+-trees, the built-in cost models for B+-trees do not estimate well the processing cost since they do not take the particular structure and partitioning of interval data into account.

Our techniques aim at the collection of statistics, the estimation of selectivity, and the prediction of I/O cost. Thereby, the optimizer of the database system is enabled to place the user-defined index at its optimal position in the query execution plan. According to [5] and [13], such a cost-based approach is preferable to rule-based approaches when referencing user-defined methods as predicates. The two main design aspects for the above mentioned functions are:

Effectiveness. The extensible optimizer uses the selectivity estimation to determine a good join order for complex SQL queries. It then evaluates the available cost models to choose the most efficient access path to the data. The objective is to keep the relative error of selectivity and cost estimations sufficiently small to rank the user-defined index accurately among alternative access methods.

Efficiency. In order to obtain an efficient execution plan for a DML operation, the optimizer framework calls the estimation functions for each contained interval predicate. To reduce the total runtime of query optimization, the execution cost for the estimation functions should be kept minimal. Furthermore, data statistics required for the estimation functions should also be efficiently collected.

The architecture of extensible optimization is analogous to extensible indexing as illustrated in Figure 1: Whereas the new methods are built on top of the relational SQL layer, they are object-relationally embedded by implementing the

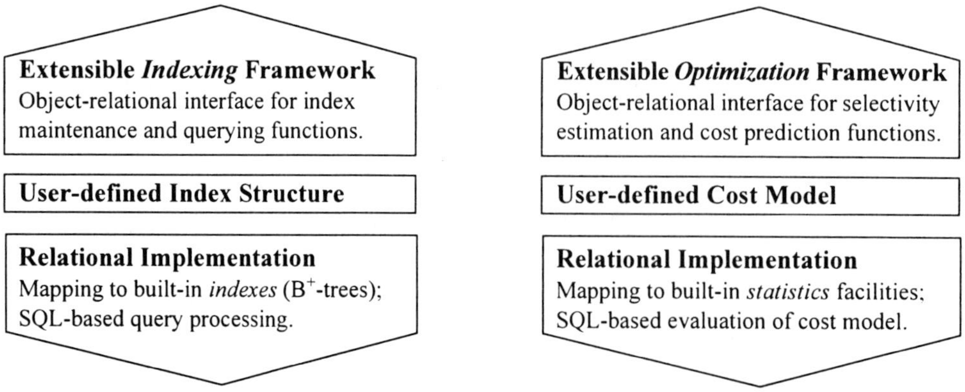

Fig. 1. Analoguous architectures for the object-relational embedding of user-defined index structures and cost models into extensible indexing and optimization frameworks, respectively.

respective interfaces of the frameworks. In case of our new cost models, we particularly propose methods to estimate the selectivity of a given range query on a database of intervals (function *getSelectivity*) and a method to predict the cost of processing that query (function *getIndexCost*). In this paper, we focus on the predicate *overlaps* (= *intersects*) which is considered to be the most important one [10]. Nevertheless, the RI-tree supports all of the PERIOD predicates as well [18] and, moreover, has been also successfully applied to spatial queries [19]. The proposed cost model can easily be extended to these kinds of queries.

The organization of the paper follows the requirements of extensible optimization frameworks and proceeds in the following way: After sketching related work on selectivity estimation and cost prediction in Section 2, Section 3 briefly recalls the RI-tree [17]. In Section 4, we propose two approaches to estimate the selectivity of intersection queries on interval data. The first approach is based on user-defined histograms, whereas the second one relies on the built-in statistics of standard database systems. Section 5 derives a cost model for estimating the I/O cost of a given query on the RI-tree. After an empirical evaluation of the presented methods in Section 6, the paper is concluded in Section 7.

2. Related Work

2.1. Selectivity Estimation

In order to determine a good estimate for the selectivity of a specific predicate without retrieving the actual results, the predicate has to be evaluated on a sufficiently accurate approximation of the data distribution. The computation of such an approximation is known as one of the most difficult problems, for instance in case of selectivity estimation of extended objects [1]. The many existing approaches fall into three different classes: parametric techniques, sampling, and statistics.

Parametric techniques approximate the given data by using a standard mathematical distribution. For databases comprising extended objects, many proposals exploit intrinsic characteristics of the stored data, including the usage of the *Correlation Fractal Dimension* on point sets by Belussi and Faloutsos [2] or the *SLED* property of real segment data proposed by Proietti and Faloutsos [24]. A limitation for parametric techniques results from the requirement of a-priori assumptions about the data distribution. In contrast, *sampling* adapts to the actual data distribution by processing a small fraction of the stored tuples. This paradigm has been pursued and evaluated by Lipton, Naughton and Schneider [21] and by Haas et al. [12]. *Statistics* are a very popular approach in database systems, as they typically can be efficiently computed and occupy only a small amount of secondary storage. For linearly ordered domains, the most commonly used statistics type in commercial database servers are quantiles of the original data. For the selectivity estimation on non-uniform distributions of extended objects, histograms are a common technique. An extensive analysis on different kinds of spatial histograms has been published by Acharya, Poosala and Ramaswamy [1]. Whereas histograms can be naturally applied to one-dimensional interval data, a quantile-based approach has to operate on a linear representation of the original intervals. In this paper, we present and evaluate techniques for interval data on both types of statistics, histograms as well as quantiles.

2.2. Cost Estimation

A wide range of cost models has been presented in the literature for various index structures for extended objects, including the technique of Kamel and Faloutsos [16] for intersection queries on packed R-trees, or the *REGAL* law for R-tree entries by Proietti and Faloutsos [25]. Recently, cost models have also been extended to handle joins of extended objects and the presence of database buffers as in the proposals of Huang, Jing and Rundensteiner [11], Leutenegger and Lopez [20], or Theodoridis, Stefanakis and Sellis [33]. Whereas previous research has mainly concentrated on the design and evaluation of cost models for stand-alone access methods, the following sections develop an approach that can be fully implemented on top of existing object-relational database systems.

3. The Relational Interval Tree

The RI-tree is a relational storage structure for interval data (*lower*, *upper*), built on top of the SQL layer of any RDBS. By design, it follows the concept of Edelsbrunner's main-memory interval tree [8] and guarantees the optimal complexity for storage space and for I/O operations when updating or querying large sets of intervals.

3.1. Relational Storage and Extensible Indexing

The RI-tree strictly follows the paradigm of relational storage structures since its implementation is purely built on (procedural and declarative) SQL but does not assume any lower level interfaces to the database system. In particular, built-in index structures are used as they are, and no intrusive augmentation or modification of the database kernel is required.

On top of its pure relational implementation, the RI-tree is ready for immediate object-relational wrapping. It fits particularly well to extensible indexing frameworks as already proposed in [32] and illustrated in Figure 1. These frameworks, which are provided by the latest object-relational database systems, including IBM DB2 Universal Database [14] [7], Informix Universal Server [15] [6], or Oracle Server [23] [29] enable developers to extend the set of built-in index structures by custom access methods in order to support user-defined datatypes and predicates without weakening the reliability of the entire system.

3.2. Dynamic Data Structure

The structure of an RI-tree consists of a binary tree of height h which covers the range $[1, 2^h-1]$ of potential interval bounds. It is called the virtual backbone of the RI-tree since it is not materialized but only the root value 2^{h-1} is stored persistently in a metadata table. Traversals of the virtual backbone are performed purely arithmetically by starting at the root value and proceeding in positive or negative steps of decreasing length 2^{h-i}, thus reaching any desired value of the data space in $O(h)$ CPU time and without causing any I/O operation. For the relational storage of intervals, the node values of the tree are used as artificial keys: Upon insertion of an interval, the first node that hits the interval

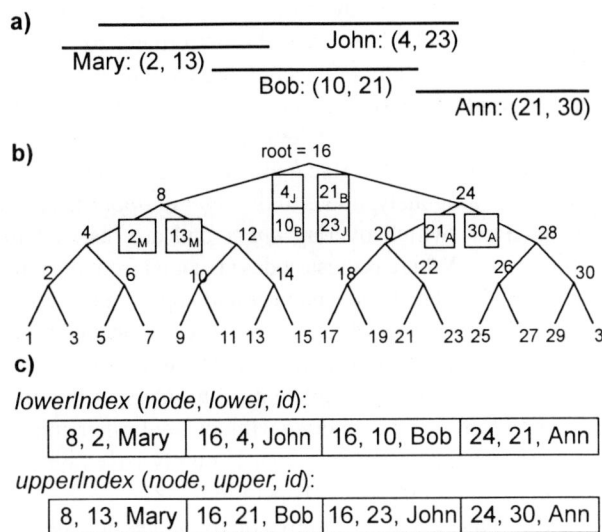

Fig. 2. Example for an RI-tree. **a)** four intervals. **b)** virtual backbone and registration positions of the intervals. **c)** resulting relational indexes *lowerIndex* and *upperIndex*

when descending the tree from the root node down to the interval location is assigned to that interval.

An instance of the RI-tree then consists of two relational indexes which in an extensible indexing environment are preferably managed as index-organized tables. The indexes obey the relational schema *lowerIndex* (*node, lower, id*) and *upperIndex* (*node, upper, id*) and store the artificial key value *node*, the bounds *lower* and *upper*, and the *id* of each interval. An interval is represented by exactly one entry in each of the two indexes, and therefore, $O(n/b)$ disk blocks of size b suffice to store n intervals. For inserting or deleting intervals, the *node* values are determined arithmetically, and updating the indexes requires $O(\log_b n)$ I/O operations per interval.

The illustration in Figure 2 provides an example for the RI-tree. Let us assume the intervals (2,13) for Mary, (4,23) for John, (10,21) for Bob, and (21,30) for Ann (Fig. 2a). The virtual backbone is rooted at 16 and covers the data space from 1 to 31 (Fig. 2b). The intervals are registered at the nodes 8, 16, and 24. The interval (2,13) for Mary is represented by the entries (8, 2, Mary) in the *lowerIndex* and (8, 13, Mary) in the *upperIndex* since 8 is the registration node, and 2 and 13 are the lower and upper bound, respectively (Fig. 2c).

3.3. Intersection Query Processing

To minimize barrier crossings between the procedural runtime environment and the declarative SQL layer, an interval intersection query (*lower*, *upper*) is processed in two steps. The procedural query preparation step descends the virtual backbone from the root node down to *lower* and to *upper*, respectively. The traversal is performed arithmetically without causing any I/O operations, and the visited nodes are collected in two main-memory tables, *left queries* and *right queries*, as follows: nodes to the left of *lower* may contain intervals which overlap *lower* and are inserted into *left queries*. Analogously, nodes to the right of *upper* may contain intervals which overlap *upper* and are inserted into *right queries*. Whereas these nodes are taken from the paths, the set of all nodes between *lower* and *upper* belongs to the so-called *inner query* which is represented by a single range query on the node values. All intervals registered at nodes from the *inner query* are guaranteed to intersect the query and, therefore, will be reported without any further comparison. The query preparation step is purely based on main memory and requires no I/O operations.

In the subsequent declarative query processing step, the transient tables are joined with the relational indexes *upperIndex* and *lowerIndex* by a single, three-fold SQL statement (Figure 3). The upper bound of each interval registered at nodes in *left queries* is compared to *lower*, and the lower bounds of intervals stemming from *right queries* are compared to *upper*. The *inner query* corresponds to a simple range scan over the intervals with nodes in (*lower*, *upper*). The SQL query requires $O(h \cdot \log_b n + r/b)$ I/Os to report r results from an RI-tree of height h since the output from the relational indexes is fully blocked for each join partner.

```
SELECT id FROM upperIndex i, :leftQueries q
    WHERE i.node = q.node AND i.upper >= :lower
UNION ALL
SELECT id FROM lowerIndex i, :rightQueries q
    WHERE i.node = q.node AND i.lower <= :upper
UNION ALL
SELECT id FROM lowerIndex   /* or upperIndex */
    WHERE node BETWEEN :lower AND :upper;
```

Fig. 3. SQL statement for an intersection query with bind variables for *left queries*, *right queries*, *lower* and *upper*

For systems which do not support transient main-memory tables as bind variables, we use set containment predicates 'i.node IN leftQueries' and 'i.node IN rightQueries'. The query sets are then composed by string concatenation. In any case, no I/O operations are required in the query preparation step.

4. Selectivity Estimation

Accurate estimations of query result sizes are a necessary input for many components of the underlying database system. In particular, the selectivity estimation for an interval intersection query can be used by the built-in optimizer

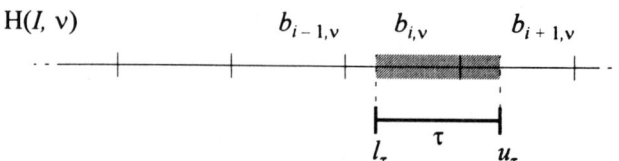

Fig. 4. Selectivity estimation on an interval histogram.

to find an efficient join order and to determine the best available access method [27] [23]. Selectivity estimation is also required to provide the user with an approximate prediction about the potential execution time of DML statements. In the following, we propose a histogram-based approach ('equi-width histograms') and a quantile-based approach ('equi-count histograms').

4.1. Histogram-Based Selectivity Estimation

In order to cope with arbitrary interval distributions, histograms can be employed to capture the data characteristics at any desired resolution. We start by giving the definition of an interval histogram:

Definition 1 *(Interval Histogram).*
Let $D = [1, 2^h - 1]$ be a domain of interval bounds, $h \geq 1$. Let the natural number $v \in N$ be the *resolution*, and $\beta_v = (2^h - 1)/v$ the corresponding *bucket size*. Let $b_{i,v} = [1 + (i - 1) \cdot \beta_v, 1 + i \cdot \beta_v)$ denote the *span of bucket i*, $i \in \{1, ..., v\}$. Let further $I = \{(l, u), l \leq u\} \subseteq D^2$ be a database of intervals. Then, $H(I, v) = (n_1, ..., n_v) \in N^v$ is called the *interval histogram* on I with *resolution* v, iff for all $i \in \{1, ..., v\}$:
$$n_i = |\{\psi \in I \mid \psi \text{ intersects } b_{i,v}\}|$$

In order to compute an interval histogram on a database I of n intervals, $O(n/b)$ disk blocks have to be touched, assuming a blocked storage of I by a page size b. The computation is performed by standard SQL and wrapped by a stored procedure that complies with the statistics collection interface of the extensible optimization framework (function *getSelectivity*). Based on $H(I, v)$, we compute a selectivity estimate by evaluating the intersection of the query interval with each bucket span $b_{i,v}$ (cf. Figure 4):

Definition 2 *(Histogram-based Selectivity Estimate).*
Given an interval histogram $H(I, v) = (n_1, ..., n_v)$ with bucket size β, we define the *histogram-based selectivity estimate* $\sigma_I(I, \tau)$, $0 \leq \sigma_I(I, \tau) \leq 1$ for an intersection query $\tau = (l_\tau, u_\tau)$ by the following formula:

$$\sigma_I(I, \tau) = \left[\sum_{i=1}^{v} \frac{overlap(\tau, b_{i,v})}{\beta} \cdot n_i \right] \cdot \left[\sum_{i=1}^{v} n_i \right]^{-1}$$

where *overlap* returns the intersection length of two intersecting intervals, and 0, if the intervals are disjoint.

Note that long intervals may span multiple histogram buckets. Thus, in the above computation, we normalize the expected output to the sum of the number n_i of intervals intersecting each bucket i rather than to the original cardinality n of the database. In order to support query intervals with a very small duration, the average length of the stored intervals could also be considered for the estimation.

4.2. Quantile-Based Selectivity Estimation

Due to the replication of intervals across bucket boundaries, the accuracy of the histogram-based selectivity estimation may deteriorate with longer interval lengths or higher histogram resolutions. In addition, the runtime required for the histogram computation is increased by the cost of barrier-crossings between the declarative environment of the SQL layer and our stored procedure. Fortunately, most ORDBMS comprise efficient built-in functions to compute single-column statistics, particularly for cost-based query optimization. Available optimizer statistics are accessible to the user by the relational data dictionary. The basic idea of our quantile-based selectivity estimation is to exploit these built-in index statistics rather than to add and maintain user-defined histograms. We start with the definition of a quantile vector, the typical statistics type supported by relational database kernels. Then, we describe its application to node values of the RI-tree.

Definition 3 *(Quantile Vector).*
Let $(S, \leq)$ be a totally ordered multi-set. Without loss of generality, let $S = \{s_1, s_2, ..., s_k\}$ with $s_j \leq s_{j-1}$, $1 \leq j < k$. Then, $Q(S, v) = (q_0, ..., q_v) \in S^v$ is called a *quantile vector* for S and a *resolution* $v \in N$, iff the following conditions hold:

(i) $q_0 = s_1$

(ii) $\forall i \in 1, ..., v: \exists j \in 1, ..., k: \quad q_i = s_j \wedge \dfrac{j-1}{k} < \dfrac{i}{v} \leq \dfrac{j}{k}$

Definition 4 *(Node Quantiles).*
Let *lowerIndex* be the relational index on *(node, lower, id)* for an instance T of the RI-tree. Let $N = \pi_{node}(lowerIndex)$ be the projected multi-set of node values. Then, $Q(N, v)$ is called the vector of *node quantiles* on T with *resolution* v.

Based on the node ordering materialized in the *lowerIndex* (or *upperIndex*), the computation of $Q(N, v)$ on an RI-tree storing n intervals has an I/O complexity of $O(n/b)$, where b is the disk block size. By using the node quantiles for an RI-tree index, we get an aggregated view on the locations of the stored intervals. In addition, we may use some knowledge about the one-dimensional durations which is given by the following definition that captures the average distances of the interval bounds to the respective fork node:

Definition 5 *(Average Node Distances).*
Let T be an instance of the RI-tree with *lowerIndex* and *upperIndex* relations. Then, the *average lower distance* $\delta_{lower}(T)$ and the *average upper distance* $\delta_{upper}(T)$ is defined as:

(i) $\delta_{lower}(T) = \text{avg}(node - lower)$

(ii) $\delta_{upper}(T) = \text{avg}(upper - node)$

The average values δ_{lower} and δ_{upper} are computed with $O(n/b)$ I/O complexity, and if possible, along with the quantile statistics. If the built-in statistics of the hosting database system comprise single-column averages on *node*, *lower*, and *upper*, then δ_{lower} and δ_{upper} can be simply derived from these existing statistics: $\delta_{lower} = \text{avg}(node) - \text{avg}(lower)$ and $\delta_{upper} = \text{avg}(upper) - \text{avg}(node)$. For interval databases with a highly skewed distribution of interval lengths, δ_{lower} and δ_{upper} can be replaced by quantiles on $\pi_{node-lower}(lowerIndex)$ and $\pi_{upper-node}(upperIndex)$.

Our goal is to compute the selectivity estimate in constant time, i.e. independent not only from the cardinality, but also from the granularity of the interval data. Instead of submitting the $O(h = \log_2 root + 1)$ node queries on the RI-tree, we evaluate the quantiles with respect to the span of nodes touched during the processing of a potential interval intersection query.

Definition 6 *(Span of Touched Nodes).*
For a given RI-tree T and an intersection query τ, the range $\theta(T, \tau) = (l_\theta, u_\theta)$ is called the *span of touched nodes*, iff l_θ is the minimal and u_θ is the maximal node on the virtual backbone that is touched while processing the query τ on T.

Lemma 1. *Let* $D = [1, 2^h - 1]$ *be the interval domain covered by an RI-tree T with root $= 2^{h-1}$. For an intersection query $\tau = (l_\tau, u_\tau) \in D$, the span of touched nodes $\theta(T, \tau) = (l_\theta, u_\theta) \in D$ is computed by the following formulas:*

(i) $l_\theta = 2^k$, $k = \log_2(l_\tau)$,

(ii) $u_\theta = 2^h - 2^k$, $k = \lfloor \log_2(2^h - u_\tau) \rfloor$.

Proof. (i) The leftmost node touched during the arithmetic traversal of the backbone is the last node before we first step into a right subtree. Following the left branch yields a 0-bit, following the right branch yields a 1-bit in the binary representation of the actual node value. Thus, the leftmost node l_θ has exactly one bit set at the first position of a 1-bit in l_τ. (ii) Analogously, the rightmost node u_θ is derived from the first 0-bit in the binary representation of u_τ by a mirrored consideration. ∎

We estimate the number of results yielded by the *inner*, *left*, and *right queries* for an intersection query $\tau = (l_\tau, u_\tau)$ based on the node quantiles $Q(N, v) = (q_0, ..., q_v)$. Figure 5 provides a graphical interpretation of the following calculations: the number of results r_{inner} from the *inner query* can

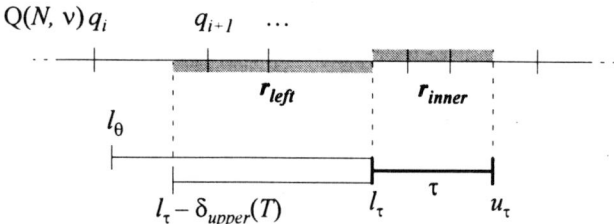

$Q(N, v)q_i \qquad q_{i+1} \quad \cdots$

r_{left}

r_{inner}

l_θ

$l_\tau - \delta_{upper}(T) \qquad l_\tau \qquad \tau \qquad u_\tau$

Fig. 5. Selectivity estimation on node quantiles.

be estimated by evaluating the overlap of τ with the quantiles (analogously to Section 4.1):

$$r_{inner} = \sum_{i=1}^{v} \left(\frac{overlap(\tau, (q_{i-1}, q_i))}{q_i - q_{i-1}} \cdot \frac{|N|}{v} \right)$$

To estimate the number of results r_{left} retrieved by the *left queries*, we only have to consider quantiles falling into the range $(leftBound_\tau, l_\tau)$, where $leftBound_\tau = \max(l_\theta, l_\tau - \delta_{upper}(T))$ and $\theta(T, \tau) = (l_\theta, u_\theta)$:

$$r_{left} = \sum_{i=1}^{v} \left(\frac{overlap((leftBound_\tau, l_\tau), (q_{i-1}, q_i))}{q_i - q_{i-1}} \cdot \frac{|N|}{v} \right)$$

The estimation of the number of results r_{right} of the *right queries* is done analogously to r_{left}. Finally, we define:

Definition 7 *(Quantile-based Selectivity Estimate).*
The quantile-based selectivity estimate $\sigma_N(I, \tau)$ of the intersection query τ on an interval database I is given by

$$\sigma_N(I, \tau) = \frac{r_{left} + r_{inner} + r_{right}}{|N|}$$

As desired, the quantile vector is a non-replicating statistics on interval data, and the data sets contributing to the results r_{left}, r_{inner} and r_{right} are disjoint. In consequence, $0 \le r_{left} + r_{inner} + r_{right} \le |N|$ holds and, thus, $0 \le \sigma_N(I, \tau) \le 1$.

5. Model for I/O Cost

In order to achieve a seamless declarative integration of the Relational Interval Tree into extensible indexing frameworks as provided by modern object-relational database systems, a cost model has to be registered at the extensible optimization framework. In this section, we present a cost model for interval intersection queries on the RI-tree, based on the estimated selectivity and the range queries generated for the underlying B^+-trees.

We assume the selectivity estimation $\sigma(I, \tau)$ for an intersection query $\tau = (l_\tau, u_\tau)$ on an interval data set I to be determined as shown above. In our derivation of a cost model to estimate the number of touched B^+-tree blocks for arbitrary

intersection queries τ, we use that expected selectivity as input for the estimation of the I/O operations.

Let us recall from Section 3.3 that the query preparation step does actually cause no I/O operations since the traversal of the backbone structure is done purely arithmetically, and the generated join partners are managed in main memory. The I/O complexity of $O(h \cdot \log_b n + r/b)$ [17] for an intersection query retrieving r results from an RI-tree of height h comprises components of the following two types:

- First, the directories of the relational indexes (built-in B^+-trees) have to be traversed in order to navigate on the disk to the first result, if any, for each join partner. Let us denote this portion of I/O operations by $join_{IO}$ and let us recall that $join_{IO} = O(h \cdot \log_b n)$.
- Second, the results for each join partner are reported by scanning contiguous leaf blocks of the relational indexes. We call this portion of I/O operations $output_{IO}$. Since the output is blocked, i.e. there are no gaps between the answers for a single range query, the complexity $output_{IO} = O(r/b)$ is guaranteed.

In contrast to the very general complexity analysis, a cost model has to compute actual numbers of I/O operations for specific interval queries. Our model relies on the following two observations:

1. In a real user environment with many concurrent queries, substantial parts of the B^+-directories typically reside in the main memory and can be managed by the built-in LRU-cache of the DBMS [22]. According to a common assumption, we count two I/O operations for each leaf-block access in order to estimate the number of blocks actually read from disk.

2. The transient join partners are processed in increasing order (*left queries*, *inner query*) or decreasing order (*right queries*) with respect to the *node* value in the composite indexes on (*node, upper, id*) and (*node, lower, id*), respectively. Due to this ordered access, pages that are read several times during query processing will rarely be displaced from the LRU cache between the accesses. We therefore assume that each leaf page is retrieved only once from secondary storage.

Based on these assumptions, we derive individual formulas for the components $output_{IO}$ and $join_{IO}$ in the following.

$output_{I/O}$. For a given RI-Tree T on a set I of intervals, let $L = $ leaf-blocks($upperIndex$) $\approx$ leaf-blocks($lowerIndex$) be the number of leaf blocks in the B^+-trees, $L = O(n/b)$, and τ be an interval intersection query performed on T. The answers retrieved from $upperIndex$ and from $lowerIndex$ are guaranteed to be disjoint, and we estimate $output_{IO}(T, \tau)$ as the fraction of L predicted by the selectivity estimate $\sigma(I, \tau)$ on T:

$$output_{IO}(T, \tau) = \sigma(I, \tau) \cdot L$$

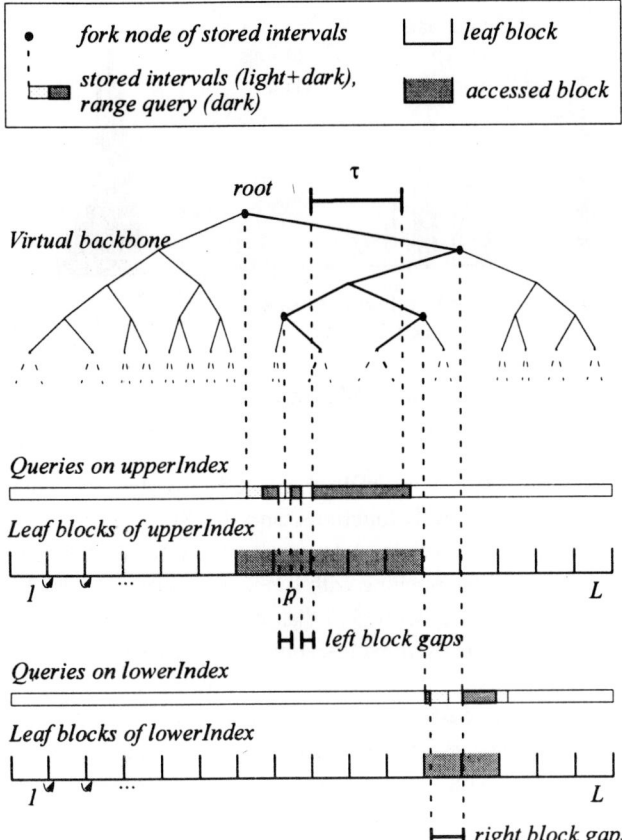

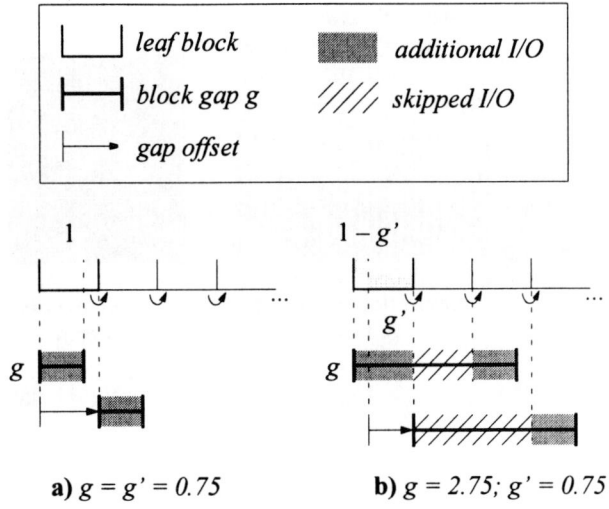

Fig. 7. Additional I/O due to block gaps g between range queries.

a) $g = g' = 0.75$ **b)** $g = 2.75;\ g' = 0.75$

Fig. 6. Touched leaf blocks and query gaps for an intersection query τ.

join$_{I/O}$. The formula for *join*$_{I/O}$ includes the number of leaf block accesses caused by the navigation in the B$^+$-tree directories for the join partners. Since the leaf blocks are read from two independent B$^+$-tree indexes, we capture the join overhead for the set of *left queries* and *inner queries* on the one hand and for the set of *right queries* on the other hand separately.

Figure 6 provides an illustration for our considerations and depicts the leaf blocks in the *lowerIndex* and *upperIndex* that are read for a query $\tau = (l_\tau, u_\tau)$. Note that the virtual backbone is drawn to the scale of the population in the indexes, and not to the original domain of $D = [1, 2^h - 1]$.

The leaf block p in the *upperIndex*, for example, is touched multiple times during query processing. According to the locality-preserving read schedules for LRU buffers, the multiple accesses to block p count for a single leaf access only. This estimation is complemented by the additional heuristics to count two physical disk accesses for a single leaf block access in order to take care of the I/O caused by traversing the index directory.

An important observation for *join*$_{I/O}$ is that the results of different join partners in general do not form a contiguous

range of entries in the leaf blocks of the indexes. Although the results are blocked for each single *left query*, *inner query*, and *right query*, there are typically gaps between the blocked result sets of different join partners. In order to model the distribution of gaps, we first determine the gaps between the node values, $NGaps_{left}(\tau)$ and $NGaps_{right}(\tau)$, for a given intersection query τ on an RI-tree T. Then, we derive the expected corresponding gaps between disk blocks, $BGaps_{left}(\tau)$ and $BGaps_{right}(\tau)$.

Estimation of Node Gaps. For the estimation of node gaps, we traverse the virtual backbone on $D = [1, 2^h - 1]$, and we collect the lengths $NGaps_{left}(\tau) = \{\zeta_1, \ldots, \zeta_l\}$ of gaps to the left of the query interval τ, i.e. in the range $[1, l_\tau]$ between consecutive nodes touched by the *left* and *inner queries*, and the lengths $NGaps_{right}(\tau) = \{\xi_1, \ldots, \xi_r\}$ of gaps to the right of τ, i.e. in the range $[u_\tau, 2^h - 1]$ between the *right queries*, respectively.

Estimation of Block Gaps. Let L_θ be the average number of nodes per leaf block in the span $\theta(T, \tau)$ of touched nodes for τ. We estimate the corresponding block gaps among the range queries for τ by the multi-sets $BGaps_{left}(\tau)$ and $BGaps_{right}(\tau)$ of real numbers:

$$BGaps_{left}(\tau) = \frac{\zeta_1}{L_\theta}, \ldots, \frac{\zeta_l}{L_\theta},\ BGaps_{right}(\tau) = \frac{\xi_1}{L_\theta}, \ldots, \frac{\xi_r}{L_\theta}.$$

The value of L_θ is easily estimated by using the persistent statistics on T along with the cardinality n and the number of leaf blocks L, similarly to Section 4. After having computed the number and extension of gaps between the blocked sections of *output*$_{I/O}$, we use this information to estimate *join*$_{I/O}$. Depending on the length and the position of each block gap g, a specific number of leaf block accesses

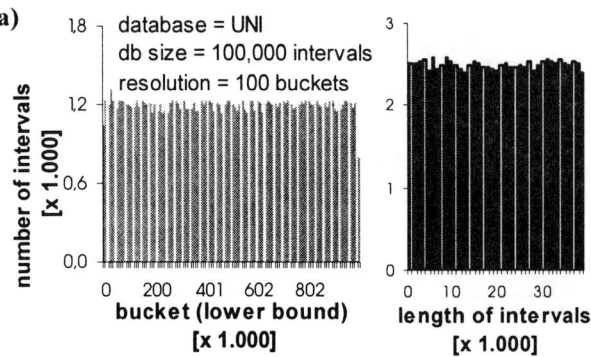

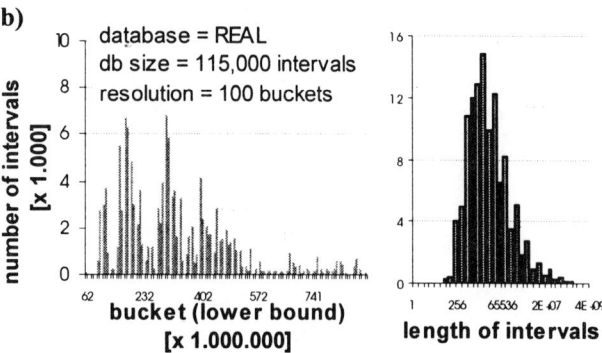

Fig. 8. Histograms of interval distributions **a)** for uniform data, **b)** for real data.

occurs. For gaps smaller than one disk block, i.e. $g \leq 1$, the I/O is increased by this very gap length g with a weight of 1 (cf. Figure 7a). According to Figure 7b, the I/O overhead for larger gaps depends on the gap offset to the leaf blocks and is restricted to blocks at the gap border. For gaps $g > 1$, our formula to estimate the contribution $gap_{I/O}(g)$ of a gap g to $join_{I/O}$ therefore focuses on the fraction $g' = g - \lfloor g \rfloor$. Since we assume a uniform distribution of gap offsets with respect to the leaf blocks in the *upperIndex* and *lowerIndex*, the mean contributions of the left and right borders of a gap $g > 1$ to $gap_{I/O}(g)$ are $1 + g'$ with weight $1 - g'$ and g' with weight g'. The overall value then sums to $(1 - g') \cdot (1 + g') + g' \cdot g' = 1$, and the distinction of cases simplifies to

$$gap_{I/O}(g) = \min(1, g).$$

With respect to I/O cost, random access to a leaf block is, therefore, only beneficial if the preceding block gap is larger than the size of a disk block. In consequence, gaps covering only fractions of a disk block could be sequentially scanned without causing any I/O overhead (this observation opens up a promising potential to further improve the performance of the RI-tree). For all gaps between the range queries on the *upperIndex* and *lowerIndex* for a given query interval τ on an RI-tree T, we estimate the additional I/O for the join processing as

$$join_{I/O}(T, \tau) = \sum_{g \in BGaps_{left}(\tau) \cup BGaps_{right}(\tau)} gap_{I/O}(g).$$

The total I/O cost for an interval intersection query τ on an RI-tree T is then summarized by

$$total\_cost_{I/O}(T, \tau) = output_{I/O}(T, \tau) + join_{I/O}(T, \tau).$$

6. Empirical Evaluation

6.1. Experimental Setup

We implemented the proposed functions for the estimation of selectivity and execution cost on the Oracle Server Release 8.1.6 using built-in methods for statistics collection, analytic SQL functions, and the PL/SQL procedural runtime environment. All experiments were performed on an Athlon/750 machine with IDE hard drives. The database block cache was set to 500 disk blocks with a block size of 8 KB and was used exclusively by one active session. The experiments for the evaluation of statistics, selectivity estimation, and cost model have been executed on various interval databases. We have used a synthetic dataset of intervals following a uniform starting point and length distribution (*UNI*) and intervals derived from a real dataset (*REAL*). For both databases UNI and REAL, Figure 8 depicts the histogram statistics. A peak in the histogram visualization denotes a high density of interval data. In case of the UNI dataset, the data space [1..1,000,000] is covered with a uniform density.

To evaluate the quality of the selectivity and cost prediction, we determined the average relative error of the estimates. This measure denotes the ratio of the absolute estimation error to the actual query result, averaged over a set of queries S. If e_i is the estimated and r_i is the actual result size of a query q_i, the average relative error of the estimated selectivity for S is defined as:

$$\text{Avg relative error (selectivity)} = \left(\sum_{q_i \in S} |r_i - e_i| \right) \bigg/ \left(\sum_{q_i \in S} r_i \right)$$

For the estimations of the actual I/O cost, the average relative error is defined analogously. This measure is a common technique to evaluate selectivity estimations and cost models, see e.g. [1]. It is undefined if all queries in a query set produce zero output or zero cost. However, this is not the case for our evaluation. As an alternative, the geometric average of relative errors could be used as in [25]. But, as the relative error of some predictions reached zero, this measure would be undefined, and, therefore, could not be used throughout our experiments. The following results show the

138

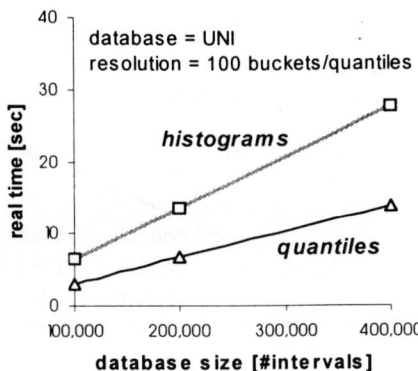

Fig. 9. Computation cost of histogram-based and quantile-based statistics.

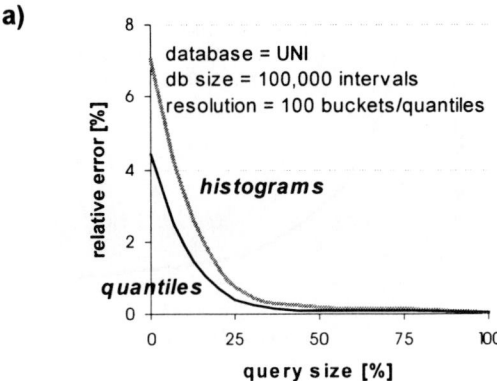

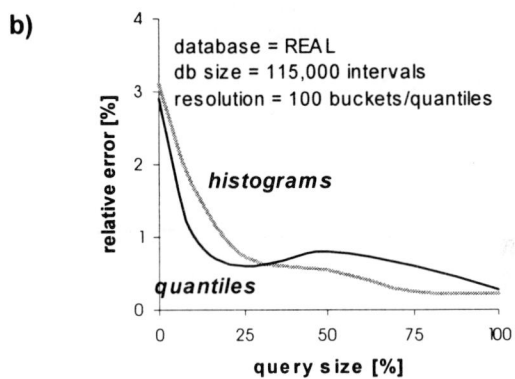

Fig. 10. Relative error of selectivity estimation for histogram-based and quantile-based statistics **a)** on uniform data, **b)** on real data.

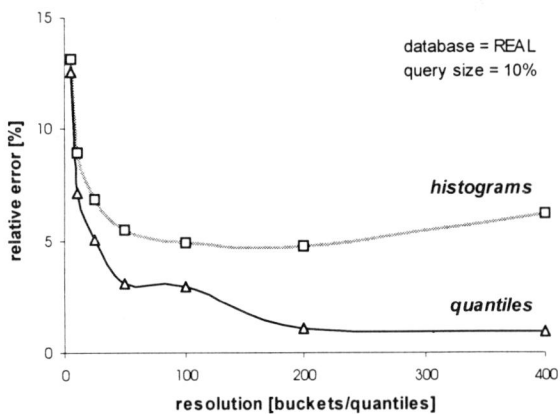

Fig. 11. Relative error of selectivity estimation for varying resolution of the statistics.

averages of, in total, 100 intersection queries for the UNI and REAL databases.

6.2. Computation of Statistics

The persistent statistics must be recomputed in order to adapt to changing data distributions. For highly dynamic data, the database administrator might even decide to trigger the computation of important statistics periodically. Therefore, a low execution cost for the creation of statistics is essential. Figure 9 compares the total runtime of computation for the histogram statistics to the quantile statistics for increasing database size, using 100% samples. Due to the overhead of barrier crossing between PL/SQL and SQL, the quantile-based approach outperforms the histogram-based approach by a factor of 2.

6.3. Selectivity Estimation

In the next set of experiments, we evaluate the average relative error with respect to the query size, i.e. the percentage of the data space covered by the query region. Figure 10 shows the relative error of the histogram-based and quantile-based statistics on the UNI and REAL database. The resulting accuracy of both, the quantile-based approach and the histogram-based approach is very high. For higher selectivities, the quantile-based approach performs slightly better, yielding estimation errors around 4.5% and 2.9% for the UNI and REAL database, respectively. This result can be explained by the fact that quantiles adapt to the local density of the data, whereas histograms partition the whole data space using buckets of identical size. The next experiment in Figure 11 depicts the average relative error for different resolutions of the persistent statistics, evaluated for a set of intersection queries having 10% average query size. As expected, the estimation error increases significantly for coarser resolutions due to the replication of intervals across

bucket boundaries (cf. Section 4.2). Beyond a global optimum at some 100 buckets, the error of the histogram-based approach increases for higher resolutions, due to the repli-

139

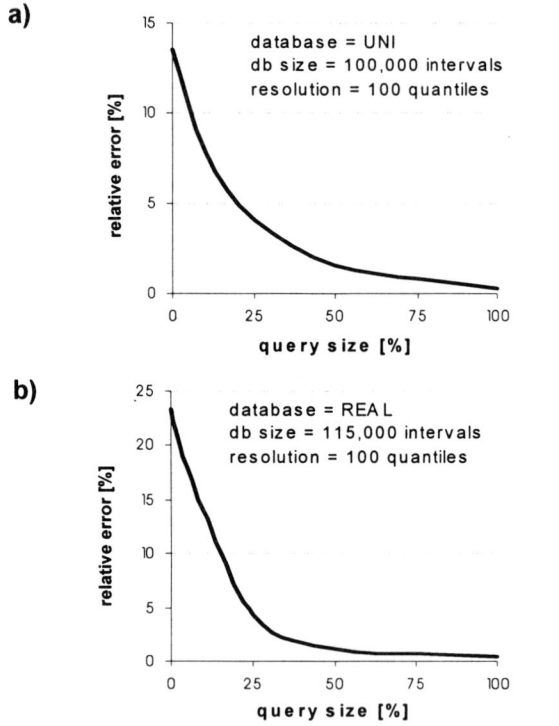

a)

b)

Fig. 12. Relative error for cost estimation **a)** on uniform data, **b)** on real data.

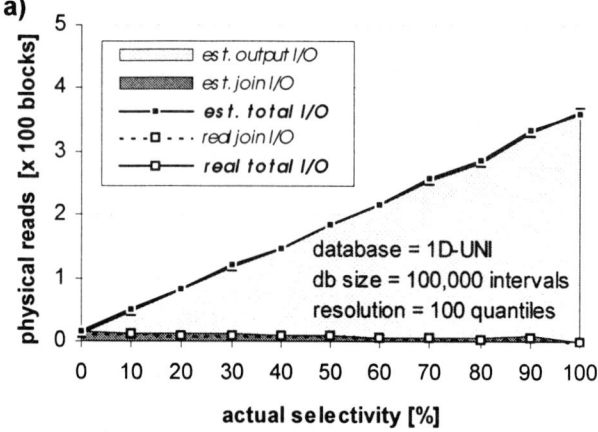

a)

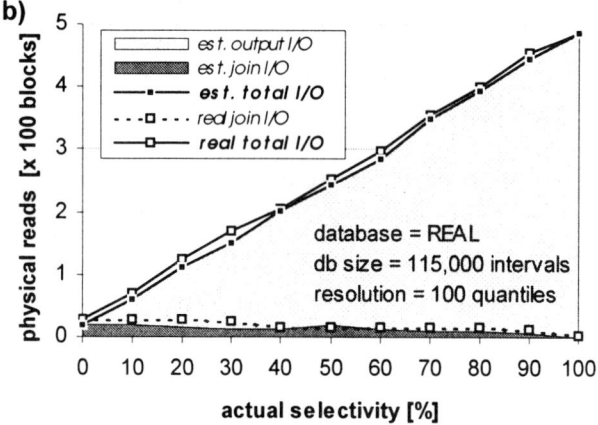

b)

Fig. 13. Output cost and join overhead for queries evaluated **a)** on uniform data, **b)** on real data.

cation of intervals spanning multiple histogram buckets. Therefore, we focus on the quantile-based approach in the following experiments, as the representation of intervals is non-redundant. Regarding the runtime, a single selectivity estimation using statistics with a resolution of 100 quantiles for the UNI and REAL databases took about 0.05 seconds on the average.

6.4. Cost Estimation

We used the estimated quantile-based selectivity of the previous section as input for the I/O cost model. The extensible query optimizer uses the resulting estimations to decide upon the usability of the RI-tree for specific queries. Figure 12 presents the relative error of the estimated cost for the UNI and REAL databases. The relative errors stay below 14% and 23%, respectively. Figure 13 compares the absolute estimations and the actual cost for the blocked output of results ($output_{IO}$). In addition, $join_{IO}$ denotes the overhead due to the nested-loop join with the transient query tables. For the sake of comparability to the analytical I/O complexity mentioned in Section 3.3, the results are shown with respect to the actual query selectivity. Our interpretation of these results is twofold: First, the real I/O cost show that the total I/O is largely determined by the cardinality of the query result, whereas the overhead for the join process-

ing remains almost constant. The relative cost of the join overhead decreases from 100% at 0% selectivity to 0.2‰ at 100% selectivity (Figure 13a). According to these empirical results, the overhead of $join_{IO}$ is negligible for higher values of the query selectivity. Second, we observe that our cost model not only yields tight estimations for the total query cost, but also reflects the distribution between the output and join cost rather accurately. Regardless of the actual query selectivity, the cost computation on the databases UNI and REAL took about 0.05 seconds for a single interval intersection query.

7. Conclusions

High quality selectivity estimation and cost prediction are the fundamentals of effective query optimization. Particularly for complex query objects and complex query predicates, the recent object-relational database servers provide extensible optimization frameworks that go along with the extensible indexing frameworks, in order to complete the seamless integration of user-defined index structures into the declarative DML. In this paper, we present an example for such an extension and focus to the relational interval tree (RI-tree) that already fits well to modern object-relational extensible indexing frameworks. We particularly propose models to estimate the selectivity of interval intersection queries and to predict the cost for query processing. With respect to the generation and management of statistics, the proposed quantile-based selectivity estimation reuses as much built-in functionality of the RDBMS as possible. According to our experimental evaluation, the computed estimations are very accurate. For highly selective queries on a database of real interval data, the relative error for the selectivity estimation was around 2.9%. The corresponding errors for the I/O cost model amount to 23% and 3.3%, respectively.

In our future work, we plan to adapt the proposed techniques to support general interval relationships as required for temporal applications [18], and, in addition, to spatial queries in GIS and CAD applications [19].

References

[1] Acharya S., Poosala V., Ramaswamy S.: *Selectivity Estimation in Spatial Databases*. Proc. ACM SIGMOD, 13-24, 1999.

[2] Belussi A., Faloutsos C.: *Estimating the Selectivity of Spatial Queries Using the 'Correlation' Fractal Dimension*. Proc. VLDB, 299-310, 1995.

[3] Böhm C., Klump G., Kriegel H.-P.: *XZ-Ordering: A Space-Filling Curve for Objects with Spatial Extension*. Proc. SSD (LNCS 1651), 75-90, 1999.

[4] Bozkaya T., Özsoyoglu Z. M.: *Indexing Valid Time Intervals*. Proc. DEXA, LNCS 1460, 541-550, 1998.

[5] Boulos J., Ono K.: *Cost Estimation of User-Defined Methods in Object-Relational Database Systems*. ACM SIGMOD Record, 28(3), 22-28, 1999.

[6] Bliujute R., Saltenis S., Slivinskas G., Jensen C.S.: *Developing a DataBlade for a New Index*. Proc. ICDE, 314-323, 1999.

[7] Chen W., Chow J.-H., Fuh Y.-C., Grandbois J., Jou M., Mattos N., Tran B., Wang Y.: *High Level Indexing of User-Defined Types*. Proc. VLDB, 554-564, 1999.

[8] Edelsbrunner H.: *Dynamic Rectangle Intersection Searching*. Inst. for Information Processing Report 47, Technical University of Graz, Austria, 1980.

[9] Faloutsos C., Roseman S.: *Fractals for Secondary Key Retrieval*. Proc. ACM PODS, 247-252, 1989.

[10] Gaede V., Günther O.: *Multidimensional Access Methods*. ACM Computing Surveys 30(2), 170-231, 1998.

[11] Huang Y.-W., Jing N., Rundensteiner E. A.: A Cost Model for Estimating the Performance of Spatial Joins Using R-trees. Proc. SSDBM, 30-38, 1997.

[12] Haas P. J., Naughton J. F., Seshadri S., Stokes L.: *Sampling-Based Estimation of the Number of Distinct Values of an Attribute*. Proc. VLDB, 311-322, 1995.

[13] Hellerstein J., Stonebraker M.: *Predicate Migration: Optimizing Queries with Expensive Predicates*. Proc. ACM SIGMOD, 267-276, 1993.

[14] IBM Corp.: *IBM DB2 Universal Database Application Development Guide, Version 6*. Armonk, NY, 1999.

[15] Informix Software, Inc.: *DataBlade Developers Kit User's Guide, Version 3.4*. Menlo Park, CA, 1998.

[16] Kamel I., Faloutsos C.: *On Packing R-trees*. Proc. ACM CIKM, 490-499, 1993.

[17] Kriegel H.-P., Pötke M., Seidl T.: *Managing Intervals Efficiently in Object-Relational Databases*. Proc. VLDB, 407-418, 2000.

[18] Kriegel H.-P., Pötke M., Seidl T.: *Object-Relational Indexing for General Interval Relationships*. Proc. SSTD (LNCS 2121), 522-542, 2001.

[19] Kriegel H.-P., Pötke M., Seidl T.: *Interval Sequences: An Object-Relational Approach to Manage Spatial and Temporal Data*. Proc. SSTD (LNCS 2121), 481-501, 2001.

[20] Leutenegger S. T., Lopez M. A.: *The Effect of Buffering on the Performance of R-Trees*. Proc. ICDE, 164-171, 1998.

[21] Lipton R. J., Naughton J. F., Schneider A. D.: *Practical Selectivity Estimation through Adaptive Sampling*. Proc. ACM SIGMOD, 1-11, 1990.

[22] Lomet D.: *B-tree Page Size When Caching is Considered*. ACM SIGMOD Record 27(3), 28-32, 1998.

[23] Oracle Corp.: *Oracle8i Data Cartridge Developer's Guide, Release 2 (8.1.6)*. Redwood Shores, CA, 1999.

[24] Proietti G., Faloutsos C.: *Selectivity Estimation of Window Queries for Line Segment Datasets*. Proc. ACM CIKM, 340-347, 1998.

[25] Proietti G., Faloutsos C.: *I/O Complexity for Range Queries on Region Data Stored Using an R-tree*. Proc. ICDE, 628-635, 1999.

[26] Ramaswamy S.: *Efficient Indexing for Constraint and Temporal Databases*. Proc. ICDT (LNCS 1186), 419-431, 1997.

[27] Selinger P. G., Astrahan M. M., Chamberlin D. D., Lorie R. A., Price T. G.: *Access Path Selection in a Relational Database Management System*. Proc. ACM SIGMOD, 23-34, 1979.

[28] Slivinskas G., Jensen C. S., Snodgrass R. T.: Adaptable Query Optimization and Evaluation in Temporal Middleware. Proc. ACM SIGMOD, 127-138, 2001.

[29] Srinivasan J., Murthy R., Sundara S., Agarwal N., DeFazio S.: *Extensible Indexing: A Framework for Integrating Domain-Specific Indexing Schemes into Oracle8i*. Proc. ICDE, 91-100, 2000.

[30] Snodgrass R. T.: *Developing Time-Oriented Database Applications in SQL*. Morgan Kaufmann, 2000.

[31] Shen H., Ooi B. C., Lu H.: *The TP-Index: A Dynamic and Efficient Indexing Mechanism for Temporal Databases*. Proc. ICDE, 274-281, 1994.

[32] Stonebraker M.: *Inclusion of New Types in Relational Data Base Systems*. Proc. ICDE, 262-269, 1986.

[33] Theodoridis Y., Stefanakis E., Sellis T.: *Efficient Cost Models for Spatial Queries Using R-Trees*. IEEE Trans. on Knowledge and Data Engineering (TKDE), 12(1), 19-32, 2000.

Efficient Techniques for Range Search Queries on Earth Science Data
(Extended Abstract)

Qingmin Shi and Joseph F. JaJa
Institute for Advanced Computer Studies,
Department of Electrical and Computer Engineering,
University of Maryland, College Park, MD 20742, USA
{qshi,joseph@umiacs.umd.edu}

Abstract

We consider the problem of organizing large scale earth science raster data to efficiently handle queries for identifying regions whose parameters fall within certain range values specified by the queries. This problem seems to be critical to enabling basic data mining tasks such as determining associations between physical phenomena and spatial factors, detecting changes and trends, and content based retrieval. We assume that the input is too large to fit in internal memory and hence focus on data structures and algorithms that minimize the I/O bounds. A new data structure, called a Tree-of-Regions (ToR), is introduced and involves a combination of an R-tree and efficient representation of regions. It is shown that such a data structure enables the handling of range queries in an optimal I/O time, under certain reasonable assumptions. We also show that updates to the ToR can be handled efficiently. Experimental results for a variety of multi-valued earth science data illustrate the fast execution times of a wide range of queries, as predicted by our theoretical analysis.

1. Introduction

Considerable amounts of spatio-temporal data sets are generated on a daily basis with the amount of remotely sensed data alone expected to exceed several terabytes per day within the next few years. The sources of geospatial data are quite diverse and include satellite imagery, geographical information systems, census data, and environmental assessment and planning. These data sets offer unprecedented opportunities for exploring associations between environmental phenomena and spatial factors, building environmental models, detecting changes and finding interesting spatio-temporal patterns and trends. In spite

of a significant progress in the development of geospatial data mining techniques, the exploration of large amounts of geospatial data by content remains quite difficult. The NASA supported Earth Science Information Partnership (ESIP) Federation, that includes all the major data centers for earth sciences, developed a list of major scenarios for which content-based retrieval techniques will be critical [3]. Almost all of these scenarios involve the fundamental problem of determining spatio-temporal regions over which a certain number of parameter values satisfy certain constraints, for example the values fall within certain ranges or increase within certain bounds over a time period. This paper develops efficient techniques for addressing the core problem of determining regions within a large scale raster geospatial data set whose parameters' values fall within specified ranges. A forthcoming paper will show how these techniques can be extended to handle a time series of such data. The techniques developed here have a strong theoretical foundation and are coupled with an extensive set of experimental results that illustrate the efficiency of these techniques.

Briefly our main contributions are:

- The development of an efficient representation of raster data sets consisting of a combination of an R-tree built around the parameter values and a decomposition of the spatial-space into regions described by their boundaries. The overall complexity to build this structure is dominated by two external sorting steps.

- The querying over arbitrary value ranges of the parameters can be done very efficiently in time that is approximately proportional to the time it takes to read the output from external memory.

- An efficient way to handle updates.

- Extensive experimental tests on remotely sensed data

confirmed the efficiency of our representation in terms of fast execution times of a wide variety of random queries.

The remainder of this paper is organized as follows. In Section 2, we define the problem and the computational model used for analyzing our algorithms. The related work is discussed in Section 3, while our data organization structure is described in Section 4. Sections 5 and 6 present the algorithms for building our structure and handling general queries. In Section 7, we briefly discuss how updates are supported. The experimental evaluation of our methods is summarized in Section 8.

2. Problem Definition and Computational Model

We assume that we are given a grid G of size $N_x \times N_y$ representing a spatial region decomposed into $N = N_x N_y$ cells. A k-tuple $(f_1^{(i,j)}, f_2^{(i,j)}, \cdots, f_k^{(i,j)})$ is associated with each cell (i,j) in G such that each parameter $f_l^{(i,j)}$ is a certain numerical attribute corresponding to cell (i,j). We assume that G is too large to fit in internal memory and the result of a query may or may not fit in memory. The problem is to develop a representation of this grid in such a way that the following query can be answered very quickly (in time proportional to reading the output from external memory):

Determine all the regions in which the parameters' values fall within specified ranges: $a_l \leq f_l \leq b_l$, *for all* $1 \leq l \leq k$ *(a* query window*).*

A region is defined as the maximal set of connected cells where the parameter values satisfy the constraints in each cell. The output to our query consists of a list of all the cells in these regions such that all the cells in the same region are assigned the same label. Our techniques will carry out to more robust definitions of regions such as density-base regions in [12].

Solving the above problem involves efficient identification of the cells whose parameters' values fall within the specified ranges and fast groupings of cells into connected regions. The main focus here is on minimizing the query time. In addition, the storage of the new structure is also important since the size of the raw data is assumed to be large. In general, we aim to achieve the following three properties:

- The size of the new representation should be comparable to the raw data.

- The construction time of the new representation should be efficient in the sense that the input data should only be scanned a few times.

- Queries should be answered very quickly in time proportional to the output size.

To analyze our algorithms, we will use the standard two-level I/O model [4] defined by the following parameters:

N: the size of the input;

M: the internal memory size; and

B: the size of a disk block.

We assume that $B^2 < M < N$. An I/O operation is defined as the transfer of one block of contiguously stored data between disk and internal memory. Hence scanning an input of size N stored contiguously on a disk takes $O(N/B)$ I/O operations in this model.

3. Related Work

A major component of our problem requires the handling of multidimensional range queries of point data. A large number of external data structures and algorithms have been proposed to deal with this problem. In contrast to the two-dimensional case where solutions that provide provable good performance exist (see for example [21, 17, 7]), most data structures for high dimensional data are aimed at achieving good practical performance (A recent survey can be found in [13]). Among them, the R-tree [14] has been widely accepted as an efficient external tree structure for handling multidimensional data sets. Many dynamic R-tree variations have appeared in the literature (see for example [20, 8, 16, 10]) and they differ mainly in the heuristics used to split or merge nodes when node overflows or underflows occur. More relevant to the work reported here is the static case where the entire data set is known beforehand. Techniques that deal with this type of data are sometimes called tree-packing or bulk-loading. Various tree packing techniques aimed at improving node utilization and minimizing the *minimum bounding rectangles (MBRs)* of nodes have been introduced. They either sort data based on some spatial orders and recursively pack them into tree nodes level-by-level from bottom up (such as in [19, 15, 18]), or recursively partition data using various heuristics (such as in [24, 9]). Techniques for efficiently constructing dynamic R-trees for static data sets have also been explored. They view the construction of an R-tree for a static data set as a batched insertion problem and use lazy buffering strategy to achieve optimal I/O complexity [11, 6].

All the above techniques report the individual points that satisfy the range query. For raster data, one needs to find regions for which the attributes fall within the query window. Very few attempts have been made to address this problem. Most past work revolves around organizing the data objects hierarchically according to their spatial locations and summarizing their parameter values at different levels. STING [23] stores statistical information about the types and parameters of distributions for subsets of data in a hierarchical grid structure. This information can be used during a query to identify *relevant* cells that are later clustered using for example density based methods [12]. Yang et al. [25] addressed the same problem as in this paper but for a single parameter, and proposed a two level hierarchical structure that uses histograms to summarize the distributions of data values. The histograms of the high level cells are then clustered. The representative histogram summarized from histograms in the same cluster is used to decide whether that cluster of cells should be checked for a given query. These previous methods provide approximate answers without any guaranteed accuracy.

In the following three sections, we will discuss how our proposed data structure is used to solve this problem.

4. Data Structure

The proposed data structure, called the *Tree-of-Regions* (ToR) is an extension of the R-tree structure, although the same technique can be used to extend other tree structures. Each node v of the R-tree defines a k-dimensional value range and we associate with it a set of spatial regions such that the cells in each region have attributes that fall within this value range. Each leaf node of a ToR corresponds to a unique k-tuple from G and contains a pointer to the regions whose parameter values are equal to the k-tuple. Each internal node occupies an entire block and has $O(B)$ children. It contains a *minimum bounding rectangle (MBR)*, which is the tightest bounding rectangle of the union of the minimum bounding rectangles of its children. For leaf nodes, MBRs reduce to k-tuples. It is clear that the spatial region induced by the MBR of an internal node is the union of the spatial regions induced by its children. Figure 1 shows a ToR with $B = 3$ for a data set with two parameters. The top part of the figure illustrates the tree structure. Distinct k-tuples and their corresponding leaf nodes are depicted as dots. Internal nodes are represented by their MBRs shown as rectangles. The bottom part consists of four spatial regions $R(a)$, $R(b)$, $R(c)$, and $R(d)$ that are associated with leaf nodes a, b, c, and internal node d, respectively.

There are two benefits of storing regions at higher level nodes. First, during a range query, if the MBR of a node is covered entirely by the query window, the regions corresponding to that node can be reported immediately, with no need to explore the descendants of that node. Second, higher level nodes tend to have larger regions. By pre-storing these

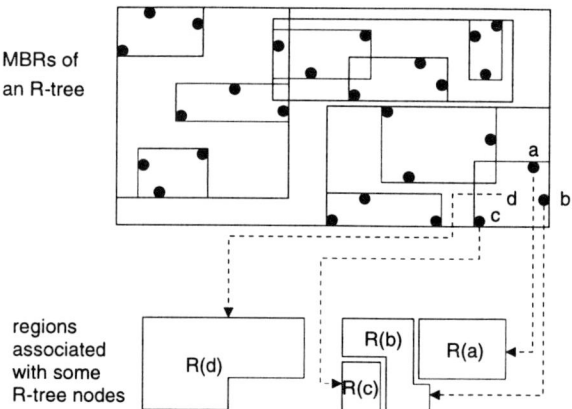

Figure 1. An example of the ToR

larger regions, we can compute the connected regions much more efficiently.

These potential benefits come at a storage cost as cells are duplicated along the path from the root to the leaf nodes. A *light* version of the ToR that associates regions only with the leaf nodes is another possible choice. However, it turns out that the query performance of the light ToR is much inferior without introducing any significant space savings. Related experimental results will be reported in the full paper.

A region associated with a ToR node is represented as a list of non-overlapping horizontal segments, each consisting of a maximal set of horizontally connected cells. Each segment is stored as a triple (y, x_left, x_right), where x_left and x_right are the x-coordinates of its leftmost and rightmost cells and y is its y-coordinate. Segments in the list are sorted using y as the primary key and x_left as the secondary key. Using segments to represent regions have several benefits. First, this representation maintains all boundary information of a region. Second, the amount of storage required is proportional to the perimeter of the region, which is much smaller than the area. Third, merging regions reduces to merging segments, which can be done quite efficiently as we will show soon. Figure 2 shows the merging of two regions represented using segment lists.

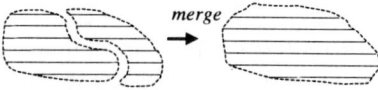

Figure 2. Merge of two regions represented by segment lists

Our overall data structure consists of three files: the *segment file*, the *leaf file*, and the *tree file*. Each file contains elements of the same size. The segment file contains the set of segment lists corresponding to the tree nodes. The list of segments corresponding to the same node are always stored contiguously on disk. The leaf file consists of the leaf nodes. The tree file is used to store the internal nodes. The reason we separate the leaf nodes from the internal nodes is that they have different structures. A leaf node does not contain the array of child pointers as does an internal node. Since we target our solution for very large data sets, we do not make the assumption that either of the three files will fit entirely in memory.

5. Tree Construction

The construction of a ToR consists of three steps:

1. The creation of the leaf nodes, where each leaf node corresponds to a distinct k-tuple of attribute values. This step includes the creation of all segments corresponding to the leaf nodes.

2. The creation of the internal nodes. Exactly how the internal nodes are constructed depends on the type of R-tree used. In this paper, we will use the packed Hilbert R-tree [15].[1]

3. The determination of the segment lists corresponding to the internal nodes.

5.1. Construction of Leaf Nodes

The objective of the first step is to find distinct k-tuples and their corresponding segment lists. These k-tuples form the leaf nodes of the tree structure. We assume that the raw data consists of a set of records, one for each cell. The record for cell (i, j) is in the form of $(f_1^{(i,j)}, f_2^{(i,j)}, \cdots, f_k^{(i,j)}, j, i)$. (Without causing confusion, we will call such record cell as well.) Reformatting is needed if the raw data are stored in a different format. Finding distinct k-tuples is achieved by sorting all cells using the key sequence $(f_1, f_2, \cdots, f_k, j, i)$. Since we are making the assumption that the data set resides on a disk, an external merge sort algorithm [22] is used. This sorting guarantees that cells having the same values are stored contiguously and, furthermore, horizontally adjacent cells that have the same values are also stored contiguously. This allows us to use a single scan through the

---

[1]There are three reasons for choosing the Hilbert R-tree. First, it is has been widely regarded as very competitive among all the R-tree variations [13]. Second, constructing such a tree structure can be done very efficiently, since it involves only one sort of the data set. Third, it has served as a base of performance comparison for many recently proposed data structures [18, 9, 6]. Note that the Hilbert R-tree can be replaced by any other static R-tree without affecting the remaining tree construction algorithm and the query algorithm.

sorted cells to create both the leaf nodes and the associated segment lists. Cells corresponding to the same k-tuple are merged into horizontal segments which are then stored contiguously in a *segment file*. Leaf nodes are created for distinct k-tuples in the same sorted order and stored in a *leaf file*. Each leaf node contains a k-tuple, an integer indicating the number of segments in the associated segment list, and a pointer to the beginning of that list in the segment file.

Clearly, this step involves the external sorting of N cells whose I/O complexity is $O(N/B \log_{M/B} N/B)$.

5.2. Construction of Internal Nodes

The Hilbert R-tree packing algorithm packs as many children into a parent node as possible while trying to make sure that children of the same parent are spatially close by using the Hilbert "space-filling" curve. The tree is constructed from bottom up. N_0 leaf nodes (at level 0) are first sorted according to their ascending Hilbert values. The first B leaf nodes in the sorted list are removed from the list and grouped under the same parent node at level 1. The next B leaf nodes are again chosen and put under another parent node. This continues until there are no leaf nodes left. After all internal nodes at level 1 are created, they are grouped similarly into nodes at level 2. The only difference is that, the internal nodes are no longer sorted based on their Hilbert values. Instead, they are grouped according to the order in which they are created. Tree nodes thus are created level by level until there is only one node that becomes the root of the R-tree.

The complexity of this step is dominated by the Hilbert sorting of the leaf nodes, which requires $O(N_0/B \log_{M/B} N_0/B)$ I/O operations. N_0 normally is much smaller than N.

5.3. Creation of Internal Segment Lists

The creation of the segment lists for internal nodes is done by recursively merging the segment lists of their children, starting from the leaf level.

Note that the segments in each list have been sorted in increasing order using keys y and x_left, and stored contiguously on the disk. Merging horizontal segments can be done in a similar way as the merging phase of the external sorting, while combining horizontally adjacent segments. Segments in a list are always brought into memory in blocks. A buffer of size B is allocated for each list. (Note that we have at most B lists per node and $B^2 < M$.) The smallest segment among the first segments of all the lists is repeatedly removed and added to the output segment list until all lists become empty. During this process, whenever a buffer is empty, another block of segments in the corresponding list is retrieved from the disk. The output segments are also buffered and added to the segment file in blocks.

Suppose the total number of segments associated with the leaf nodes is S, then the creation of the segment lists for the internal nodes just above the leaf nodes requires $O(S/B)$ I/O operations. As a result, the I/O complexity of the entire process of creating internal segment lists could be $O(S/B \log_B N_0/B)$, which may seem to be worse than the external sorting. However, in practice both S and N_0 are much smaller that N. Furthermore, the number of segments often decreases rapidly as the tree level gets higher. As a result, this step is normally dominated by the previous two steps.

6. Range Queries

Given a query window w, an *allocation node* in the ToR is a node whose MBR is covered entirely by w and whose parent is not an allocation node. Figure 3 shows the allocation nodes, depicted as dashed rectangles for internal nodes and gray dots for leaf nodes, for the dotted window describing a range query.

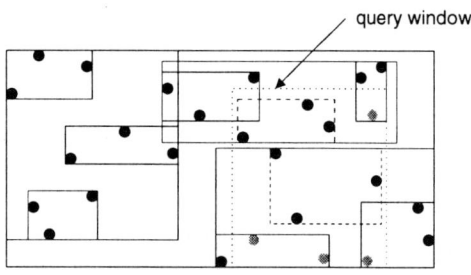

Figure 3. Allocation nodes

Answering a range query consists of determining the set of allocation nodes followed by merging the segment lists of these allocation nodes horizontally. Finally, this list of segments is merged vertically to create the output regions.

6.1. Identifying Allocation Nodes

The search for the allocation nodes starts from the root with the set of allocation nodes initialized as empty. If a node has no intersection with w, then no action is taken. If a node is fully covered by w, then it is identified as an allocation node and added to the set of allocation nodes. Otherwise, if the node intersects w, the same procedure is repeated for each of its children.

Note that this process is similar to the search in an R-tree. The difference is that a search path stops at an allocation node instead of continuing until a leaf is reached. As a result, the number of tree nodes accessed is much smaller than

the number of distinct k-tuples that fall in the query window. This number is even smaller comparing to the number of cells in the output regions since many cells may share the same allocation node. As we will show in our experimental results, the execution time of this step turns out to be only a very small portion of the total query time.

6.2. Merging Segments Horizontally

After the allocation nodes are determined, their associated segments are merged so that horizontally connected segments are combined into a single segment. A segment list merging algorithm similar to the one used in the tree construction can be used here. There is one difference, however. Since the number of allocation nodes f could be larger than M/B, multiple iterations might be needed as follows. In each iteration, every M/B segment lists are merged into a single list. There will be $O(\log_{M/B} f)$ iterations. Let F be the total number of segments associated with the allocation nodes. The I/O complexity for each iteration is $O(F/B)$. The total complexity for the horizontal merge is then $O(F/B \log_{M/B} f)$. We denote the list of segments after the horizontal merge as L and its cardinality as T.

6.3. Merging Segments Vertically

To identify the connected regions, we need to assign a label to each output cell such that cells from different connected regions have different labels. Using the sorted list L, finding connected regions can be done very efficiently, in fact in $O(T/B)$ I/O time.

Note that horizontally connected output segments have already been merged in the horizontal merge phase. What remains to be done is to merge the segments vertically to create regions.

We first use $O(T/B)$ I/O operations to scan L once to partition it into T_y sublists, T_y being the number of different y-coordinates of these segments. Each sublist contains segments with the same y-coordinate. This is possible since segments in L have already been sorted using their y-coordinates as the primary keys.

If L fits in internal memory, then we can apply an internal merging algorithm as follows. First, a graph H is created, whose vertices correspond to the segments in L. If two segments are adjacent to each other vertically, their corresponding vertices are connected by an edge in H. H is represented as a set of adjacency lists, one for each vertex. Second, a connected components algorithm based on depth-first search is used to find the regions.

Each sublist with y-coordinate y has one sublist *above* (*below*) it if there exists a sublist with the y-coordinate equal to $y - 1$ ($y + 1$). To construct the adjacency list for the first segment s in a sublist L_s, we scan the sublists above and below until all segments vertically adjacent to s are found and the first such segments are recorded in s's adjacency list. Then we continue to scan the same two sublists for the next segment in L_s and keep doing so until the adjacency lists for all the segments in L_s are created. Figures 4(a) and (b) give a simple example of a segment list L consisting of five segments and its corresponding adjacency list. Given a segment, its adjacent segments can be found by scanning the corresponding *above* and *below* sublists, starting from the segments recorded in its adjacency list.

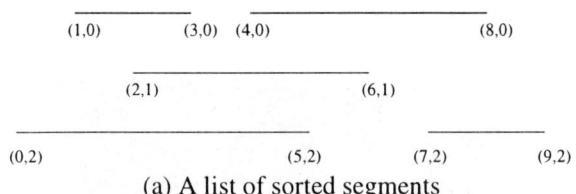

(a) A list of sorted segments

	segment	above	below
0	$(0,1,3)$	NULL	2
1	$(0,4,8)$	NULL	2
2	$(1,2,6)$	0	3
3	$(2,0,5)$	2	NULL
4	$(2,7,9)$	NULL	NULL

(b) The corresponding adjacency lists

Figure 4. Adjacency lists

If L does not fit in internal memory, we determine the connected regions as follows. We read as many sublists as the internal memory size allows, starting from the one with the smallest y value. We call this set of sublists a *stripe*. The internal merging algorithm described above is then applied to label all the segments in that stripe. These labeled segments are then written back to disk. Next, we again read as many sublists as possible, but starting from the lowest sublist in the last stripe (we call this sublist the *lower boundary* of the last stripe and the *upper boundary* of the current stripe). Since the segments in that boundary (the *boundary segments*) have already been labeled, their labels are propagated to other segments connected to them. New labels will be assigned to segments that do not connect with any of these boundary segments. If, during the labeling process, we find out that two segments from the upper boundary with different labels are actually connected then the label of one of them is changed. This change is kept in an *label-update table (LUT)* for that boundary. LUT is also written to disk after the current stripe is processed. The same process continues as we read the sublists stripe by stripe with two con-

tiguous stripes sharing a boundary until all segments are labeled.

After the downward labeling process, an upward updating operation is performed as follows. We repeatedly read a stripe and the LUTs of its upper and lower boundaries, starting from the stripe that is just above the lowest stripe. For each stripe, we update the labels of the segments in it using the label changes maintained in the lower LUT. These changes are also used to update the upper LUT. This process continues until the labels of the segments in the first stripe are updated. Details will appear in the full paper.

Under the reasonable assumption that the size of each of the T_y sublists is less than $O(M)$, we can make sure that a stripe and its upper and lower LUTs can be loaded into memory simultaneously, thus guaranteeing that the operations described above are possible. It is obvious that both the downward labeling and the upward updating processes require $O(T/B)$ I/Os for reading and writing the segments. The additional cost for reading and writing the LUTs is clearly less than $O(T/B)$ because each LUT is only accessed $O(1)$ times and the total size of the LUTs is less than the total size of the boundaries, which is less than T.

7. Handling Updates

In this section, we describe briefly how to handle the case where the parameters' values of a certain cell (i,j) are updated. Let the old values of cell (i,j) be $(f_1, f_2, \cdots, f_k)$ and the new values be $(g_1, g_2, \cdots, g_k)$.

Updating the value of cell (i,j) is done by deleting (i,j) from the segments associated with the R-tree nodes whose MBRs contain the old values and adding it to the R-tree nodes whose MBRs cover the new tuple. To delete (i,j), we first find the leaf node v that corresponds to the k-tuple $(f_1, f_2, \cdots, f_k)$. If the leaf node has more than one cells associated with it, then we only remove (i,j) from all the lists of segments associated with the nodes in the path from the root to v. Similarly, to insert cell (i,j) back into the ToR structure, we first locate the leaf node that corresponds to $(g_1, g_2, \cdots, g_k)$. If such a leaf node exists, then (i,j) is added to its associated segment list and to that of its ancestors. Otherwise, a new leaf node is created and inserted into the data structure. Deleting or adding an R-tree node can be done in a way similar to updating an ordinary R-tree (or an R*-tree) with the additional effort to update the segment lists of the nodes affected. We shoud note that only a small portion of the ToR structure, i.e. a constant number of nodes and their associated segment lists, needs to be updated. In the case where the values of many cells need to be modified, the I/O cost per update can be further reduced by using batched update methods such as the buffer tree [5].

147

8. Experimental Results

We tested our new approach on a number of raster data sets generated from satellite data. We describe here two types, a global coarse resolution and a fine resolution. The first type consists of the standard AVHRR (Advanced Very High Resolution Radiometers) data products that form a 1-degree by 1-degree of global coverage generated from 10 day composites. Geophysical parameters contained in the data set include: Normalized Difference Vegetation Index (NDVI), two reflectance channels (channels 1 and 2), three brightness temperature channels (channels 3, 4, and 5), and date and hour of observation [2]. NDVI is the ratio of the contrast between the response of the two reflectance channels. We used three of these parameters (NDVI, channel 1 and channel 4). The total number of cells in each AVHRR data set is 64K. The second type is the TM (Thematic Mapping) data [1]. Each TM data is a 7200-by-8192 grid representing a region with 30 meter resolution. Each cell has seven parameters (bands), of which we used five (bands 1, 2, 4, 5, and 7) on a 1000×1000 grid. A total of 44 data sets are used in our experiments. 24 of them are AVHRR global 1-degree by 1-degree data and the remaining 20 are the TM data.

The tree construction and query answering algorithms were coded in C. All the experiments were conducted on a Pentium III 550Mz machine with 1GB Memory running Linux 2.2.19. The page size B was chosen to be 8192 bytes.

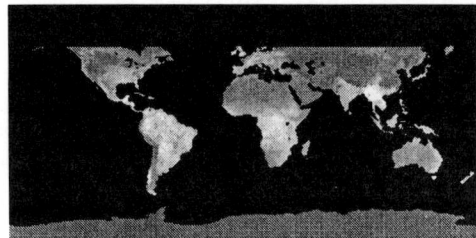

(a) NDVI values, 10-day composite
Jan. 01, 1989

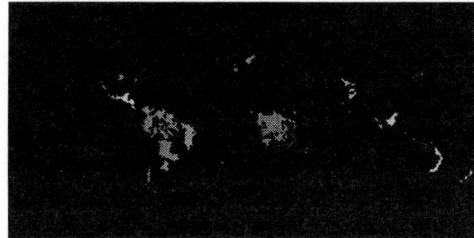

(b) regions where NDVI ≥ 0.4 and
channel 3 brightness temperature ≥ 260.0 *Kelvin*

Figure 5. Sample query results (AVHRR)

8.1. Sample Query Results

Figures 5 and 6 give two sample query results. Figure 5(a) is the global 1-degree by 1-degree NDVI map. Figure 5(b) shows the areas with high temperature and high NDVI values, which approximately correspond to the rain forests and the wooded grasslands that mainly locate in Central America, Central Africa, South Asia, and the east coast of Australia. Different colors are used to denote different connected regions. Figure 6(a) is band 7 for part of a TM scene in Columbia. Figure 6(b) shows the query result that largely corresponds to *nonforests*, which typically have high values in bands 4 and 7.

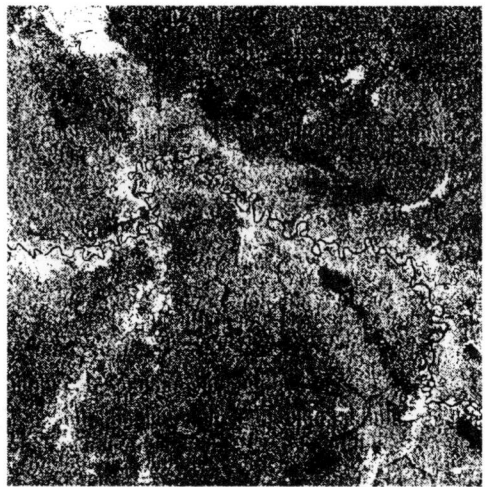

(a) Band 7, Part of Columbia (Path/Row: 6/66)
Oct 16, 1996

(b) regions where band 4 ≥ 20 and band 7 ≥ 20

Figure 6. Sample query results (TM)

8.2. Query Performance

We will first examine the overall query performance and then focus on the main components of the query algorithm. We will also compare the number of output segments and output cells to demonstrate the importance of representing regions using segment lists.

For each ToR, we generated query windows of 5 different sizes ranging from 5% to 25% of the size of the MBR of the root node. For each window size, 30 query windows were randomly and uniformly generated within the root MBR. All performance numbers are averaged over these queries. We will report the experimental results for the TM data, which is much larger than the AVHRR data.

Figure 7 shows the overall query execution time as contributed by the three main steps. We can see that the amount of time it takes to identify the allocation nodes is very small comparing to the horizontal and vertical segment merge times. The horizontal merge step takes up most of the execution time, while the vertical merge step was done much faster. Overall, it can be seen that the queries are handled extremely fast, within 5 seconds for the TM data even for queries with large windows. Furthermore, the query time is proportional to the output size as had been indicated by our earlier analysis.

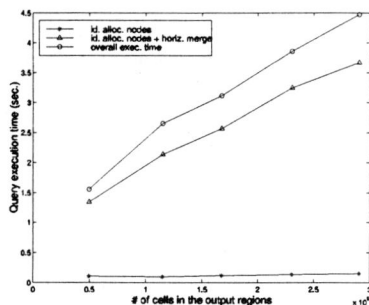

Figure 7. Overall query performance

Figures 8(a) and (b) show the comparison of the theoretical bounds and the observed bounds in terms of execution time and number I/O operations. The X-axis represents the theoretical complexity $F/B \log_{M/B} f$, where f is the number of allocation nodes and F is the number of segments associated with them. Since B and M do not change in our experiments, this theoretical bound differs from $F \log_2 f$ by only a constant. Thus, using the latter will not affect the shape of the curves. These two figures demonstrate that our experimental results and the theoretical results are quite consistent. The performance of the vertical segment merge is shown in Figure 9. The results are consistent with our theoretical complexity analysis as well.

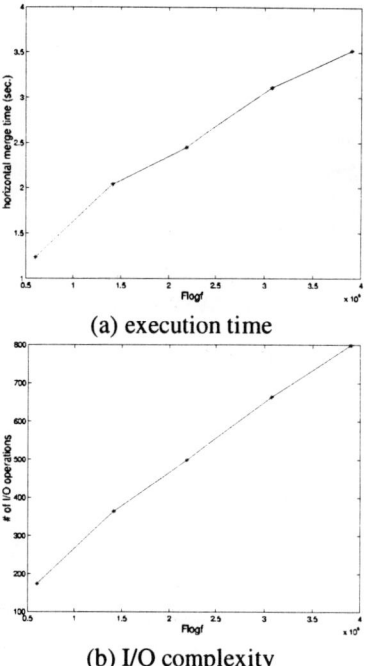

(a) execution time

(b) I/O complexity

Figure 8. Complexity of Horizontal Merge

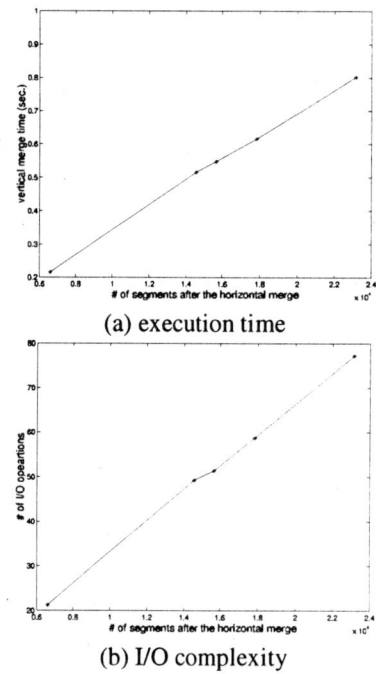

(a) execution time

(b) I/O complexity

Figure 9. Complexity of Vertical Merge

Finally, we compare the number of output cells and output segments to demonstrate the effectiveness of the seg-

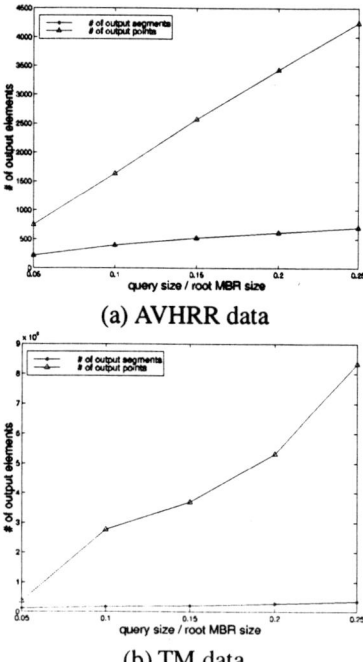

(a) AVHRR data

(b) TM data

Figure 10. Number of output segments v.s. number of output cells

ment representation of output regions. Figure 10 shows the average number of output segments and output cells for both AVHRR and TM data. Note that while the number of output cells increases quite fast as the size of the query window increases, the number of output segments increases very slowly in both cases. This has enabled the various steps in our query answering algorithm to be carried out quite fast.

References

[1] Landsat thematic mapper data. http://edc.usgs.gov/glis/hyper/guide/landsat_tm.

[2] Goddard DAAC NOAA/NASA Pathfinder AVHRR Land (PAL). http://daac.gsfc.nasa.gov/ REFERENCE_DOCS/dataset_references/pal_summary.html, 1999.

[3] Content-based search and data mining. http://www.esipfed.net/clusters/content_based/ sci_scen.html, 2000.

[4] A. Aggarwal and J. S. Vitter. The input/output complexity of sorting and related problems. *Communications of the ACM*, 31(9):1116–1127, Sept. 1988.

[5] L. Arge. The buffer tree: A new technique for optimal I/O-algorithms. In *Proceedings of the 4th International Workshop on Algorithms and Data Structures (WADS'95)*, pages 334–345, Kingston, Ontario, Canada, Aug. 1995.

[6] L. Arge, K. Hinrichs, J. Vahrenhold, and J. S. Vitter. Efficient bulk operations on dynamic R-trees. In *Proceedings of the 1st Workshop on Algorithm Engineering and Experimentation*, pages 328–348, Baltimore, MD, Jan. 1999.

[7] L. Arge, V. Samoladas, and J. S. Vitter. On two-dimensional indexability and optimal range search indexing. In *Proceedings of the Eighteenth ACM SIGACT-SIGMOD-SIGART Symposium on Principles of Database Systems*, pages 346–357, Philadelphia, PA, May 1999.

[8] N. Beckmann, H.-P. Kriegel, R. Schneider, and B. Seeger. The R*-tree: An efficient and robust access method for points and rectangles. In H. Garcia-Molina and H. V. Jagadish, editors, *Proceedings of the 1990 ACM SIGMOD International Conference on Management of Data*, pages 322–331, Atlantic City, NJ, May 1990.

[9] S. Berchtold, C. Böhm, and H.-P. Kriegel. Improving the query performance of high-dimensional index structures by bulk-load operations. In *Proc. 6th Int. Conf. Extending Database Technology, EDBT*, pages 216–230, Mar. 1998.

[10] S. Berchtold, D. A. Keim, and H.-P. Kriegel. The X-tree : An index structure for high-dimensional data. In *VLDB'96, Proceedings of 22nd International Conference on Very Large Data Bases*, pages 28–39, Mumbai (Bombay), India, Sept. 1996.

[11] J. V. den Bercken, B. Seeger, and P. Widmayer. A generic approach to bulk loading multidimensional index structures. In *VLDB'97, Proceedings of 23rd International Conference on Very Large Data Bases, August 25-29, 1997, Athens, Greece*, pages 406–415, 1997.

[12] M. Ester, H.-P. Kriegel, J. Sander, and X. Xu. A density-based algorithm for discovering clusters in large spatial databases with noise. In *Proceedings of the Second International Conference on Knowledge Discovery and Data Mining (KDD-96)*, pages 226–231, Portland, OR, Aug. 1996.

[13] V. Gaede and O. Günther. Multidimensional access methods. *ACM Computing Surveys*, 30(2):170–231, 1998.

[14] A. Guttman. *R*-trees: A dynamic index structure for spatial searching. In *Proceedings of the ACM SIGMOD International Conference on Management of Data*, pages 47–57, Boston, MA, June 1984.

[15] I. Kamel and C. Faloutsos. On packing R-trees. In *In Proceedings of the Second International Conference on Information and Knowledge Management*, pages 490–499, 1993.

[16] I. Kamel and C. Faloutsos. Hilbert R-tree: An improved R-tree using fractals. In *Proceedings of the Twentieth International Conference on Very Large Databases*, pages 500–509, Santiago, Chile, 1994.

[17] P. C. Kanellakis, S. Ramaswamy, D. E. Vengroff, and J. S. Vitter. Indexing for data models with constraints and classes. *Journal of Computer and System Science*, 52(3):589–612, 1996.

[18] S. T. Leutenegger, J. M. Edgington, and M. A. Lopez. STR: A simple and efficient algorithm for R-tree packing. In *Proceedings of the 13th International Conference on Data Engineering (ICDE'97)*, pages 497–507, Apr. 1997.

[19] N. Roussopoulos and D. Leifker. Direct spatial search on pictorial databases using packed R-trees. In *Proceedings of*

ACM-SIGMOD 1985 International Conference on Management of Data, pages 17–31, Austin, TX, Dec. 1985.

[20] T. K. Sellis, N. Roussopoulos, and C. Faloutsos. The R+-tree: A dynamic index for multi-dimensional objects. In *Proceedings of the 13th Interntational Conference on Very Large Data Bases*, pages 507–518, Brighton, England, Sept. 1987.

[21] S. Subramanian and S. Ramaswamy. The P-range tree: A new data structure for range searching in secondary memory. In *Proceedings of the ACM-SIAM Symposium on Discrete Algorithms*, pages 378–387, 1995.

[22] J. S. Vitter. External memory algorithms. In *Proceedings of the Seventeenth ACM Symposium on Principles of Database Systems*, pages 119–128, New York, NY, USA, June 1998.

[23] W. Wang, J. Yang, and R. Muntz. STING: a statistical information grid approach to spatial data mining. In *Proceedings of the Twenty-Third International Conference on Very Large Data Bases*, pages 186–195, Athens, Greece, Aug. 1997.

[24] D. A. White and R. Jain. Algorithms and strategies for similarity retrieval. Technical Report VCL-96-101, Visual Computing Laboratory, University of California, San Diego, CA, 1996.

[25] R. Yang, K.-S. Yang, M. Kafatos, and X. Wang. Value range queries on earth science data via histogram clustering. In *First International Workshop on Temporal, Spatial, and Spatio-Temporal Data Mining*, pages 62–76, Lyon, France, Sept. 2001.

Scientific Multimedia Data Management

Access Support Tree & TextArray:
A Data Structure for XML Document Storage & Retrieval

Dieter Scheffner   Johann-Christoph Freytag

Department of Computer Science
Humboldt-Universität zu Berlin, Germany
{scheffne | freytag} @dbis.informatik.hu-berlin.de

Abstract

The characteristics of XML documents require new ways of storing and querying such documents. Queries on both textual content and structural aspects must be supported efficiently. For this reason, we examined existing work on both document storage approaches and models for querying documents to derive requirements that are essential for the storage of XML documents. As a result of our study, we designed the Access Support Tree and TextArray (AST/TA) data structure. The important idea of the AST/TA data structure is the separation of the (logical) structure of a document from its "visible" text content. The latter is represented as a single contiguous string. At the same time the AST/TA data structure provides a tight integration to guarantee consistent changes. We introduce the AST/TA data structure formally by its abstraction, namely the AST/TA model and compare requirements of our AST/TA approach with those found in the current literature. Finally, we describe the advantage of the AST/TA model based on the AST/TA design principles.

1. Introduction

XML has become widely accepted for both data representation and exchange of information over the Internet. The amount of such data is rapidly growing. Thus, the demand increases for systems to manage large amounts of XML documents or data, which greatly vary in structural and access-related requirements, in an efficient and reliable way. Conventional DBMSs, however, are designed for structured rather than for semi-structured data or XML documents. For this reason, we decided to design a management system for XML documents from scratch—the XML Query Execution Engine (*XEE*). We build our *XEE* system based on a data structure suitable for providing efficient operations on XML documents, in particular with respect to updates and to increased flexibility for many applications.

In this paper, we introduce the first step of our ongoing research on physical storage implementation for XML documents, namely the *Access Support Tree* (*AST*) and *Text-Array* (*TA*) data structure. The AST/TA data structure has been implemented in main memory and provides our base concept for maintaining XML documents on persistent storage. For convenience, we introduce an abstraction of our data structure, which we refer to as the *AST/TA model*.

The AST/TA model takes advantage of merging well-known and established concepts from different fields to support search and update operations on large collections of XML documents. The requirements for the physical storage design are mainly influenced by the requirements of our *XEE* system as follows:

(1) Query evaluation is supported by integrating both the concept as database query languages and the concept as information retrieval, thus supporting such concepts as, e.g., the *Information Retrieval Query Language* (*IRQL*) [2] and the *Proximal Nodes Model* [5].

(2) The idea of separating the layout from the structure and content of a document is extended to the separation of structure and content, thus taking over the basic DBMS concept of partitioning data and meta data.

(3) The structure of documents to be stored is not necessarily constrained by any schema, i.e., it must be possible to store such *generic* documents even if they do not come with any schema information.

(4) Efficient operations on documents must be supported, especially while updating the structure and the content of documents.

Two of our goals are to support information retrieval and structural queries equally, and to accomplish a separation of structure and content. For this reason, we decided to keep the entire "visible" text content of an XML document as a single contiguous string maintaining the original order of the text in the document. The "visible" text content is the text a user may have in mind when formulating a query. Consequentially, the text content is neither fragmented nor interspersed with markup. Consider the following sample query put to an electronic shop: *"Retrieve documents containing* `Our Price: $11.96`*"*. Assuming that the price is tagged with markup, e.g., `... Our Price: <price>$11.96</price> ...`, in our desired XML document, we recognize the following advantages: (1) The search string matches directly the presentation string—no filtering is needed, since the storage representation is not stained with markup. Furthermore, the distances in the "visible" text of documents match their distances on storage. (2) By not fragmenting the text content, we need at most one access to fetch the complete string `Our Price: $11.96` from storage. Consequently, we are able to implement efficient search and browse operations on the "visible" text.

In addition, queries on document structure only, e.g., *"Are there any priced books?"*, may be handled without looking at the text content at all. We experienced that the predominant part of XML documents—far more than 50%—consists of text content only. Thus, the separation of structure and content makes it possible to present the navigation structure of documents more densely. That is, with respect to an implementation of this concept on persistent storage, we expect to map document structures to fewer pages than complete documents require. For this reason, navigating document structures becomes more efficient, due to shorter access paths within the page presentation of such structures.

By maintaining (unstructured) text content as a single contiguous string, we are able to provide operations that refine the structure of a document in a state of flux. That is, we add new "markups" to the structure rather than insert real markups into the text content of the document. For example, a shop administrator realizes that it might reasonable to make the price of a product explicitly available to customers by tagging it. Such a change only affects the structure of the document. Thus, the impact on the overall system is much smaller. Technically, this means, our approach supports *tag insertion* which is addressed as the *tag insertion problem* and is an important part of text mining [16].

Another advantage of referring to the text content as a single contiguous string is that we may build multiple or independent hierarchies [5] of logical structures on top of the same text content. For example, when we consider the sentence `"Max wears a hat."`, we may add structure as in `<txt><name>Max</name>wears a hat.</txt>` or as in `<txt><subj>Max</subj><pred>wears</pred><obj>a hat</obj>.</txt>`.

Our data structure also supports different views of text efficiently, i.e., text as a sequence of characters and as a sequence of words [9]. Considering text as a sequence of characters is a very simple and flexible way for manipulating and accessing texts. Nowadays, well-known representatives of such texts are descriptions of, e.g., protein sequences in molecular biology. However, it is advantageous for natural languages, such as German or English, if we consider text as a sequence of words, because we may avoid repetition of characters which only indicate word boundaries and layout. For this purpose, we take into account text normalization similar to the normalization in PAT [7], thus facilitating a more efficient access to words.

Eventually, our approach may be useful for replacing or supplementing existing storage structures, respectively. The latter alternative means, that those parts of XML documents that conventional systems cannot handle efficiently are stored in our data structure. That is, the AST/TA data structure might be plugged into such systems as an "XML data type".

The concept we use in our approach, i.e., considering the text content as a contiguous string and representing the logical structure of documents in a separate hierarchy, has already been introduced earlier. To the best of our best knowledge, this was done only twice, i.e., in the framework of the *Structured Multimedia Document DBMS (SMD DBMS)* [1] and in the *Proximal Nodes Model* [5]. With regard to SMD DBMS, only few implementation issues are discussed in [6]. The Proximal Nodes Model is, according to its authors, a purely logical model for querying document databases. Therefore, it does not mandate any specific implementation. Neither of these approaches refers to XML documents directly. Hence, we are convinced, it is worthwhile to investigate deeper implementation details based on this concept, in particular with regard to XML documents.

2. Related Work

XML or SGML documents are often treated as objects and thus stored and managed by an OODBMS. Alternatively, documents are "forced" into relational or object-relational DBMSs by managing documents as data in one or more tables. Moreover, we experienced that, according to the *ANSI/X3/SPARC* architecture model, most of these approaches more or less rely on the conceptual level rather than on the physical level, thus leaving the physical organization to the DBMS.

In this section, we briefly examine storage and management concepts of documents in the *XML Extender* (IBM® DB2® Extenders™) [3], in *HyperStorM* (Hypermedia Document Storage and Modeling) [10], in *Structured Multimedia Document DBMS* [1], and in NATIX (Native XML Repository) [4].

XML Extender. Based on the DB2 object-relational DBMS, the XML Extender manages XML documents in

Table 1. Overview of criteria and approaches

criterion	XML Extender		HyperStorM	SMD DBMS	NATIX	AST/TA
	XML Column	XML Collection				
support of update operations	✓	✓	✓	✓	✓	✓
contiguous text content	✓	–	to a degree (hybr. approach)	✓	–	✓
storage of text without markup	–	✓	to a degree (hybr. approach)	✓	✓	✓
separation of doc. structure & content	–	✓	–	✓	–	✓
support of cont.-based search	✓	–	✓	✓	–	✓
support of generic document storage	✓	–	–	–	✓	✓
support of text normalization	–	–	–	–	–	✓
operational support of text mining	–	–	–	–	–	✓

✓ : criterion applies to the approach.　　— : criterion does not apply.

two different ways, namely as *XML Column* and as *XML Collection*. XML Column supports the storage of complete and marked up documents in a single table column. *User Defined Functions* give access to documents and parts of documents. To search the structure of a document efficiently, the user must create indexed side tables referring to well-chosen elements and element attributes of the document. In contrast to an XML Column, an XML Collection is based on the idea of "dismantling" documents. Elements or element attributes are mapped into columns of one or more tables. The text content of elements and the values of element attributes become table values. *Document Access Definitions* take care of the structural glue of documents. Dismantling and reconstructing documents is accomplished with the help of *Stored Procedures*.

HyperStorM. The objective of HyperStorM was to build an application database framework for storing structured documents in a system coupling an object-oriented DBMS and an information retrieval system [10]. Their approach distinguishes three strategies to represent SGML documents in an object-oriented database: (1) a completely structured database-internal representation of documents, i.e., each logical document component corresponds to a database object, (2) documents are stored as BLOBs in the database, and (3) the hybrid approach of (1) and (2), i.e., some "non-flat" elements are represented by individual database objects in an object hierarchy while "flat" elements represent parts of the document in their native form (text interspersed with markup). It is an administration task to decide which element becomes a "flat" or a "non-flat" element.

Structured Multimedia Document DBMS. Within the framework of the Structured Multimedia Document DBMS (SMDDBMS), an object-oriented MMDBS was developed for storage of SGML documents [1] in the presence of DTDs. The text content of documents is stored as a contiguous text string as a whole rather than being fragmented [6]. The logical structure of a document is represented by an object hierarchy of which the objects that refer to text content reference to "their text" with so-called *annotations*. An annotation is a logical reference that indicates the first and the last character of the substring to which the concerning object refers.

NATIX. NATIX is a repository for the storage and management of large tree-structured objects, preferably XML documents [4]. Essentially, the idea of NATIX is to map the logical structure of an XML document directly into the corresponding physical structure. Unlike the logical structure, the physical tree structure is equipped with additional nodes that help manage large trees, such that the tree might be split up among several pages in storage. Since the materialized tree is the direct mapping of the logical structure of the document, the text content is fragmented and is interspersed with structural data like element and attribute names, etc. NATIX may store generic documents that do not depend on any schema.

Comparing Approaches. Table 1 summarizes the properties of the approaches introduced in the previous paragraphs and compares them with the requirements for our approach.

3. The Access Support Tree/TextArray Model

3.1. Motivation

The AST/TA model is the abstraction of the AST/TA data structure that is relevant and is used for representing XML documents in physical storage. The AST/TA model describes and summarizes the components of XML documents and their relationship to each other with regard to their storage representation rather than to their conceptual models as provided by query models like the *Document Object Model* [11], the *XQuery 1.0 and XPath 2.0 Data Model* [15] or the *Proximal Nodes Model* [5]. The AST/TA model provides a base concept for maintaining XML documents on persistent storage. Based on the concept of an persistent AST/TA model, we intend to implement the DOM, the XQuery 1.0 and XPath 2.0 Data Model, or the Proximal Nodes Model on secondary storage.

Well-formed XML documents represent the actual instances of documents. They actually consist of only an XML **[document element]**[1] each. XML declarations and XML document type declarations are optional and provide additional information constraining XML document elements. Since *XML Schema* [14] provides means for specifying schemata in the form of XML documents themselves, XML document type declarations may be reduced to just references to other XML documents. The AST/TA model is designed to support *generic* XML documents, i.e., documents having no constraints other than well-formedness. For this reason, the AST/TA model devotes to XML document elements only.

3.2. The static AST/TA Model

An important goal of our approach is to separate meta data from data for efficient management of XML documents on storage. Therefore, we distinguish the Access Support Tree from the TextArray of a document element. The entire logical structure of the document element, including element attributes, comments etc., is incorporated in an ordered tree—the AST—whereas the TA keeps the entire text content as a single contiguous string with regard to its original order in the document. Thus, the XML document element is represented by an AST/TA pair in physical storage.

The relationship between AST and TA is established by logical references called *text surrogate values*. A text surrogate value is a vector indicating the start position and the length of a text segment referenced to in a TA. Every vertex of the logical structure receives a text surrogate value, thus text segments might be reached from any vertex and ASTs become indexes. If we apply this linking concept to documents that do not change their text content in length, text surrogate values alone might be sufficient. We introduce *offsets* that adjust the text surrogate values after updates changing the length of the text in TAs. Hereby, the number of vertices to be modified is kept as small as possible. Otherwise, all values that span or follow the location of update in the logical structure would have to be modified, making such updates inefficient. The offset concept is also applied to text surrogate values by specifying the length rather than the end position of text segments. That is, when updating a surrogate value, in most cases, either the start position or the length must be updated only.

An AST/TA Example. Figure 1 shows an example of an AST/TA pair with its corresponding document element. This example is based on a *normalized* TA. The symbols ␣ and # act as substitutes for word separators. We address the normalization of TAs in the next section.

The sample AST consists of seven vertices (r, ..., x) representing their corresponding document components, i.e., elements, a comment, and text components. For example, vertex r labeled with `header` refers to the document element itself. Vertex t represents a comment. We label such vertices with their textual content only, e.g., `check year`, leaving out syntactic additives as `<!--` and `-->`. On the contrary, text vertices such as v and x have no labels. They act as place holders for the text components they represent in the AST tree. In addition to labels, element vertices may have an attribute set assigned to. Thus, vertex s (`author` element) comes with the refinements of `from="1832"` and `to="1908"` representing the attributes `from` and `to` with their corresponding values `"1832"` and `"1908"`.

Edges (solid lines) link vertices of the tree. They refer to the immediate part-of relationship of the document components. For example, the `author` element is a direct part (child) of the `header` element. Thus, the AST tree matches the logical structure of the document element.

Expressions in parenthesis, e.g., `(14,14)`, depict text surrogate values. That is, the vertices u and x both refer to the text segment that starts at position `14` in the TA and has a length of `14`. Arrows give a notion of text surrogate values in the figure pointing to the start of corresponding text segments in the TA.

For an efficient implementation of update operations, all vertices do have a defined start position in their surrogate values, thus aligning vertices to text segments. Therefore, the comment vertex t has the defined start position of `14` in its surrogate value. Vertices representing processing instructions and comments and vertices that stand for *true empty elements* acquire a length of -1 in their surrogates. Addressing the problem of empty elements, we must distinguish elements that do not span over any text content in a TA—*true empty elements*—and "PCDATA elements" that

---

[1]We use the notation introduced in XML Information Set specification [13].

158

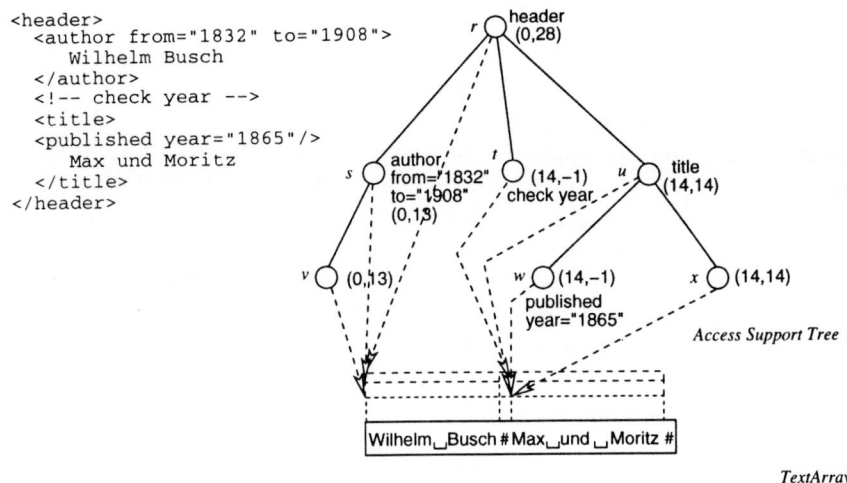

```
<header>
   <author from="1832" to="1908">
      Wilhelm Busch
   </author>
   <!-- check year -->
   <title>
   <published year="1865"/>
      Max und Moritz
   </title>
</header>
```

Figure 1. AST and normalized TextArray

refer to empty word content and, therefore, acquire a length of 0 in their surrogates. Providing these different measures of length, the AST model satisfies the different semantics of such elements.

We left out the offsets in our example, since we assume the document not being updated yet. Therefore, all offsets are equal to 0.

Definitions. The Definitions 1, 2, and 3 summarize our AST/TA model. The identifiers we use in our definitions, namely `element`, `PI`, `Comment`, `CharData`, `CDSect`, `EntityRef`, `Attribute`, and `Name` refer to the left-hand sides of the corresponding productions in the *XML Standard* [12]. As character references are only used for overcoming limitations of the character encoding of documents and for "escaping" characters that would be otherwise interpreted as markup, we postulate character references to be expanded. Furthermore, we assume that validating XML parsers expand entity references, whenever it is possible. Otherwise, entity references are handled in their native form, e.g., `&entity-ref;`, as "normal" text of type #PCDATA. In Def. 3 we use the term *textual content* to refer to processing instructions and comments leaving out the syntactical additives `<--`, `-->`, `<?`, and `?>`.

Formal definitions of our AST/TA data structure and its relationship to the well-known *XML Information Set* [13] can be found in [8].

Definition 1 (TextArray) *A TextArray is the contiguous sequence τ of characters in document order that represents all parts of an XML document referring to the type set $T_{PCDATA} := \{$CharData, CDSect, EntityRef$\}$. τ_i is the ith character of τ, where $i \in \{0, \dots, length(\tau) - 1\}$.*

The following properties are valid for TextArrays: (1) all character references are expanded in TextArray segments that refer to the type of `CharData`, *(2) CDATA sections are stored as specified by the type of* `CDSect` *leaving out the syntactic additives* `<![CDATA[` *and* `]]>`, *and (3) entity references are stored as specified by the type of* `EntityRef` *referring to their native form.* □

Definition 2 (Normalized TextArray) *A normalized Text-Array is a TextArray with the following additional properties. Sequences of white spaces (#x20, #x9, #xD, #xA) are reduced to a single space (#x20 – word separator) or they are removed if they appear next to markup boundaries or next to the beginning of the TextArray. Markup boundaries are represented by the character #x0 (separator). We may use the character #x0 for separators, since it is not part of any XML document [12]. For implementation reasons, we make markup boundaries explicit at the end of the TextArray, however, not at its beginning.* □

Definition 3 (Access Support Tree) *An Access Support Tree is an ordered tree $(\mathcal{V}, \mathcal{E})$. A vertex $v \in \mathcal{V}$ represents the corresponding structure component of an XML document element. An edge $e \in \mathcal{E}$ expresses the relationship between parent and child vertices in $\mathcal{V}$, thus $\mathcal{E} \subseteq \mathcal{V} \times \mathcal{V}$.*

Siblings of one parent vertex are ordered with respect to a successor relation for any two neighbor siblings. Therefore, they may be counted from left to right starting with 0.

Every vertex has a type $\vartheta \in T$ assigned to it. We may separate T in two subsets: $T := T_{labeled} \cup T_{PCDATA}$ with $T_{labeled} := \{$element, PI, Comment$\}$ and $T_{PCDATA} := \{$CharData, CDSect, EntityRef$\}$.

There is exactly one root vertex of type `element`. *All*

vertices, except the root *vertex, have a parent vertex of type* element.

Every vertex of type $\vartheta \in T_{labeled}$ *receives a label* $\lambda \in L$ *representing the* Name *or the* textual content *of the corresponding component. Vertices of type* $\vartheta \in T_{PCDATA}$ *carry the empty word* ε *as label. Vertices* v *of type* element *may have a set of corresponding* Attributes a_v, *including their names and values, assigned to it.*

Every vertex obtains a text surrogate value $\sigma \in \mathbb{N}_0 \times \mathbb{Z}$. *Text surrogate values are vectors consisting of the two components* σ_p *and* σ_l, *which refer to the start* position *and the* length *of the text segment in a TextArray referenced by a vertex. Thus, text surrogate values are logical references. Vertices of type* $\vartheta \in T_{PCDATA}$ *reference their corresponding character sequence in the TextArray that they represent in a document. Vertices of type* element *reference the text segment that is the "concatenation" of all text segments to which their descendant vertices of type* $\vartheta \in T_{PCDATA}$ *refer.*

Offsets $\omega \in \mathbb{Z}$ *adjust the* position σ_p *of a text surrogate value, such that vertices reference their text segments via their text surrogate values correctly after updates.* $\square$

In the following, we discuss issues of integrating our proposed text normalization into the AST/TA model. For the integration of other relevant XML features such as handling of entities and XML namespaces, we refer to our technical report [8].

Text Normalization. It is easy to handle text content as a sequence of characters, because the text remains in its original form. However, when considering text content as a sequence of words, we must be aware of *separators* and *word separators*. If we disregard separators, words would merge in a TA. The following example shows this effect—the symbol "$\sqcup$" indicates single word separators:

```
Instance:   <author>Wilhelm Busch</author
            ><title>Max und Moritz...
TextArray:  Wilhelm␣BuschMax␣und␣Moritz...
```

Busch and Max merged into BuschMax within the TA, thus Max may not be recognized as an independent word any longer.

Furthermore, we have to take into consideration different semantics of separators and word separators in a TA, if we intend to remove character sequences that cross markup boundaries. Let us take a look at the following example—for better reading, the symbol "#" replaces the #x0 character here, indicating separators:

```
Instance:   <author>Wilhelm Busch</author
            ><title>Max und Moritz...
TextArray:  Wilhelm␣Busch#Max␣und␣Moritz...
```

There is no problem in removing the string helm␣Bu, since the removal happens directly within the author element. However, if we want to remove, e.g., sch#Ma, that crosses markup boundaries, we expect the markup boundary to be maintained:

```
Instance:   <author>Wilhelm Bu</author
            ><title>x und Moritz...
TextArray:  Wilhelm␣Bu#x␣und␣Moritz...
```

The separator must remain in the TA as long as the markup </author><title> remains part of the instance.

To handle these and other challenges which notably arise in text normalization, we encapsulate the access to TextArrays on behalf of both views of text in our implementation. Thus, we provide a generic interface for the operations that the AST/TA model specifies.

3.3. Operations of the AST/TA Model

The AST/TA model provides two levels of operations: (1) operations for *generating* an AST/TA pair from an XML document and for *restoring* an XML document from an AST/TA pair and (2) *search*, *insert*, and *delete* operations that are based on the AST/TA pair. Operations of level (1) may be considered as *Data Definition* operations of DBMSs. Whereas (2) represents the level of *Data Manipulation* operations.

Generating and Restoring XML Documents. We assume that the XML processor *generates* both the AST and the TA from an XML document at the same time. Furthermore, we expect an XML processor to perform any necessary normalization, e.g., the carriage-return and line-feed and the attribute-value normalization as described in the *XML Standard* [12]. Additionally, the processor may perform text normalization as proposed in Definition 2. During the process of generating an AST/TA pair, all character references are expanded; if schema information is available, entity references are expanded as well. Beyond that, text surrogate values are calculated and all offsets are set to 0.

The reverse operation of generating AST/TA pairs is *restoring* XML documents from AST/TA pairs. This process is mainly ruled by the serialization of the tree structures coming up with AST/TA pairs. With performing different kinds of normalization while generating AST/TA pairs, the restored XML document may differ from its original source.

Manipulating TextArrays. Searching the TextArray means retrieving an arbitrary sequence of characters or words from the "visible" text content of an XML document. The TA enables random access to arbitrary segments of the

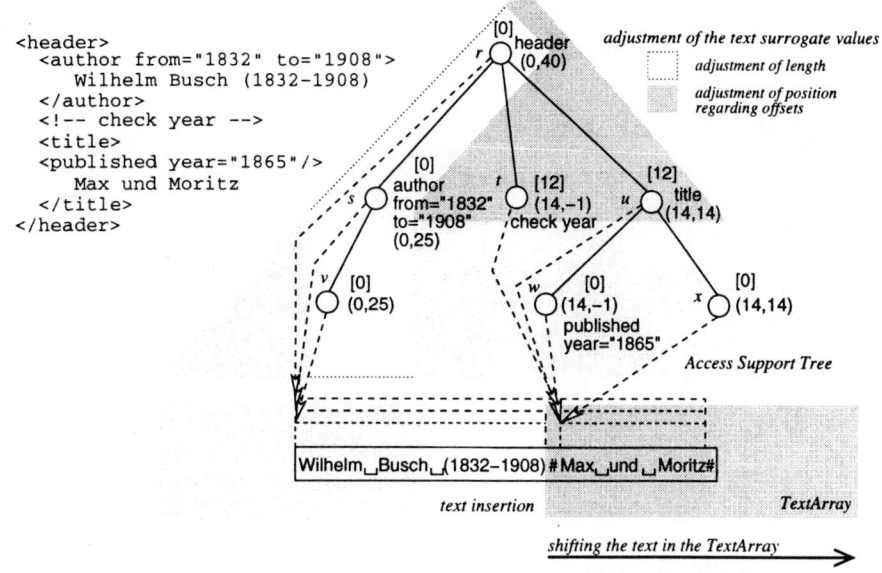

Figure 2. Inserting text into the TextArray

text content. For this purpose, we need to know the length of the desired text segment and its start position in the TA. Alternatively, the TA allows to scan its text.

By representing the text content of an XML document as a single contiguous string, the TA provides efficient operations both for random access to and for scanning the text. As we do not need to rearrange the text, random access is performed in constant time, whereas scanning needs time depending on the length of the TextArray.

The TA facilitates insertions and deletions of arbitrary sequences of characters or words respectively, to change the text content of an XML document. Finding an insertion or deletion position is done in constant time. However, inserting or deleting a sequence of characters or words may take much more time, because all text that follows the corresponding position in a TA must be shifted, i.e., $length(\tau)/2$ bytes on average. For this reason, we implement TAs as *positional B*-trees* to improve the performance. Insertions into or deletions from normalized TAs are additionally more complex, because the normalized state of those TAs must be maintained.

Deletions and insertions are similar to each other. In contrast to insertions, we must specify a valid deletion length referring to TAs. For example, we may delete up to 20 characters beginning at position 7 in the TA of Figure 1. If we delete 14 characters, the remaining TA is `Wilhelm#Moritz#`. Again, normalized TAs must be left in a normalized state afterwards. That is, if we specify deleting 13 characters, 14 characters must actually be deleted; otherwise, the TA is in an ill-formed state: `Wilhelm#␣Moritz#`. The separator sequence "`#␣`" violates the normalization. Adjusting the deletion length is integrated into our deletion algorithm.

After changing the length of TAs, it is necessary to adjust text surrogate values and offsets in ASTs. Figure 2 shows an example in which the text "`(1832-1908)`" was inserted after the word `Busch`. The example refers to a normalized TextArray. Thus, we must insert an additional space at the beginning of the text insertion causing an effective insertion length of 12. With respect to the example in Figure 2, we perform the AST adjustment as follows: We start in the root vertex r and increment its text surrogate length by 12. In the next step, we search for the child vertex of r that spans over the text insertion and increment its length also by 12. The offsets of all siblings vertices (t and u) that follow to the right of that child vertex (s) are incremented by the insertion length of 12. (In Figure 2 we use square brackets to constitute offsets, e.g., `[12]`.) This procedure continues until we reach the text vertex v (the leaf) that directly spans over the insertion. By using offsets rather than none to adjust text surrogate values, we do not need to visit all of the vertices that follow the location of an update within an AST. Thus, if C is the maximum number of child vertices and d is the maximum depth in an AST, we must visit no more than $C \times d$ vertices, using this approach.

As for deletions, updates of offsets and text surrogate values are analogous to updates after insertions. However, we need a more sophisticated algorithm when deleting character or word sequences that go beyond markup boundaries. Unlike at insertions, the adjustment procedure for ASTs may reach a non-leaf vertex of which some children span only over parts of the deleted text. Therefore, this subtree

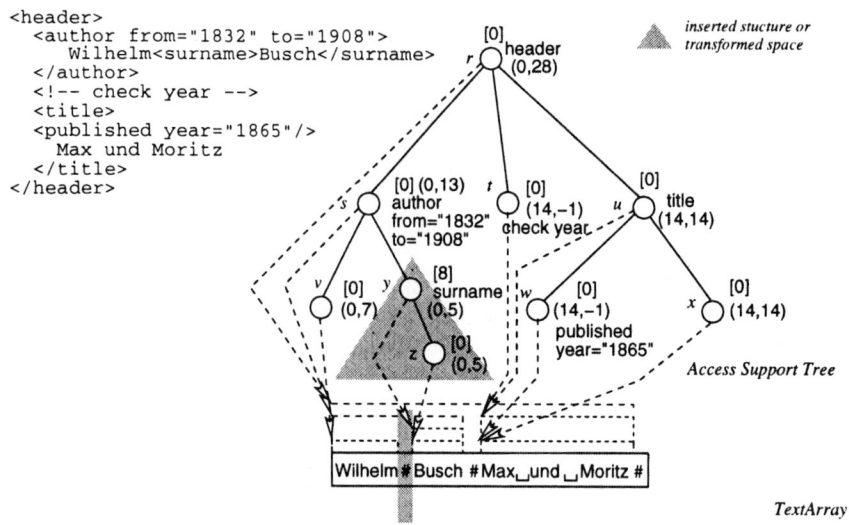

```
<header>
   <author from="1832" to="1908">
      Wilhelm<surname>Busch</surname>
   </author>
   <!-- check year -->
   <title>
   <published year="1865"/>
      Max und Moritz
   </title>
</header>
```

Figure 3. Inserting structure into the AST

must be traversed to adjust text surrogate values and offsets of the vertices according to their contributions.

Manipulating ASTs. The manipulation of ASTs refers to the structure of XML documents. For searching the structure of a document, the AST provides operations navigating the tree to find vertices having specific characteristics. For example, we are able to search for vertices with specific element names, attribute names, attribute values, specific positions to siblings, or references to specific text segments according to their corresponding text surrogate value and offsets.

Insertions referring to ASTs are operations that refine the structure of documents only, therefore enhancing the structure of ASTs. For example, we might apply some refinement to our sample document, such that `<author from="1832" to="1908" >Wilhelm Busch</author>` becomes `<author from="1832" to="1908">Wilhelm<surname>Busch</surname ></author>`. Such an insertion of a logical structure into a document is a *tag insertion* that might be the result of a text mining task [16]. Figure 3 illustrates the previous example based on a normalized TextArray. The AST receives two additional vertices y and z with y of type `element` representing the new `surname` markup. Vertex y is the parent of z which references the refined text segment `Busch`. Since the underlying TA is normalized, the former word separator (␣) prefixing the word `Busch` must be converted into a separator (#), making the new markup boundary explicit in the TA. This is not necessary with non-normalized TAs, because there are no separators to consider.

For convenience, we focus in the following only on triv-

ial structural insertions as in our example, which refine texts referenced by single text vertices rather than higher level structures in ASTs. We note that we may perform more sophisticated structural insertions. For such trivial insertions, we perform the following steps: After having found our desired sequence of characters or words that is to be marked by a new tag, we know the position and the length of this sequence. We search the AST—beginning from the root—for the text vertex that spans over this sequence. As all vertices of an AST come with text surrogate values, we follow exactly one path to the target vertex. This text vertex must be "split" in such a way that the AST integrates its additional structure correctly. That is, the root vertex of the new structure becomes a sibling of this text vertex and the text surrogate values and offsets must be adjusted accordingly. With respect to Figure 3, the text vertex v receives a new text surrogate length of 7 and the offset of the root vertex y is set to 8, thus aligning all text surrogate values of the newly inserted structure.

We note that structural insertions do not have any impact on the overall AST; changes in the AST are rather limited to the location of insertion. Beyond that, if TextArrays hold sequences of characters rather than of words, there is no need to access TAs at all.

We also allow the user to delete structure from an AST. That means, vertices may be removed from ASTs. Removing a vertex is simple. All children of the vertex to be removed become children of its grand parent. Only, if the corresponding TextArray is normalized, separators might have to be exchanged by word separators.

Structural updates of ASTs also involve updates of element names, attribute names, attribute values etc. These

162

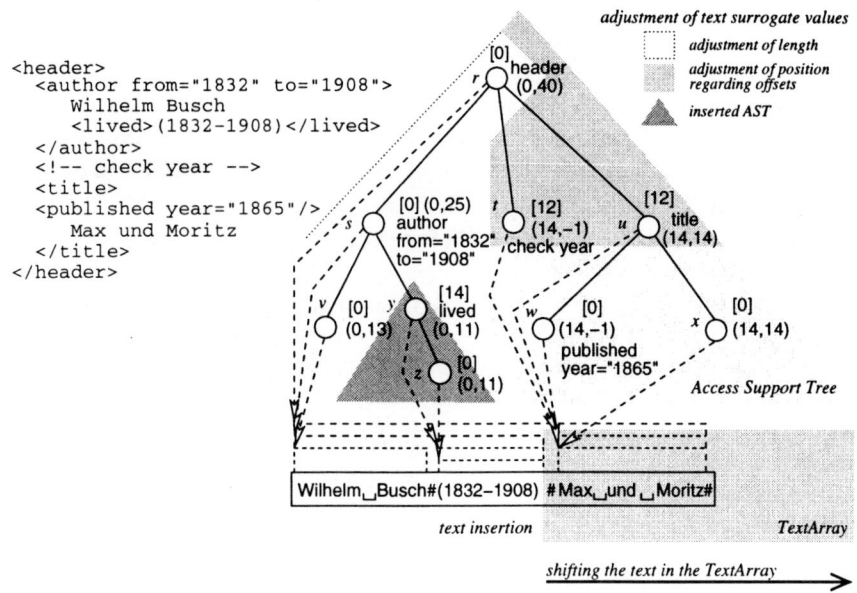

```
<header>
    <author from="1832" to="1908">
        Wilhelm Busch
        <lived>(1832-1908)</lived>
    </author>
    <!-- check year -->
    <title>
        <published year="1865"/>
        Max und Moritz
    </title>
</header>
```

Figure 4. Insertion of an AST/TA pair into an AST/TA pair

operations are simple operations mainly based on structural search, therefore we do not address them here.

Manipulating AST/TA pairs. In the previous sections, we considered the manipulation of ASTs and TAs separately. We now focus on manipulating ASTs and TAs jointly.

A simple operation on AST/TA pairs is the search for the minimal enclosing element of an arbitrary text segment of a TA. For example, in Figure 1, the minimal enclosing element of the string Busch#Max is the header element. Based on text surrogate values, this search needs to follow exactly one path from the root vertex to the desired element vertex in the AST. If C is, again, the maximum number of child vertices and d is the maximum depth in an AST, at most $C \times d$ vertices must be processed.

Inserting an AST/TA pair into another AST/TA pair is equivalent to inserting one XML document into another. The position of such insertions depends on different characteristics of the AST/TA pair in which to insert. On the one hand, insertion positions may depend on text positions. For example, consider the query: *"Insert y into <a>xz</a> after x"* that results in <a>xyz</a>. On the other hand, insertion positions may depend on structural properties as in the query: *"Insert y into <a>xz</a> as the first child of a"* resulting in <a>yxz</a>.

Figure 4 describes an example of the latter type of insertions. Here, the AST/TA pair representing the XML element <lived>(1832-1908)</lived> is inserted as last child into vertex s (author element). The algorithm for non-normalized TextArrays works as follows: First, the

vertex s in which the insertion is to occur is searched. During the search the length parameter of the surrogate values of r and s and the offsets of the siblings that follow the search path are incremented by the length of the additional text content "(1832-1908)". Finally, the root vertex y of the AST to be inserted is placed as the last child of vertex s; the new text is added to the TA. This kind of insertion needs to pass through the path from the root to the destination vertex exactly once.

Unfortunately, the insertion requires a second walk through the search path, when we insert into AST/TA pairs that are built on normalized TAs as shown in Figure 4. We do not know the real insertion length in the TA at the beginning; it might change, because of considerations of separators. Thus, we cannot update the text surrogate values while searching the destination vertex. Therefore, the corresponding text surrogate values and offsets must be updated after the insertion of text into the TA requiring a second walk through the search path.

As for deletions, we provide operations on AST/TA pairs similar to insertions.

4. Conclusion

The AST/TA model introduces our approach for storing and retrieving *generic* XML documents. The basic idea of the AST/TA data structure is the separation of meta data from data, namely the separation of document structure from content. We designed two physically independent storage units: the Access Support Tree for taking the meta data and the TextArray for taking the data of XML

163

documents, in which TextArrays maintain the data in "one lump" in document order. Hereby, the AST/TA data structure enables efficient access to documents based on both content and structure, facilitating information retrieval and database-like querying.

Additionally, our approach refers to relevant aspects of XML; it incorporates aspects specific to XML that cannot be ignored, e.g., the *XML Information Set* [13], white spaces handling, and handling of XML namespaces and entities. Due to space limitations, we cannot discuss such issues here in detail, we therefore refer the interested reader to [8]. The AST/TA data structure supports some of these aspects directly, namely by our proposed scheme to handle processing instructions and comments and by the text normalization we proposed. Moreover, the AST/TA data structure facilitates additional operations such as, e.g., efficient *tag insertions*—we are able to add structure or metadata to documents efficiently.

We have implemented the AST/TA data structure in main memory, notably to prove our design concept. We experience the following results: (1) We are able to provide the AST/TA data structure with efficient update operations, although we support the additional feature of viewing the document content as sequence of words. (2) In case, we consider the document content as a sequence of characters only, we may build more than one AST on top of a TA. (3) It is worthwhile to implement TAs as, e.g., positional B^*-trees. (4) The number of vertices processed in textual insertions is reduced considerably by using offsets for text surrogate values. (5) The number of offsets to be updated is still too much high and must therefore be reduced when implementing the AST/TA data structure on secondary storage.

Currently, we move our main memory implementation of AST/TAs to secondary storage within the framework of our XML Query Execution Engine (*XEE*). We would like to provide persistent AST/TAs for handling large scale XML document collections. We face the following main challenge when implementing the AST based on secondary storage: The AST tree mapped onto disk pages appropriately, such that the number of offsets to be changed are minimized for updates. Based on such appropriate mapping of ASTs into secondary storage, the AST/TA data structure is a good candidate for implementing, e.g., a persistent DOM efficiently.

Eventually, we carefully review the interdependence between the AST and the TA for possible improvements and the design of a suitable interface that enables convenient access to XML documents represented by AST/TA pairs. Furthermore, we intend to map query languages to the AST/TA data structure. The integration of both the concept of database query languages and the concept of information retrieval plays an important role with respect to this mapping. In addition, we take a look at optimizations with respect to AST/TAs and plan, therefore, to apply various kinds of indexes to both the AST and the TA to improve the performance. For this purpose, we take advantage of the indexes we proposed, e.g., for XML namespaces, entities etc. in [8]. The *XEE* system provides the testbed for the AST/TA data structure implemented on secondary storage.

References

[1] Database Systems Research Group (University of Alberta). *Multimedia Data Management.* www.cs.ualberta.ca/~database/multimedia/multimedia.html, 1998.

[2] A. Heuer and D. Priebe. IRQL – Yet Another Language for Querying Semi-Structured Data? Technical Report Preprint CS-01-99, Universität Rostock, 1999.

[3] International Business Machines Corporation. *XML Extender (Administration and Programming).* IBM, 2000.

[4] C.-C. Kanne and G. Moerkotte. Efficient Storage of XML Data. In *Proceedings of ICDE, San Diego, California.* IEEE Computer Society, 2000.

[5] G. Navarro and R. A. Baeza-Yates. Proximal Nodes: A Model to Query Document Databases by Content and Structure. *Information Systems*, 15(4):400–435, 1997.

[6] M. T. Özsu, D. Szafron, G. El-Medani, and C. Vittal. An Object-Oriented Multimedia Database System for a News-on-Demand Application. *Multimedia Systems*, 3(5-6):182–203, 1995.

[7] A. Salminen and F. W. Tompa. PAT Expressions: An Algebra for Text Search. *Acta Linguistica Hungarica*, 41(1-4):277–306, 1992-93.

[8] D. Scheffner. Access Support Tree & TextArray: Data Structures for XML Document Storage. Technical Report HUB-IB-157, Humboldt Universität zu Berlin, 2001.

[9] F. W. Tompa. Views of Text. Digital Media Information Base (DMIB '97), November 1997.

[10] M. Volz, K. Aberer, and K. Böhm. An OODBMS-IRS Coupling for Structured Documents. *Data Engineering Bulletin*, 19(1):34–42, 1996.

[11] World Wide Web Consortium. Document Object Model (DOM) Level 2 Core Specification, Version 1.0. Technical Report REC-DOM-Level-2-Core-20001113, W3C, November 2000.

[12] World Wide Web Consortium. Extensible Markup Language (XML), Version 1.0 (Second Edition). Technical Report REC-xml-20001006, W3C, October 2000.

[13] World Wide Web Consortium. XML Information Set. Technical Report REC-xml-infoset-20011024, W3C, October 2001.

[14] World Wide Web Consortium. XML Schema Part 1: Structures. Technical Report PR-xmlschema-1-20010330, W3C, March 2001.

[15] World Wide Web Consortium. XQuery 1.0 and XPath 2.0 Data Model. Technical Report WD-query-datamodel-20010607, W3C, June 2001.

[16] S. Yeates and I. H. Witten. On Tag Insertion and its Complexity. In *Proceedings of PRICAI 2000: International Workshop on Text and Data Mining, Melbourne, Australia,* pages 52–63, 2000.

Image Data Model for an Efficient Multi-Criteria Query: A Case in Medical Databases

R. CHBEIR, S. ATNAFU, L. BRUNIE

LISI – INSA de Lyon, 20 Avenue A. Einstein
F-69621 VILLEURBANNE - FRANCE
Phone: (+33) 4 72 43 88 99 - Fax: (+33) 4 72 43 87 13
E-mail: {rchbeir, satnafu, lionel.brunie}@lisi.insa-lyon.fr

Abstract

Since the last two decades, image database management has been practiced using different image representation methods. In the literature, images are represented using two paradigms: the metadata-based and the content-based representations. Image retrieval using the metadata is done using the traditional database operations. However, image retrieval by its low-level features requires similarity-based operations. Practice has shown that both types of operations are needed for an efficient image database management system. Particularly in Medical Image databases, such a mixed form of retrieval is very important. In this paper, we first present a global image data model that supports both metadata and low-level descriptions of images. We illustrate our work with real examples in the medical domain. Then, using an original image data repository model, we show how relational and similarity-based operations can be integrated. Both image and salient object are considered in our model. A prototype called MIMS (Medical Image Management System) has been realized to validate the main aspects of our approach.

1. Introduction

During the last two decades, a lot of work has been done in information technology in order to integrate image data in the standard data processing environments of different applications [3, 6, 10, 13]. The two different approaches used for the representation of images are: the metadata-based and the content-based approaches. The metadata-based representation uses alpha-numeric attributes to describe the context and/or the content of an image. This metadata representation of images is made with human assistance. Metadata are stored as accessory information to the visual objects. Retrieval by metadata representation follows the traditional techniques [14, 19]. However, it is mostly, difficult or not possible to fully or adequately describe an image using metadata representation. Though a complete description of images using metadata is not possible, so far physicians use the metadata retrieval as a primary method in the domain of medicine [25].

The other approach for representation of images is using its low-level contents such as its color, texture, and shape [1, 2, 16]. The representations using low-level features are derived through feature extraction algorithms. Image retrieval using these features is done by methods of similarity and hence is a non-exact matching. The research efforts exerted in the area of Content-Based Image Retrieval (CBIR), has made this technique of retrieval promising and an area of high importance [2, 5, 6, 10, 12]. This method is being adopted in different application areas. For example, content-based medical image retrieval can be used to assist training, enhanced image interpretation, clinical decision-making, automated archiving, etc. However good the content-based approach is, we cannot ignore the necessity of using metadata as a supplementary means of description and retrieval. The issue of developing an efficient medical image retrieval system is as important as the need of understanding and treating a particular malady to cure ones health.

The current trend is then, towards systems that use both metadata- and content-based image retrieval. There are a number of efforts for such mixed systems of query [4, 23, 24]. A combined use of the two techniques can produce a means of data retrieval that can satisfy different application areas. Often, in image databases, users need multi-criteria queries. A multi-criteria query is a query that considers different levels of abstraction of image descriptions such as metadata, colors, shapes, textures, salient objects features, etc. For instance, consider the following query:

Query 1: *Retrieve all brain X-rays taken between 01/01/2000 and 12/31/2000 where an anomaly is positioned as the image on the screen (upper-left part of the left lobe), identified as hypervascularized tumor, with a dark gray dominant color, and a volume greater than 30 mm³.*

Another important type of query is a query that requires an image-oriented join operation (i.e. a join on the low-level image representation components of image tables). In such a query, a similarity-based join operation is required other than the common similarity-based selection operation. Figure 1 shows medical image database tables, where the Patient table contains personal data of patients without any metadata description of the images and the other table contains images and their fully annotated thesaurus of all anomaly cases. If physicians or researchers need to get more information on patients having similar anomalies that exist in the thesaurus table (Query 2), they need to make similarity-based join on the image content descriptions:

Query 2: *Retrieve all personal data of patients having similar anomalies of lung X-rays with its corresponding treatments in the Thesaurus image table.*

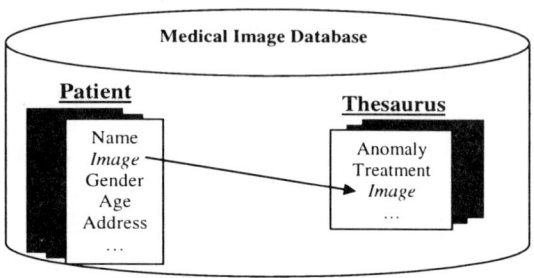

Figure 1: Similarity-based join operation on image contents description.

Though, such multi-criteria queries are more adequate in different areas of applications, there are only few systems that address all the issues of:

- adequate description of images using both metadata and content-based representations,
- using both relational and similarity-based system of operations in a query,
- exploiting the capabilities of the current commercial data management systems, rather than developing specialized database systems.

In this paper, we address these issues and present a novel image data model that effectively considers both multi-criteria image descriptions and retrieval for an image data management. We also introduce our original image data repository model with which we can apply a combination of relational and similarity-based system of database operations. These proposals are made in a way that can be integrated with the current widely used data management systems. Our proposal is supported with an application in the medical domain. We have implemented

a prototype called MIMS (Medical Image Management System) that shows most of our proposals here.

The rest of this paper is organized as follows. Section 2 summarizes related works. In Section 3, we present our novel image data model for managing image data and give an application example in medical domain. We also present our image data repository model, and discuss how multi-criteria operations can be applied on. Query examples using MIMS are presented in Section 4. Finally, conclusions are given in Section 5.

2. State of the Art

The metadata and the content-based descriptions of image data are the two common practices of image representations [11]. The metadata oriented approach has been practiced since the last two decades in different fields of applications (medical, security, etc.). Due to the nature of the images acquired for different purposes, the need for a specialized treatment is found commendable in many cases. This makes the idea of a generalized approach for metadata annotation difficult. Since subjectivity, ambiguity and imprecision are usually associated with specifying the content of images, metadata descriptions are considered as incomplete and domain-dependent [14]. First, describing the content and the semantic of each image object is difficult, because probable descriptions are numerous and each person may describe the image differently. Secondly, image description based on salient object position and relationships between objects (spatial facet) has proven to be imperfect at retrieval process where translation, scaling, perfect and multiple rotations, or any arbitrary combination of transformations is applied. For instance, the spatial content in terms of relationships in surgical or radiation therapy of brain tumors is decisive because the location of a tumor has profound implications on a therapeutic decision [22]. Thirdly, there is a great waste of important information when describing an image by metadata.

On the other hand, the work on content-based image analysis, representation and retrieval attracted a large number of researchers for more than a decade. As a result, a promising level of work for an effective representation and content-based retrieval of image data by the low-level features of color, texture, shape, etc. has been performed [6, 10, 12]. A large number of content-based image retrieval prototype systems has been developed and tested [1, 3, 6, 12, 16]. However, these systems give less or no emphasis to the role of metadata-based image retrieval. Using content-based representation of images as the only means lacks the means of describing semantic interpretation of images. Hence, a combined use of metadata and content-based is indispensable.

To support content-based image retrieval in the standard DBMS, a number of initiatives exist both in the research and commercial environments (QBIC in DB2 [3], VIR Image Engine in Oracle [9, 12], The Excalibur Image DataBlade module in Informix [8]). DISIMA is an object oriented system that even considers salient objects of images in the query system [17, 18]. However, none of the above does support all the necessary operations one needs for an effective image database management. For instance, operations such as the "similarity-based join" are not supported by the current systems. The commonly practiced feature of these systems is that, given a query image, they search its most similar images from a list of images using their respective content-based image retrieval engines. That is, the attempts so far did not exceed from these one-to-many content-based image retrieval operations and are limited in supporting complex similarity-based operations and mixed queries involving both content- and relational operations.

The efficiency of image retrieval is strongly related to the representation model. The better the features of the image data are represented, the more the image retrieval is able to satisfy complex queries. In the literature, several image data models have been proposed [11, 22, 24]. However, these models lack an appropriate representation of various necessary image related data for different applications. The work in [11] for example, doesn't consider content and semantic representations of salient object related data and the relationship between salient objects. The works in [22, 23, 24] are restricted because they do not allow the integration of various types of low-level features. A convenient image data model that supports most of the necessary operations on image content is a primary requirement. Moreover, since an image is a complex entity that may be composed of several objects of interest, DBMSs should consider the management of salient objects and operations on them. When salient image objects are supported, systems need to consider spatial (metric, directional, topological) operators.

3. Integrating Image Related Data in DBMS

In this section, we propose an image data representation model that integrates both metadata and content-based image description in a DBMS. Based on a novel image data model and an original data repository model, our representation also supports salient object related data. Our proposals here are illustrated with a practical application in the area of medical domain. Though we emphasized here on medical domain, our representation can be applied to different areas of applications under an Object-Relational DBMSs paradigm.

Consider a case of a patient data in medical domain:

"Mrs Danielle Lee, is a 44 years old patient, manifests low back pains, worsening shortness of breath, hemoptysis, and a weight loss of 10 lbs over the last three weeks. She has a smoking history of 2 packets of cigarettes per day for 27 years. Recently, she has decreased to 1 packet per day of "light" cigarettes. She claims that she only consumes alcohol on weekends".

The medical file contains two images (Figure 2):

- Lungs radiography, taken at 19/10/1998, shows a tumor in the inferior part of right lung and the shifting of the trachea to the right.
- Brain MRI, taken at 05/11/1999, shows a metastasis at the parietal left lobe.

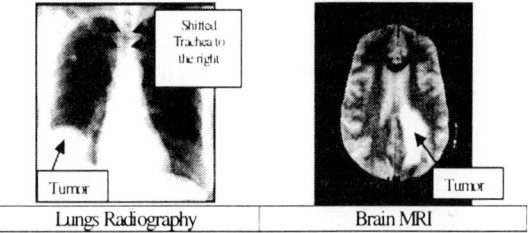

Figure 2: Medical Images registered in the medical file of Mrs. Lee.

This example will be used to show how our representation model captures various required image related information, and to illustrate how both metadata and content-based image operations are possible with the consideration of salient objects.

3.1 Image Data Model

The model that we propose here describes the image data in several levels of abstraction. Our image model has two main spaces: the external space and the content space (Figure 3). A short description of the different components of this model is given below.

3.1.1 The External Space (ES)

The external space captures the information associated to the image data that are not related to its content. The data in the external space are all alphanumeric data (Figure 4). The External Space has three components or subspaces:

- *The context-oriented subspace*: contains application-oriented data that are completely independent of the image content and have no impact on the image description. For example, in a medical application, it contains information such as the hospital name, the physician identity, the patient name, patient's age, his gender, etc.

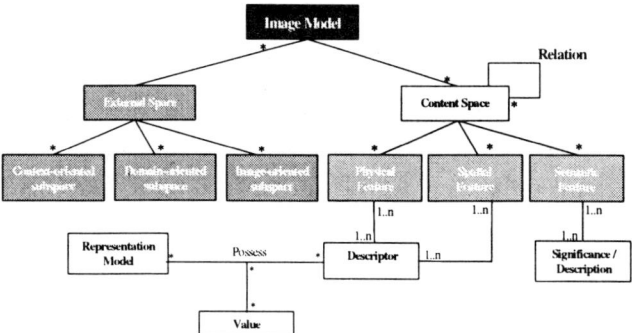

Figure 3: An image data model in UML.

- *The domain-oriented subspace*: consists of the data that are directly or indirectly related to the image. This subspace is very important in that it allows one to highlight several associated issues. For example, in medical image domain, it contains information like, the medical doctor's general observations, previous associated diseases, etc. The domain-oriented subspace can also assist in identifying associated medical anomalies.

- *The image-oriented subspace*: corresponds to the information that is directly associated to the image creation, storage, and type. For example, in medical domain, we need to distinguish the image compression type, the format of image creation (radiography, scanner, MRI, etc.), the incidence (sagittal, coronal, axial, etc.), the scene, the study (thoracic traumatism due to a cyclist accident), the series, image acquisition date, etc. These data can significantly help the image content description.

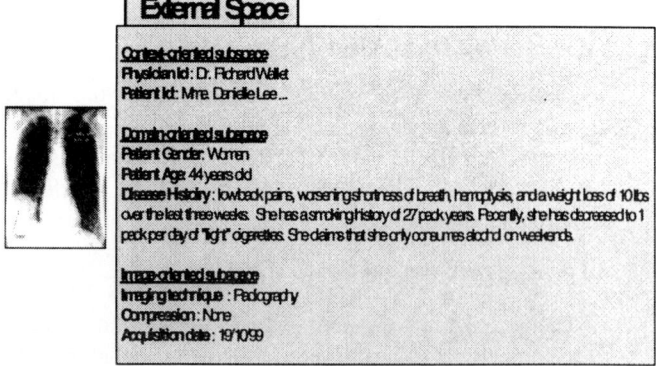

Figure 4: Example of External Space description

3.1.2 The Content Space (CS)

The content space describes the content of the image not only using content-based representation, but also using metadata description. It consists of: the physical, the spatial and the semantic features. This format of content representation is inherited by the salient object descriptions. The content space maintains relations between the salient objects, and the salient objects and the image.

- *The Physical Feature*: describes the image (or the salient object) using its low-level features such as color, texture, etc. The color feature, for instance, can be described *via* several descriptors such as color distribution, histograms, dominant color, etc. These data are in most cases pre-calculated automatically or semi-automatically and stored systematically so as to speed-up the content-based image retrieval (Figure 5). The use of physical features allows to respond for non-traditional queries in medical systems such as: "Find lung x-rays where they contain objects that are similar (by color) to a salient object SO2".

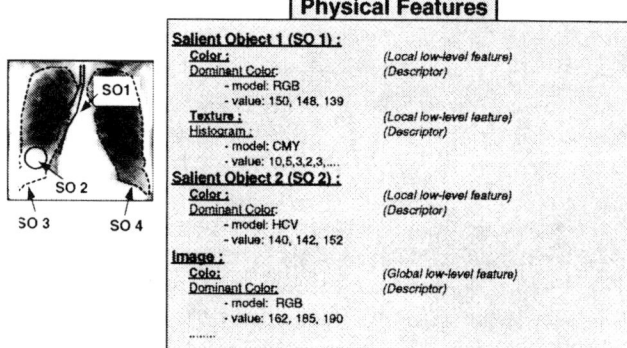

Figure 5: Example of Physical Features description.

- *The Spatial Feature*: is an intermediate (middle-level) feature that concerns geometric aspects of images (or salient objects) such as shape and position. Each spatial feature can have several representation forms such as: MBR (Minimum Bounding Rectangle), bounding circle, surface, volume, etc. The spatial feature is used to identify the relations between salient objects such as metrical (near, far, etc.), directional (right, left, above, front, etc.), and topological (touch, disjoint, overlap, equal, etc.) relations. The use of spatial features allows to respond for queries in medical systems such as: "Find lung x-rays where an object SO1 is above object SO2 and their surfaces are disjoint". Figure 6 shows an example of spatial features.

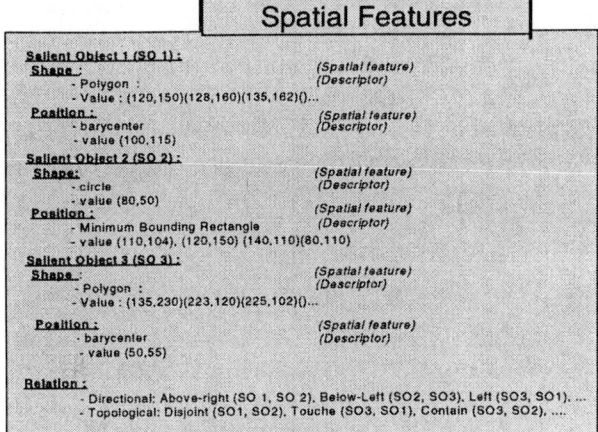

Figure 6: Example of Spatial Features description

• *The Semantic Feature*: integrates high-level descriptions of image (or salient-objects) with the use of an application domain oriented keywords. In the medical domain, for example, terms such as name (lungs, trachea, tumor, etc.), states (inflated, exhausted, dangerous, etc.), and semantic relations (invade, attack, compress, etc.) are used to describe medical image content (Figure 7). The objective of the semantic feature is to increase the expression power of medical users in a manner that usual medical terms can be used to describe images and use the same for retrieval purpose. Analyzing the semantic features of an image requires a human intervention, since explicit specialized object identification, states, and relations must be recognized. The use of semantic features is important to respond for traditional queries in medical systems such as: "Find lung x-rays where hypervascularized tumor is invading the left lung".

In this manner, the user has all the necessary descriptions of an image data and the relevant images can be retrieved by any of the facets (external, physical, spatial and semantic).

3.2 Image Data Repository Model

Modeling the image repository is a fundamental requirement for an effective storage, retrieval and a convenient integration of image data into popular DBMSs. We therefore present here an extension of the image data repository model (which we also refer it as *image table*) presented in [15]. With this extension, the model is capable to support operations on salient objects. Furthermore, it can conveniently handle both metadata and content-based image data retrieval under an Object Relational (OR) paradigm. The OR paradigm is a system

that can effectively support multimedia data in a DBMS. M. Stonebraker et al. [7] have widely elaborated the OR paradigm as the next wave of DBMSs. We use an image data repository model (or an image table model) of five components **M(id, O, F, A, P)**, where:

id   is a unique identifier of an instance of M,

O   is a reference to the image object itself which can be stored as a *BLOB* internally in the table or which can be referenced as an external *BFILE* (binary file),

F   is a feature vector representation of the object O,

A   is an attribute component that may be used to describe the object using key-word like annotations, where, *A* may be declared as an object- or a set of object types,

P   is a data structure that is used to capture pointer links to instances of other tables associated by a binary operation.

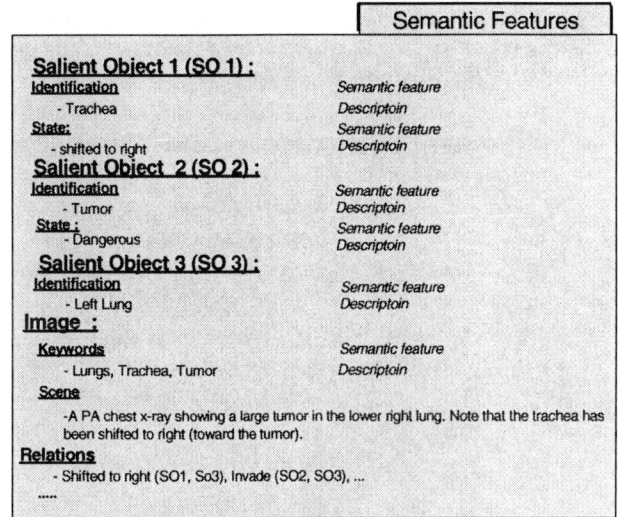

Figure 7: Example of Semantic Features description.

The three components "O", "F", and "A" can be used to capture sufficiently the context-, semantic, and the content-based information of an image. The F component of M is the part that captures physical and spatial features of an image that is primarily required to perform similarity-based operations. "P" is a column whose content is a data structure that can store links to instances of other tables during binary operations such as similarity-based join. "P" has a value "*null*" in the base tables. It has a non-null value in intermediate tables during binary operations [20].

Efficient similarity-based image retrieval in medical applications requires considering salient objects of the images. Salient objects are not separate images, but parts of an image that are of particular interest. Hence, we need to have a structure to capture salient object related data. Such a structure can be deduced from the general structure of an image repository model. We do not need to consider the components 'O' and 'P' of M in a salient object data repository model, but make a sort of liaison or link to the main image. We therefore propose a structure, $S(id_s, F_s, A_s)$, to capture the salient object related data, where:

id_s is the identifier of a salient object. An image may have more than one salient object. Each object is identified with this unique identifier,

F_s is the feature vector extracted to represent the content of the salient object,

A_s is an object attribute that can be used to capture all semantic features of the salient object.

The F_s component is the part that captures the low-level (physical and spatial) features of the salient objects. It is this F_s that is used for similarity-based operations on the salient objects. The A_s component of S contains the salient object related semantic feature data Relational operations and comparisons can be performed on this component. Figure 8 shows the content of S and its liaison with M. Only the feature vector representations of the salient objects are stored in F_s The tumor is an object of interest that is extracted from the image. The technique of extracting salient objects from an image is purely a work in image processing and is not considered in this paper.

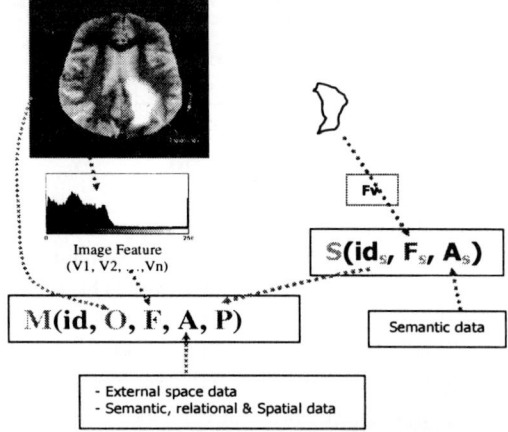

Figure 8: Managing Salient Objects in association with their source images

Integrating the image data model proposed in Section 3.1 with the image data repository model presented in this sub-section involves organizing the data models for an image and for a salient object. Since an image may consist of several salient objects, the spatial relations between the salient objects are captured in the component A of M. We discuss below the schematic structure of an image database. Considering the image repository model **M(id, O, F, A, P)**, we describe below the contents of the components F and A of M.

F(Descriptor, Model, Value):
- *Descriptor:* is the type of representation (such as Color Histogram, Color distribution, Texture Histogram, etc.),
- *Model:* is the description format (such as RGB, RHV, etc.),
- *Value:* is the content descriptor. This component contains both the Physical and Spatial Feature data;

A(ES, Sem_F, R):
- *ES:* is the External Space descriptions (consisting of Context-Oriented, Domain-Oriented, and Image-Oriented sub-spaces) as indicated in the image data Model,
- *Sem_F:* is the Semantic Feature of the Content Space of M that tells the significance and interpretation (keywords, legend, etc.) of the image, and
- *R:* is the component that captures the relations between either two salient objects or a salient object and the image;

Sem_F(Type, Description):
- *Type:* defines the type of the semantic feature (keyword, scene, etc.),
- *Description:* is a textual representation.

R(id_s, id, Relation):
- *ids:* identifies the identifier of a salient object,
- *id:* is the identifier of either an image or a salient object,
- *Relation:* represents the spatial (directional, metrical, topological) or semantic relations between them.

For the salient object repository model **S(id_s, F_s, A_s)**, the contents of the components F_s and A_s are described below.

F_s(Descriptor, Model, Value):
- *Descriptor:* is the type of representation (such as Color Histogram, Color distribution, Texture Histogram, etc.),
- *Model:* is the description format (such as RGB, RHV, etc.),.

- *Value:* is the content descriptor. This component contains both the Physical and Spatial Feature data;

A_s(Type, Description):

- *Type:* defines the type of the semantic feature (name, state, etc.),
- *Description:* is a textual representation.

With this model, we support a mixture of relational and similarity-based operations. When a query deals with relational operations, it operates on the alphanumeric attributes of M and/or S that can be treated in the traditional manner. For similarity-based queries, operations are performed on the F component of M and/or on the F_s component of S. In this paper, we only present the commonly used spatial and similarity-based operators. Further details on this and other relevant operators can be found in our previous works [15, 22].

3.3 Operations

Image management needs to consider various operations. The two major categories are the spatial and the similarity-based operations. Spatial operations are often performed using spatial relations that exist either between two salient objects or a salient object and the image. In the literature [26], the following three types of spatial relations are considered:

- *Metric relations:* determined in function of proximity that express the closeness of the objects such as near, far, etc.
- *Directional Relations:* generally determined on the basis of the direction between objects such as right, left, north, east, etc.
- *Topological relations:* determined the position of the salient objects on the basis of their shapes such as disjoint, touch, overlap, etc.

In our model, all spatial relations are captured in R of A. Hence, operations are treated in the traditional manner.

The other category of operations is the similarity-based operations. In the literature, there are two major methods for similarity-based retrieval: the k-NN search and the range query search[1]. For our operations and for the purpose of facilitating similarity-based query optimization, we choose that our operators based on the range query search method. Discussion and more details about our choice and similarity-based query optimization are found in [15]. We briefly present below only the major similarity-based operators:

---

[1] $R^\epsilon(S,q) = \{o' \in S \mid \|o' - q\| \leq \epsilon\}$, where $\|o' - q\|$ denotes the distance between o' and q.

The Similarity-Based Selection Operator:

Given an image query object *x*, an image table M and $\epsilon > 0$, a similarity-based selection operation denoted by $\delta\epsilon x(M)$ is a unary operator on an image table M performed on the component F that is given by:

$$\delta\epsilon_x(M) = \{(id,o,f,a,p) \in M \mid o' \in R^\epsilon_x(M,x)\},$$ where R^ϵ_x (M,x) denotes the range query of object x with respect to M and ϵ.

The Similarity-Based Join Operator:

Let $M_1(id_1,o_1,f_1,a_1,p_1)$ and $M_2(id_2,o_2,f_2,a_2,p_2)$ be two image tables and let ϵ be a positive real number. The similarity-based Join operator, denoted by $M_1 \otimes^\epsilon M_2$, is a binary operator on image tables M_1 and M_2 given by:

$$M_1 \otimes^\epsilon M_2 = \{(id_1,o_1,f_1,a_1,p'_1) \mid (id_1,o_1,f_1,a_1,p_1) \in M_1$$
and $p'_1 = p_1 \cup (M_2, s\_id_\epsilon(M_2,o_1))\}$, where $s\_id_\epsilon(M_2,o_1) = \Pi_{M2}.id(\delta^\epsilon o_1(M_2))$. (i.e. the ids contained by the projection on the id component of the associated instances of M_2).

The possibility of using these operators with traditional relational operators has been found possible. Some example queries are given in section 4. The similarity-based operators defined on M can also be applied on S because we adopt the same structure.

4. Multi-Criteria Queries in Image DBMS

To show the the feasibility of our models, we extended the MIMS (Medical Image Management System) prototype [21]. MIMS has been developed as a web-based image management system using Java. In MIMS, icon- and hypermedia-based interfaces are used to store and retrieve images. MIMS has given very satisfactory first phase results [21]. The user is able to formulate muli-criteria queries (external, physical, spatial and semantic features). To formulate a query, for example, for Query 1 of Section 1, the user proceeds as follows (Figure 9):

1. Choose the appropriate medical organ
2. Determine its right incidence (axial, sagital, coronal, etc.)
3. Specify the external space data (image type, acquisition date, patient name, etc.)
4. Click inside a region of the organ. This displays possible anomalies of that region. Anomalies are represented by a letter or an icon,
5. Position the anomaly inside the region that is previously clicked,
6. Specify the content space features of both image and salient objects (physical, spatial, and semantic).

171

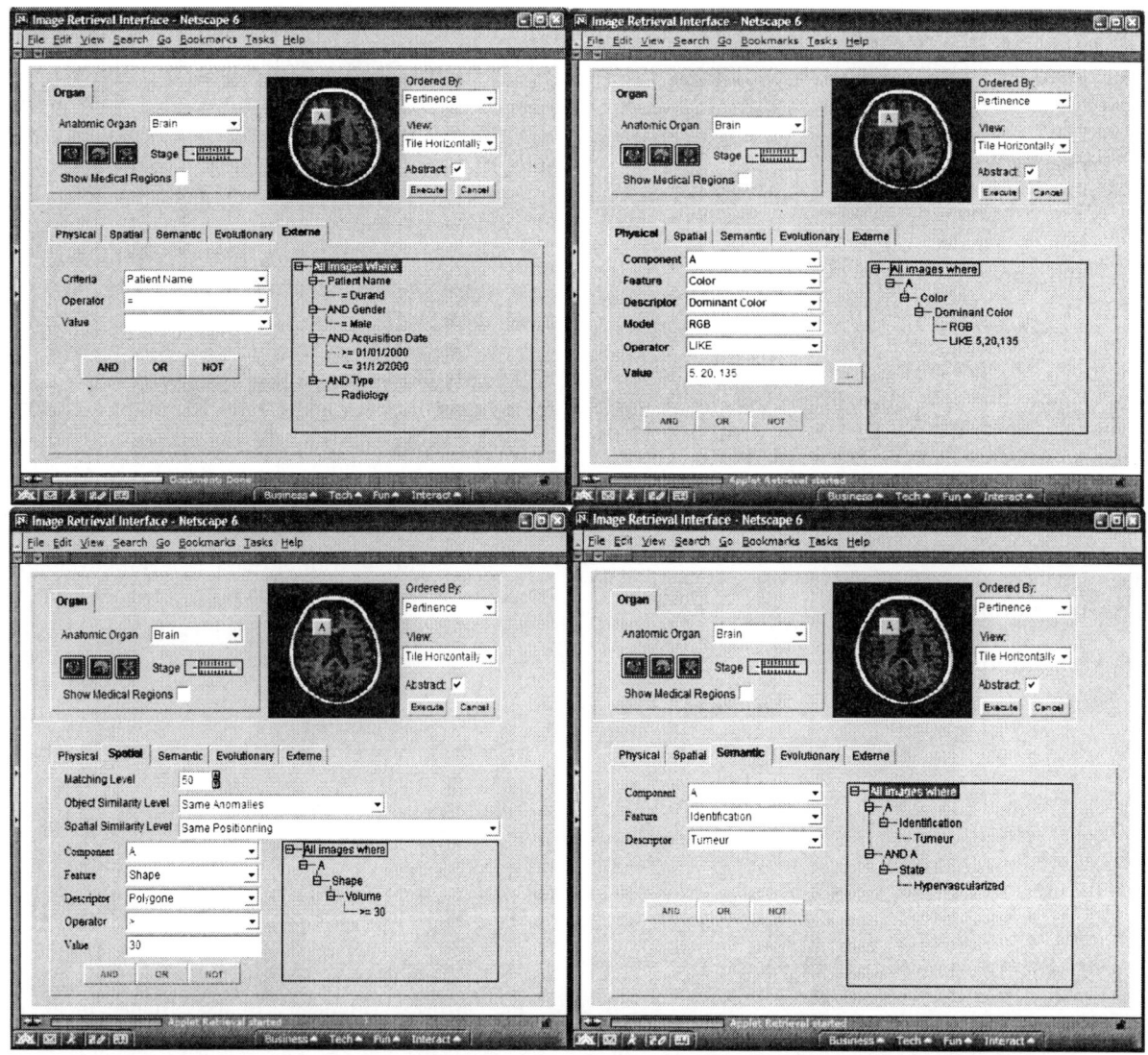

Figure 9: Visual Content of the MIMS user interface formulated for Query 1 of Section 1.

The algorithm for this formulation of query can be illustrated using the traditional OQL like language as follows.

S10: **SELECT * FROM** S
WHERE (S.A$_s$.Type = "name")
AND (S.A$_s$.Description = "tumor")

S11: **SELECT * FROM** S10
WHERE (S.A$_s$.Type = "State")
AND (S.A$_s$.Description = "Hypervascularized")

S12: **SELECT * FROM** S11
WHERE (S.F$_s$ **LIKE** ("Dominant Color", "RGB", [5, 20, 135]))

S': **SELECT * FROM** S12
WHERE (S.F$_s$. >= ("Volume", *, 30))

SELECT M.O **FROM** M
WHERE (M.A.ES.IOS.Type = "X-ray")
AND (M.A.ES.IOS.Date **BETWEEN** [01/01/00, 12/31/00])
AND (M.A.Sem_F.Type = "Organ")
AND (M.A.Sem_F.Description = "Brain")
AND (M.A.R.id$_s$ = S'.id$_s$)
AND (M.A.R.id = M.id)
AND (M.A.R.Relation = "up-left")

Notations are given as in Section 3: COS for Context-Oriented Subspace, IOS for Image-Oriented Subspace, etc.

172

Query 2 of Section 1 involves a similarity-based join operation. For this query, there are two image tables M_P and M_T for Patient and Thesaurus. Each of them contains salient objects. Hence, associated salient object tables, S_P and S_T are also considered. The algorithm for the formulation of this query is presented as follows:

SELECT	$M_P.A.ES_P.COS.Pat\_id$,
	$M_P.A.ES_P.COS.Address$,
	$M_P.A.ES_P.DOS.Gender$,
	$M_P.A.ES_P.DOS.Age$, $S_T.A_s.Description$
FROM	M_P, M_T
WHERE	$(M_P.A.ES_P.IOS.Type = \text{"X-ray"})$
AND	$(M_T.A.ES_T.IOS.Type = \text{"X-ray"})$
AND	$(S_T.A_s.Type = \text{"Treatment"})$
AND	$(S_T.F_s.Value \ \textbf{SIMILAR}^2 \ S_P.F_s.Value)$

At this level of our research, MIMS supports only the traditional and spatial operators. The similarity-based operators are in the process of integration.

5. Conclusion

An image data model that can capture the various relevant image related data is very important for an efficiently mage data description. With such a model, a multi-criteria query that considers the different features of images can be used in different application areas such as in medicine. However, the existing image data models and image retrieval proposals lack to adequately consider one or the other of the important components.

In this work, we addressed these issues and presented a novel image data model that can represent the wide range of information associated to an image data in both metadata- and content-based descriptions of images. This is the first issue that needs to be considered for an effective image data management. In order to support image data in the standard DBMS, we have proposed an original image data repository model on which we can apply a combination of relational and similarity-based database operations. We extended MIMS and presented here its query interface. We demonstrated the use of the multi-criteria operations with queries in medical domain.

Future works are envisaged to extend MIMS by integrating the similarity-based operations, and developing a query optimization model with our novel operators. The study and possible integration of the temporal and evolutionary content of image data for different application domains (such as in medicine, GIS, etc.) is another important issue to be considered.

6. References

[1] J.K.Wu and A.D. Narasimhalu and B.M. Mehtre and C.P. Lam and Y.J. Gao, CORE: A Content-Based Retrieval Engine for Multimedia Information Systems, Multimedia Systems, 1995, Vol. 3, pp. 25-41.

[2] S. Berchtold and C. Boehm and B. Braunmueller and D. A. Keim and H. P. Kriegel, Fast Parallel Similarity Search in Multimedia Databases, SIGMOD Conference, AZ, USA, 1997, pp. 1-12.

[3] A. Yoshitaka and T. Ichikawa, A Survey on Content-Based Retrieval for Multimedia Databases, IEEE Transactions on Knowledge and Data Engineering, Vol. 11, No. 1, 1999, pp.81-93.

[4] V.Oria and M.T. Özsu and L. Liu and X. Li and J.Z. Li and Y. Niu and P.J. Iglinski, Modeling Images for Content-Based Queries: The DISMA Approach, VIS'97, San Diago, 1997, pp.339-346.

[5] Jian-Kang Wu, Content-Based Indexing of Multimedia Databases, IEEE TKDE, 1997, Vol. 9, No. 6, pp.978-989.

[6] Y. Rui and T.S. Huang and S.F. Chang, Image Retrieval: Past, Present, and Future, Journal of Visual Communication and Image Representation, 1999, Vol. 10, pp.1-23.

[7] M. Stonebraker and P. Brown, Object-Relational DBMSs, Mogan Kaufmann Pub. Inc, 1999, San. Francisco, ISBN 1-55860-452-9.

[8] Excalibur Image Datablade Module User's Guide, Informix Press, March, 1999, Ver. 1.2, P. No. 000-5356.

[9] Oracle8i, Visual Information Retrieval Users Guide & Reference, Oracle Press, 1999, Release 8.1.5, A67293-01.

[10] William I. Grosky, Managing Multimedia Information in Database Systems, Communications of the ACM, 1997, Vol. 40, No. 12, pp. 72-80.

[11] William I. Grosky and Peter L. Stanchev, An Image Data Model, Advances in Visual Information Systems, Visual-2000, 4th International Conference, Lyon, France, 2000, pp. 14-25, LNCS 1929, Springer Verlag.

---

^{2} SIMILAR represents the similarity-based join operator defined in the previous section.

[12] John P. Eakins and Margaret E. Graham, Content-Based Image Retrieval: A Report to the JISC Technology Applications Program, January, 1999, Inst. for Image Data Research, Univ. of Northumbria at Newcastle.

[13] A.W.M. Smeulders and T. Gevers and M.L. Kersten, Crossing the Divide Between Computer Vision and Databases in Search of Image Databases, Visual Database Systems Conf., Italy, 1998, pp. 223-239.

[14] Amit Sheth and Wolfgang Klas, Multimedia Data Management: Using Metadata to Integrate and Apply Digital Media, McGraw-Hill, 1998, San Francisco.

[15] S. Atnafu and L. Brunie and H. Kosch: Similarity-Based Operators and Query Optimization for Multimedia Database Systems International Database Engineering & Applications Symposium (IDEAS'01), July 16-18, 2001, Grenoble, France; IEEE Computer Society Press, pp. 346-355.

[16] Remco C.Veltkamp, Mirela Tanase, Content-Based Image Retrieval Systems: A Survey, October, 2000, Technical Report UU-cs-2000-34, Department of Computer Science, Utrecht University.

[17] V. Oria and M.T. Özsu and P. Iglinski and B. Xu and L.I. Cheng, DISMA: An Object Oriented Approach to Developing an Image Database System, ICDE 2000, February, 2000, 16th Int. Conf. on Data Engineering, San Diego, California.

[18] V. Oria and M.T. Özsu and P. Iglinski and S. Lin and B. Yao, DISMA: A Distributed and Interoperable Image Database System, SIGMOD 2000, May, 2000, In Proc. of ACM SIGMOD Int. Conf. on Management of Data, Dallas, Texas.

[19] James S. Duncan and Nicholas Ayache; Medical Image Analysis: Progress over Two Decades and the Challenges Ahead; IEEE Transactions on Pattern Analysis and Machine Intelligence, Vol. 22, No. 1, January 2000.

[20] S. Atnafu and L. Brunie and H. Kosch, Similarity-Based Operators in Image Database Systems, WAIM'2001, LNCS, July, 2001, Xi'an, China, pp. 14-25.

[21] R. Chbeir, Y. Amghar, A. Flory: A Prototype for Medical Image Retrieval, International Journal of Methods of Information in Medicine, Schattauer, Issue 3, 2001.

[22] R. Chbeir, F. Favetta: A Global Description of Medical Image with a High Precision; in IEEE International Symposium on Bio-Informatics and Biomedical Engineering IEEE-BIBE'2000, IEEE Computer Society, Washington D.C., USA, (2000) November 8th-10th, pp. 289-296.

[23] Chu, W. W., Hsu, C.C., Cárdenas, A.F., and Taira, R. K., "Knowledge-Based Image Retrieval with Spatial and Temporal Constraints," IEEE Transactions on Knowledge and Data Engineering, Vol. 10, No. 6, November/December 1998, P. 872-888.

[24] M. Mechkour, "EMIR2. An Extended Model for Image Representation and Retrieval", Database and EXpert system Applications (DEXA), Sep. 1995, P. 395-404.

[25] Trayser G., "Interactive System for Image Selection", Digital Imaging Unit Center of Medical Informatics University Hospital of Geneva, http://www.expasy.ch/UIN/html1/projects/isis/isis.html

[26] Egenhofer M., Frank A., Jackson J., "A Topological Data Model for Spatial Databases", Design and Implementation of Large Spatial Databases, First Symposium SSD '89 Proceedings, Berlin, West Germany, 1990, P. 271-286.

Similarity Searching for Multi-attribute Sequences*

Tamer Kahveci   Ambuj Singh   Aliekber Gürel
Department of Computer Science   Department of Mathematics
University of California, Santa Barbara, CA 93106
{tamer,ambuj}@cs.ucsb.edu, aliekber@math.ucsb.edu

Abstract

We investigate the problem of searching similar multi-attribute time sequences. Such sequences arise naturally in a number of medical, financial, video, weather forecast, and stock market databases where more than one attribute is of interest at a time instant. We first solve the simple case in which the distance is defined as the Euclidean distance. Later, we extend it to shift and scale invariance. We formulate a new symmetric scale and shift invariant notion of distance for such sequences. We also propose a new index structure that transforms the data sequences and clusters them according to their shiftings and scalings. This clustering improves the efficiency considerably. According to our experiments with real and synthetic datasets, the index structure's performance is 5 to 45 times better than competing techniques, the exact speedup based on other optimizations such as caching and replication.

1 Introduction

Time series or sequence data sets arise naturally in many real world applications like stock market, weather forecasts, video databases, sensor-based controls, and medicine. There is a frequent need to understand the information content of this data in order to respond better to common trends, to provide corrective emergency steps, or to predict the future evolution based on past records. Some examples of queries on such datasets include finding the companies which have similar profit/loss patterns, finding similar motion patterns in a video database, finding similar patterns in medical sensor data in order to respond to patient health problems, to predict infrastructure usage by comparison with past trends, or to predict common genetic functionality by study of gene expression patterns over time.

Time series data is said to have d attributes if d values are stored for each time point. Stock market data, which is formed by storing the closing prices of a company is an example of a 1-attribute sequence. If we include the P/E ratio and the number of shares sold, this becomes a 3-attribute sequence. The trajectory of an object moving on a plane is a 2-attribute sequence, because two values (i.e. X and Y coordinates) are stored for each discrete time point. Medical data is usually multi-attribute since a single sensor is seldom sufficient to record the health of a patient: such a sequence can be obtained by using the blood pressure values, the heart beat rate, the amount of calcium, and other parameters recorded periodically from a patient.

There are many ways to compare the similarity of two time sequences. One approach is to define the distance between two sequences to be the Euclidean distance, by viewing a sequence as a point in an appropriate multi-dimensional space [1, 4, 7, 11, 19].

Range searches and nearest neighbor searches for *whole matching* and *subsequence matching* [1] have been the principal queries of interest for time series data. *Whole matching* corresponds to the case when the query sequence and the sequences in the database have the same length. Agrawal et al. [1] developed one of the first solutions to this problem. The authors transformed the time sequence to the frequency domain by using DFT. Later, they reduced the number of dimensions to a feasible size by storing the first few frequency coefficients. Chan and Fu [4] used Haar wavelet transform to reduce the number of dimensions and compared this method to DFT. The authors found that Haar wavelet transform performs better than DFT. However, the performance of DFT could be improved using the symmetry of Fourier Transforms [17]. In this case, both methods gave similar results.

1.1 Adopting new metrics for distance

Non-Euclidean metrics have also been used to compute the similarity for time sequences. Agrawal, Lin, Sawhney, and Shim [2] use L_∞ as the distance metric. Another distance metric for multi-attribute time sequences is D_{norm} [13]. Although this metric has a high recall, it allows false dismissals.

Defining the distance as some norm of the difference between two time sequences may be insufficient if the sequences can be made closer by linear transformations. The most important transformations are scaling and shifting. Scaling is needed because of the need to compare time sequences recorded on devices with different calibrations or different units.

One emerging area of applications for shift and scale invariant comparison of time sequences is data from genome microarrays. By choosing cells from an organism under different stages of development, or under different physical conditions, valuable information can be obtained about the expression of genes, their relationship to one another, and genetic pathways. Using the absolute values of the measurements may be misleading because of variation in the physical conditions like data quality and quantity, scan-

*Work supported partially by NSF under grants EIA-0080134, EIA-9986057, IIS-9877142, ANI-0123985, and NSFIIS98-17432.

175

ner quality, glass quality, hybridization conditions and post-hybridization washing. In order to allow comparisons under different experimental conditions, shift and scale invariant comparisons of the resulting sequences can be useful.

Das, Gunopulos and Mannila [6] showed that the similarity between two sequences after eliminating the outliers, and scaling and shifting can be determined in $O(n^6)$ time using *longest common subsequence* (LCSS) technique, where n is the length of the strings. However, the LCSS can be found in $O((n+m)^3\delta^3)$ time [20] if only shift invariance is involved, where m and n are length of sequences and δ is the error.

Rafiei and Mendelzon [16] developed algorithms for answering similarity queries under a set of user-specified linear transformations. When these transformations are *safe*, the queries can be answered efficiently by applying the specified transformation to an index MBR. Multiple transformations can also be applied collectively to the database sequences [15]. The problem we study here is different in that we consider all possible scalings and shiftings, not just a set of user-specified transformations.

Goldin and Kanellakis [8] propose a technique based on *normalization* for the comparison of the time sequences. A *normalized* time sequence has a zero mean and a unit standard deviation. Given a query Q, the authors present a search algorithm that finds all database sequences S such that the *normalized* Euclidean distance [1] $D_N(Q, aS + b) \leq \epsilon$ for some $a > 0$ and b. Chu and Wong [5] considered the asymmetric formulation $D_2(aQ + b, S) \leq \epsilon$. They use a transformation to map the data sequences onto the *Shift Eliminated Plane*. Both of these formulations of distance are inherently asymmetric in its treatment of query and database sequences. We focus on the symmetric notion of the distance in this paper. The restriction to only positive scalings ($a > 0$) also appears artificial.

The Landmark model [14] stores the turning points of the time sequences. The distance between time sequences here is invariant with respect to 6 transformations: shifting, uniform amplitude scaling, uniform time scaling, uniform bi-scaling, time warping, and non uniform amplitude scaling. However, the landmarks of different time sequences may correspond to different time points.

1.2 Our contribution

In this paper, we consider the similarity search problem for multi-attribute sequences. We first solve the simple case in which the distance is basically defined as the Euclidean distance. Later, we extend it to handle shift and scale invariance. We point out the problems with current methods for scale and shift invariant distance computations, and propose a new symmetric notion of distance: the distance between two time sequences is defined to be the smallest Euclidean distance after scaling and shifting either one of the sequences to be as close to the other. We define two

models for comparing multi-attribute time sequences: in the first model, the scalings and shiftings of the component sequences are dependent, and in the second model they are independent.

We propose a novel index structure called *CS-Index* (Cone Slice) for shift and scale invariant comparison of time sequences. As a part of this technique, the sequences in the database are first mapped to the shift eliminated plane [5]. The transformed points are then clustered in hierarchical cone slices. These slices are stored on disk according to an *in-order* traversal, and a pointer to each slice along with angle and spatial extent information is maintained in memory. Given any query, it is first mapped onto shift eliminated plane. The shift and scale invariant distance between the query and the slices are computed in memory to obtain a set of candidate slices. The hierarchical construction of the index structure allows early pruning. Finally, the candidate slices are read from disk in a single seek, and false hits are eliminated.

Experimental results show that the CS-Index structure performs 50 to 100 times faster than the R*-tree index structure and 5 to 10 times faster than sequential scan. The efficiency of the index structure can be further improved by selectively replicating or caching parts of the index structure.

The rest of the paper is organized as follows. We discuss the simple case in which the Euclidean distance is used as the dissimilarity measure in Section 2. We define the problem of shift and scale invariant search of multi-attribute time sequences in Section 3. In Section 4, we propose the *CS-index* structure for range queries and nearest neighbor queries. We present experimental results on a number of synthetic and real datasets in Section 5. We end with a brief discussion in Section 6.

2 Multi-attribute time sequences

A *d-attribute* time sequence is formed by storing d values at each time point. If v is a d-attribute time sequence of length l, then it can be represented as a vector $v = (v_1, v_2, ..., v_l)$, where $v_i = (v_{i,1}, v_{i,2}, ..., v_{i,d})$ for $1 \leq i \leq l$ ($v_{i,j}$ are scalars.). Figure 1(a) depicts a two-attribute sequence of length four.

The Euclidean distance, D_2, between d-attribute time sequences u and v of length l is defined as

$$D_2(u,v) = \sqrt{\sum_{1 \leq i \leq l} \sum_{1 \leq j \leq d} (u_{i,j} - v_{i,j})^2}.$$

2.1 Whole matching for multi-attribute sequences

If the length of the sequences in the database and the length of the query sequence is equal, similarity searching is called *whole matching*. Whole matching is a well studied problem for 1-attribute sequences (e.g. [1, 4, 7, 17]). Whole matching problem for multi-attribute sequences can be solved using any of the existing techniques after transforming multi-attribute sequences to 1-attribute sequences.

---

[1]This corresponds to an unbounded query in the authors' terminology.

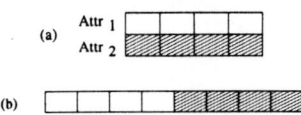

Figure 1. (a) Two-attribute sequence of length four. (b) One-attribute representation of the same sequence.

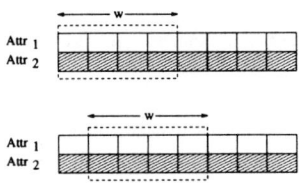

Figure 2. A sliding window of length four on a two-attribute sequence.

This transformation is performed by appending the attributes consecutively. This idea is depicted in Figure 1. Figure 1(b) shows the 1-attribute representation of the 2-attribute sequence given in Figure 1(a).

Once the multi-attribute time sequences are transformed to 1-attribute sequences, they can be viewed as points in a multi-dimensional space. For example, the sequence in Figure 1 is considered as a point in 8-dimensional space. The dimensionality of this data can be reduced using any energy preserving dimensionality reduction technique (e.g. DFT, SVD, or wavelets). These points are then indexed using any multi-dimensional index structure (e.g. R-tree [3, 9]). The distance is defined as the Euclidean distance in this multi-dimensional space.

2.2 Subsequence matching for multi-attribute sequences

Searching similar subsequences of database sequence to query sequence is called *subsequence matching*. Subsequence matching is more difficult than whole matching since the similar subsequences can be located at any location of the database sequences. Current subsequence matching techniques use sliding window based schemes to construct an index structure prior to search [7, 10, 11]. All of these techniques consider 1-attribute sequences.

Multi-attribute sequences can be handled by sliding the window over the multi-attribute sequence as in Figure 2. Each of the subsequences in these windows is then transformed into 1-attribute sequences as shown in Figure 1. Hence, each window maps to a point in a multi-dimensional space. For example, each window in Figure 2 maps to an 8-dimensional point. Later, the dimensionality of these points is reduced using a dimensionality reduction technique (e.g. DFT). The points are then indexed using the MR index structure [11].

3 Shift and scale invariant search

Simple Euclidean metric may not be sufficient for multi-attribute time sequences. In this section, we discuss the idea of shift and scale invariant distance metric. We begin with 1-attribute time sequences and later generalize to multi-attribute time sequences.

3.1 1-attribute time sequences

Consider the 1-attribute time sequences $v_1 = (2, 6, 4, 10)$ and $v_2 = (5, 7, 6, 9)$. Although the Euclidean distance between these two sequences is large, they can be made identical by scaling and shifting: $v_1 = 2 \cdot v_2 - 8 \cdot N$, where $N = (1, 1, 1, 1)$ is the normal vector. We begin by considering the existing notion of distance [5, 8].

A time sequence x can be normalized (also called *z-normalization*) as $x' = (x - mean(x))/std(x)$ [8]. That z-normalization does not minimize the Euclidean distance under all scalings and shiftings can easily be seen from the following example: Let $u = (0, 0, 1, 1)$ and $v = (3, 2, 1, 0)$, then $D_2(u', v') = 1.94$. On the other hand, $D_2(-0.5 \cdot v + 1.25, u) = 0.5$. is much less than the normalized distance.

Let v be a 1-attribute time sequence of length l, then the set of all possible scalings of v is defined as $SC(v) = \{c \cdot v | c \in \mathcal{R}\}$. The set of all scalings of v forms a line in l-dimensional space that passes through origin and v. Similarly, the set of all possible shiftings of v is defined as $SH(v) = \{v + c \cdot N | c \in \mathcal{R}\}$. The set of all shiftings forms a line in l-dimensional space which passes through v and which is parallel to the vector N. The plane which passes through the origin and is perpendicular to the normal vector N is called the *shift eliminated plane* (SE-plane). The projection of a sequence v onto the SE-plane is a point represented by $TSE(v)$. The minimum distance $d(u, v)$ between v and all scalings and shiftings of u can be computed as the distance between the projections of $SC(u)$ and $SH(v)$ onto the SE-plane. We establish the following lemma.

Lemma 1 *Given two 1-attribute time sequences u and v of the same length,*
$$d(u, v) = ||TSE(v)||_2 \cdot \sqrt{1 - (\frac{TSE(v).TSE(u)}{||TSE(v)||_2 \cdot ||TSE(u)||_2})^2},$$
where $u.v$ is the dot product of the vectors u and v, and $||v||_2$ is the second norm of v.

Though efficiently computable, the above distance formulation by Chu and Wong [5] and Goldin and Kanellakis [8] is not symmetric. That $d(u, v) \neq d(v, u)$, for some u, v, can be seen from the following example. Let $u = (0, 0, 1, 1)$ and $v = (3, 2, 1, 0)$, then
$d(u, v) = D_2(-2 \cdot u + 2.5, v) = 1$. On the other hand,
$d(v, u) = D_2(-0.5 \cdot v + 1.25, u) = 0.5$.
The scaling and shifting coefficients for this counter example are computed as in [5]. The absence of symmetry in the definition of $d(u, v)$ can lead to counter-intuitive results. For example, consider two sequences u and v from the database. It is possible that u is in the result set when

we perform a range query using v as the query sequence, but not vice-versa. It follows from Lemma 1 that the distance function $d(u, v)$ is symmetric, i.e. $d(u, v) = d(v, u)$, if and only if $||TSE(u)||_2 = ||TSE(v)||_2$. Since this condition is quite restrictive, we modify the definition of distance in order to make it symmetric.

Definition 1 *Given two 1-attribute time sequences u and v of the same length, the distance between these sequences is defined as*

$$dist(u, v) = min\{d(u, v), d(v, u)\}.$$

We could have used other functions such as *max*, or *average* in the above definition, but it is the *min* function that captures the notion of distance more adequately. We will see later that our index structure works as well for other functions.

3.2 Multi-attribute time sequences

The attributes of a multi-attribute time sequence can be *independent*, i.e., allowing independent scalings and shiftings, or *dependent*.

Let $N_{k,l}$ be the $k \times l$ matrix which is composed of all 1's. Define I_k to be the identity matrix of k dimensions. Let v be a k-attribute time sequence of length l.

Definition 2 *If v is a k-attribute dependent time sequence of length l, then the set of all possible scalings of v is defined as*

$$SC(v) = \{c \cdot v | c \in \mathcal{R}\},$$
and the set of all possible shiftings of v is defined as
$$SH(v) = \{v + c \cdot N_{k,l} | c \in \mathcal{R}\}.$$

Definition 3 *If v is a k-attribute independent time sequence of length l, then the set of all possible scalings of v is defined as*

$$SC(v) = \{C_k I_k v | C_k \in \mathcal{R}^k\},$$
and the set of all possible shiftings of v is defined as
$$SH(v) = \{v + C_k I_k N_{k,l} | C_k \in \mathcal{R}^k\}.$$

Given the above definitions of scalings and shiftings of multi-attribute time sequences, the distance function given in Definition 1 can be used for both dependent and independent multi-attribute time sequences. For example, let $u = ((0, 0), (1, 1))$ and $v = ((3, 2), (1, 0))$ be two attribute time sequences of length two. If u and v are dependent sequences, then $dist(u, v) = 0.25$ (scale v with -0.5, and shift v by 1.25.). If u and v are independent sequences, then $dist(u, v) = 0$ (scale v with 0, and shift the first attribute of v by 0 and the second attribute of v by 1.).

4 The CS-index structure

In this section, we propose a new index structure which clusters the data sequences according to both their scaling and shifting lines. We call this index structure *CS-index*

```
/*Let f be the fanout and p be the page capacity.*/
Algorithm CS-INDEX-BULKLOAD(S)
/*Let S be the set of time sequences in the database. */
  1. For all v ∈ S,      v := TSE(v).

  2. Choose a random sequence v.

  3. Sort all sequences in S in ascending order of angular distance to v.
     Let A_S be this order.

  4. Sort all sequences in S in ascending order of their distance to the
     origin. Let D_S be this order.

  5. SPLIT(S, A_S, D_S).
```

Figure 3. CS-index bulk-loading algorithm

(Cone Slice) for it resembles slices of hierarchically ordered cones. The idea is to project both the scaling lines and the shifting lines of the data sequences on to the SE-plane, and cluster the sequences whose projected shifting lines are close and for whom the angles between the projected scaling lines are small. Consideration of both the shifting line and the scaling line of data sequences is prompted by Definition 1. If we consider only the shifting lines of the data sequences, and use an R*-tree or any other similar index structure merely based on spatial closure as in [5], then there are several disadvantages: a)The angular distance between scaling lines of the data sequences clustered within a disk page may be larger than the angular distance between scaling lines of the data sequences in different pages. This may result in a large number of false hits. b) Disk I/O's for reading the candidate sequences involve random seeks, resulting in a high I/O overhead.

We will first describe the construction of the *CS-index* structure for the 1-attribute time sequences. Later we will extend the idea to the multi-attribute case.

4.1 CS-index for 1-attribute time sequences

Let q be a query sequence and let v be a time sequence in the database. Let $\theta_{q,u}$ be the angle between the vectors $TSE(q)$ and $TSE(u)$. Using Lemma 1 and Definition 1, we conclude that $dist(q, u) = min\{||TSE(q)||_2, ||TSE(u)||_2\} \cdot sin\theta_{q,u}$. This means that the distance between a query and a database sequence is based on the lengths of the projections of the two sequences on the SE-plane and the angle between the projections. Therefore, a good index structure must cluster radially, i.e., based on the distance from the origin as well as the angular distance.

The CS-index structure consists of a set of data pages stored on disk, and summary information stored in memory about the data pages. The data pages are organized in a tree structure defined by two parameters: page capacity, denoted p, and fanout, denoted f. The tree structure is virtual in the sense that it is used only for clustering of data; there are no physical index pages. The rationale for choosing the appropriate fanout will be explained later.

The data sequences are bulk-loaded into the index struc-

```
Algorithm SPLIT(S, A_S, D_S)
  • if |S| > p /* The number of points are more than page capacity */
      1. if |S| < p · f then f := |S|/p. /* Reduce fanout if there is not
         sufficient points left for current fanout */

      2. Using A_S, partition the sorted set S into subsets
         S_1, S_2, ..., S_f of size |S|/f.

      3. Obtain A_{S_i} and D_{S_i}, for each i, i = 1, ..., f, by restricting
         A_S and D_S to S_i.

      4. for i := 1 to f
           (a)  P := p sequences closest to the origin in S_i.
           (b)  store P as the next page.
           (c)  S_i := S_i − P.
           (d)  update A_{S_i} and D_{S_i}.
           (e)  SPLIT(S_i, A_{S_i}, D_{S_i}).

      5. end for

  • end if
```

Figure 4. Split Algorithm. This algorithm recursively splits the data points first according to the angles and then according to their distances from the origin.

ture as shown by algorithms in Figures 3 and 4. First, the shifting lines of all the sequences are projected onto the SE-plane (Step 1 of procedure *CS-INDEX-BULKLOAD*). Later, these sequences are sorted based on angular distance to a randomly chosen sequence (Step 3), and based on distance to the origin (Step 4). After that, procedure *SPLIT* is invoked. This procedure carries out a clustering based on angles and distances from origin. Using the angular distances, the set of sequences is partitioned into f subsets $S_1, S_2, ..., S_f$ (Step 2). Later, the angular and spatial orderings for these subsets are obtained by restricting the original orderings A_S and D_S (Step 3). In the second splitting phase (Step 4), a piece of size p (Steps 4.a, 4.b, and 4.c). is chopped from each cone by intersecting it with a sphere centered at the origin. Each of these pieces is called a *slice*. Later, the angular and spatial orderings are updated for the remainder of the subset (Step 4.d). Rest of the points are then recursively split (Step 4.d). Steps 1, 2, 3, 4.a-4.d require $O(n)$ time, where n is the size of the input set S. As a result, the time complexity, $T(n)$ of the *SPLIT* algorithm satisfies the recurrence $T(n) = fT(n/f) + O(n)$. This leads to a time complexity of $O(n \log n)$ for the *SPLIT* algorithm. Since the two sorting steps in the *CS-INDEX-BULKLOAD* algorithm also have $O(n \log n)$ complexity, the entire index construction requires $O(n \log n)$ time.

Figure 5 shows different steps of the index construction algorithm on a sample dataset when $f = 2$. The data points are projected to the SE-plane. The dataset is partitioned into f equal sized sets based on their angular distance in Figure 5(a) (Step 2 of *SPLIT*). Later, a data page is determined by clustering p closest points in one of these sets, based on Euclidean distance to the origin in Figure 5(b) (Step 4.a-4.b

of *SPLIT*). The *SPLIT* algorithm is then invoked recursively for the rest of the points in the reduced set. The reduced set is again partitioned into f sets, based on angles in Figure 5(c). Figure 5(d) presents the final index structure.

Each slice s can be viewed as the intersection of a cone with a ring that is determined by two radii, r_s and R_s, where r_s is the minimum Euclidean distance between the origin and a point on the slice and R_s is the maximum Euclidean distance between the origin and a point on the slice.

The index structure in Figure 5(d) has three levels. The root level contains two slices 4, and 11. There is a total of 14 slices, each containing at most p points. A slice s_i is said to be the *ancestor* of a slice s_j (and similarly, slice s_j is said to be a *descendant* of s_i) if the cone corresponding to s_j is contained entirely in the cone corresponding to s_i. For example, in Figure 5(d), slice 4 is the ancestor of slices 1, 2, 3, 5, 6 and 7.

The distance between a query sequence q and a slice is defined as the minimum scale and shift invariant distance between q and any point on the slice. The formal definition is as follows:

Definition 4 *Let s be a slice in the CS-index and q be a query sequence on the SE-plane. Let $\theta_{q,s}$ be the angle between q and a point in s for which $\sin\theta_{q,s}$ is the minimum. The distance between q and s is defined as:*
$$dist(q,s) = min\{||q||_2, r_s\} \cdot sin\theta_{q,s}.$$

Using this definition, we have the following theorems:

Theorem 1 *Let s be a slice in the CS-index, v be a data sequence contained in s, and q be a query sequence on the SE-plane, then*
$$dist(q,s) \leq dist(q,v).$$

Theorem 2 *Let s_1 and s_2 be two slices in the CS-index such that s_1 is the parent of s_2. Let q be a query sequence on the SE-plane, then*
$$dist(q,s_1) \leq dist(q,s_2).$$

Some important observations about the CS-index are as follows. 1) If the distance between a slice and a query sequence is greater than ϵ (Theorem 1), then the distance between the query sequence and any data sequence contained in that slice is greater than ϵ. In other words, given a range query, a slice may contain candidate sequences only if the distance between the query sequence and that slice is less than the search range. 2) If the distance between a slice and a query sequence is greater than ϵ, then the distance between that query sequence and all children of that slice is greater than ϵ (Theorem 2). 3) If the distance between a slice and a query sequence is less than ϵ, then the distance between that query sequence and all the ancestors of that slice is less than ϵ (Theorem 2).

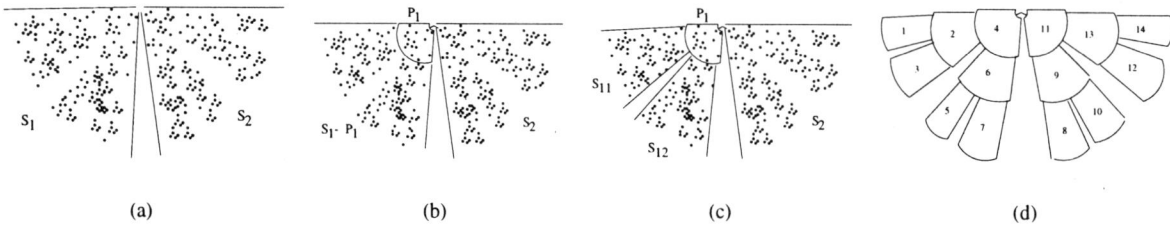

Figure 5. Construction of the CS-index when $f = 2$. The data points are transformed onto shift eliminated plane. (a) Angular partitioning, (b) spatial partitioning, (c) recursive invocation of angular partitioning, (d) final index structure.

```
Algorithm RANGE-QUERY(q, ε)
  1. C := ∅              /* the set of candidate pages*/
  2. S := ∅              /* stack for navigation*/
  3. Push all the root slices s₁, s₂, ..., s_f on to S.
  4. While S ≠ ∅
     (a)  s := Pop(S)
     (b)  if dist(q, s) ≤ ε
          i.  C := C ∪ {s}
          ii. Push all the children of s on to S
  5. Read the candidate pages in set C and perform postprocessing to
     eliminate false retrievals.
```

Figure 6. Range search algorithm

4.2 Similarity search on the CS-index structure

The range query algorithm on the CS-index is presented in Figure 6. A range query is performed in two phases: an in-memory candidate generating phase followed by a disk-based post-processing step. In the in-memory phase, the search starts from the root pages and proceeds downwards. If the distance between the query sequence and a slice is less than the query range (Step 4.b), then the method marks this slice as a candidate and expands the query to all the children of that slice. If the distance between the query sequence and a slice is greater than the search range, then the algorithm prunes that slice and all its children. Once the candidate sets are determined, the disk-based processing step begins. Using the disk placement information, a sequential scan is used to read all the candidate pages. Data points that are not in the range are pruned to ensure no false retrievals. Since the set of pages read from disk during the post-processing step is a subset of all the pages, our method performs no worse than sequential scan. In fact, it performs much better since clustering reduces the range of the sequential scan considerably. As a consequence of Theorem 1 and Theorem 2, we have the following corollary:

Corollary 1 *The range query algorithm in Figure 6 does not incur any false drops.*

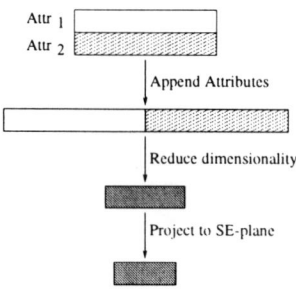

Figure 7. Transformation of a 2-attribute dependent time sequence to the SE-plane.

We present a detailed discussion of the nearest neighbor queries on the CS-index structure in the technical report version of this paper [12].

4.3 Multi-attribute CS-index

The CS-index structure, so far defined for 1-attribute sequences, can be extended easily to multi-attribute sequences. We consider the multi-attribute extension for dependent and independent time sequences separately.

4.3.1 Case 1: dependent attributes

If the multi-attribute time sequences in the database are k-attribute dependent time sequences of length l, then the problem reduces to the 1-attribute case by simply transforming the sequences into 1-attribute time sequences of length $k \cdot l$ as in Figure 1. This is justified because all the entries are scaled and shifted by the same amount. Figure 7 depicts how the dependent attributes are handled. These sequences are considered as points in a $(k \cdot l)$-dimensional space. The dimensionality of these points are then reduced using a dimensionality reduction technique (e.g. DFT). Later, the CS-index structure is constructed on these points as explained in Figures 3, 4, and 5. Range queries are performed as in Figure 6.

4.3.2 Case 2: independent attributes

If the database consists of k-attribute independent time sequences of length l, then all the attributes must be consid-

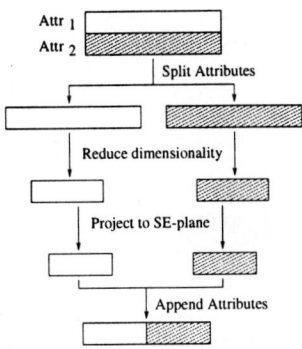

Figure 8. Transformation of a 2-attribute independent time sequence to the SE-plane.

ered separately. This is because different attributes may be scaled or shifted by different amounts. Figure 8 presents how the CS-index structure handles independent attribute sequences. We first split each time sequence into k 1-attribute time sequences of length l. This is like splitting a $k \cdot l$-dimensional space into k non-overlapping l-dimensional spaces. As a result, each k-attribute time sequence corresponds to k points in l-dimensional subspaces. We determine the SE-planes of these l-dimensional subspaces. We project the 1-attribute time sequences onto their corresponding SE-planes, and concatenate the vectors corresponding to these k projections to construct a 1-attribute $k \cdot (l - 1)$-dimensional point. We construct the CS-index on these points as described in Section 4.1. Every slice of the constructed index can be projected into k different subspaces; these projections are called *subslices*. Another choice would be to maintain k separate CS-index structures for 1-attribute $(l - 1)$-dimensional points. However, this would require additional post-processing.

For a given range query or a nearest neighbor query of k attributes, we split the query into k 1-attribute subqueries, one for each attribute. For each subquery, we obtain its distance to a slice by considering the subslice corresponding to that attribute. The distance between a query and a slice is defined to be the square root of the sum of squares of the k different subquery subslice distances. Once these distances are obtained, pruning and post-processing proceeds as in the single-attribute case. Range queries are performed similar to Figure 6. The only difference is that the distance function is computed for each attribute separately, and the results are accumulated to find the distance for the independent case.

4.4 Improving post-processing performance

The candidate slices for a range query or a NN-query are determined using an in-memory search. The postprocessing step uses one sequential scan to read the candidate slices. The performance of the index structure is therefore determined by how closely clustered the candidate slices are on

disk [2]. The number of non-candidate slices placed between the first and the last candidate slice on disk should be minimized. A clustering of candidate slices can be achieved by three techniques: carrying out a more effective pruning in the in-memory phase, optimizing the placement of pages on disk, and caching/replication of disk pages. We elaborate on these ideas next.

4.4.1 Fanout selection

We noted earlier that the fanout f is an independent parameter of the bulk-loading algorithm that is not affected by the size of disk pages. For a given dataset, a large fanout leads to thick slices that span a smaller angle, whereas a small fanout leads to thin slices that span a larger angle. The success of the pruning procedure depends on both the thickness and the angular span of the slices: a thick or a wide slice is less likely to be pruned. The right choice of fanout ensures that slices are not too thick and not too wide. This can be determined either experimentally or theoretically if the data distribution is known. In our experiments, the optimal value for fanout varied between 5 and 7.

4.4.2 Disk placement

The second parameter that improves the post-processing performance is the placement of pages on disk. Note that if the distance between a query sequence q and a slice s is less than the given search range ϵ, then the distance between q and the parent of s is also less than ϵ. Therefore, if a slice s is in the candidate set, then all its ancestors are also in the candidate set. For example, if slice 3 of the CS-index in Figure 5 is a candidate, then slices 2 and 4 are also guaranteed to be in the candidate set. In order to reduce I/O cost, slices 1, 2, and 4 should be stored contiguously on disk. In general, the slices should be placed on disk in a manner that minimizes the distance between a slice and all its descendants. This means that the slices belonging to a subtree should be stored contiguously; it does not help to interleave a subtree with slices from a sibling subtree. The second condition that minimizes the parent-child distance is that a root node should be linearized in the middle of its subtree. These two conditions imply an *in-order* placement: a tree with $2k$ subtrees is linearized by an *in-order* traversal of its first k subtrees, placement of the root slice, and an *in-order* traversal of the remaining k subtrees. This linearization is used to place the slices on the disk.

4.4.3 Replication and caching of pages

The final parameter that improves the post-processing performance is the degree of caching and replication of disk pages. Both caching and replication can reduce the number of redundant pages that are read.

Replicating a page means that we maintain a copy of the page with all its subtrees on disk. Replicating k levels of the index structure means that we replicate the pages at the first k levels of the CS-index at their children. The advantage of

---

[2]For simplicity we assume a 1-d disk model.

Dataset	Size	R-tree Size	CS-index Size
stock market dataset	2.5M	160K	25K
motion dataset	8M	393K	41K

Table 1. The size of the R-tree and the CS-index structure for stock market dataset and motion dataset.

replication is that it can reduce the distance between a page and its ancestors. Replication works best if the queries are sufficiently narrow so that all candidate pages belong to a subtree and its ancestors. Otherwise, it can lead to redundant pages being read.

Caching k levels of the index structure means to keep the pages at the first k levels of the CS-index in memory and to place the in-order linearizations of the subtrees at level k on disk. Unlike replication, caching can *only* improve the performance of our index structure.

4.5 Extending the CS-index structure to other symmetric distance functions

The CS-index structure can also be used when the distance function is obtained by using *max* or *avg* functions instead of *min* function as follows:

Case 1. $dist(q, u) = max\{d(q, u), d(u, q)\}$.

In this case, one can prove that

$$dist(q, u) = max\{||TSE(q)||_2, ||TSE(u)||_2\} \cdot sin\theta_{q,u}.$$

Similar to the *min* function, this distance function also uses a Euclidean distance and an angular distance in its computation. The CS-index structure will work well for the *max* function since it clusters time sequences based on the distance from the origin as well as the angular distance.

Case 2. $dist(q, u) = avg\{d(q, u), d(u, q)\}$.

$$dist(q, u) = \frac{(min\{d(q,u), d(u,q)\} + max\{d(q,u), d(u,q)\})}{2}.$$

Hence, $dist(q, u) = \frac{||TSE(q)||_2 + ||TSE(u)||_2}{2} \cdot sin\theta_{q,u}$. Similar to *min* and *max* functions, *avg* function is also based on the distance from the origin and the angular distance. As a result of this, the CS-index structure will work well for the *avg* function too.

5 Experimental results

We carried out several experiments to test the performance of the CS-index structure. We used three different datasets in our experiments:

1) The first dataset is a stock market dataset, obtained again from *chart.yahoo.com*. The time sequences in this dataset consist of 2 attributes. The first attribute is the closing price, and the second attribute is the volume. There are 20,000 time sequences of length 32 in this dataset.

2) The second dataset is obtained synthetically by considering four different kinds of object trails in a 2-attribute sequence. The motions that we consider are: bouncing ball, circular motion, billiard ball moving within the confines of a rectangular table with perfect carom and elasticity, and a

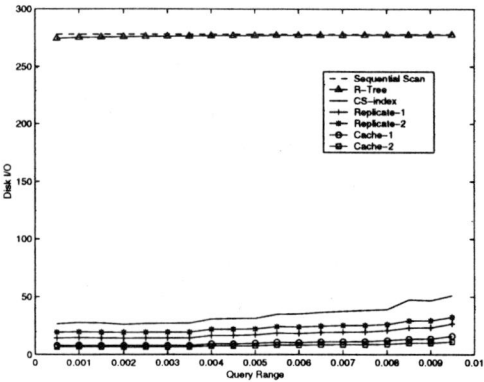

Figure 9. I/O overhead of CS-index versus R-Tree and sequential scan for the 2-attribute dependent stock market dataset.

periodic motion along a sine curve. This dataset contains 2^{15} time sequences, distributed evenly among the four different motion types. The length of the time sequences is 32.

3) The third dataset is a multi-attribute dataset of 1, 2, 4, and 8 attributes. The time sequences are obtained synthetically by adding four sine signals of random frequencies and amplitudes and some amount of random noise. Corresponding to each choice of attributes, we have 2^{15} time sequences of length 32.

We compressed the time sequences in these datasets to 4 dimensions using DFT, and then built the CS-index structure on the compressed data. Since both DFT and TSE are distance preserving transformations, there are no false drops in the resulting index structure. We also built R-Tree index as proposed by Chu and Wong [5] for comparison. Since the CS-index structure uses bulk loading, we used the VAM-Split implementation of R-Trees [21]. Table 1 displays the size of the CS-index structure and R-tree for the first two datasets. The size of the CS-index structure is much smaller than R-tree. This is because the CS-index structure disk does not store pointers to individual time sequences in the database since slices of the CS-index structure corresponds to continuous pages on the disk. In our query model, we considered range queries with 20 different values of ϵ in the range $(0.001, 0.01)$. For every value of ϵ, 1000 sequences in the database were chosen at random for querying. We assume that the page size is 4K in our experiments.

The first experiment considers the dependent attributes. 2-attribute stock market dataset. Figure 9 shows the I/O overhead for sequential scan, R-tree, and the CS-index structure. For the CS-index structure, we also present results for replicating one level, replicating two levels, caching one level, and caching two levels. The results show that the R-Tree index structure accesses almost all the pages. This can be explained as follows. The angular distance between two points may be large even if the Euclidean distance between them is small. Since R-Tree index

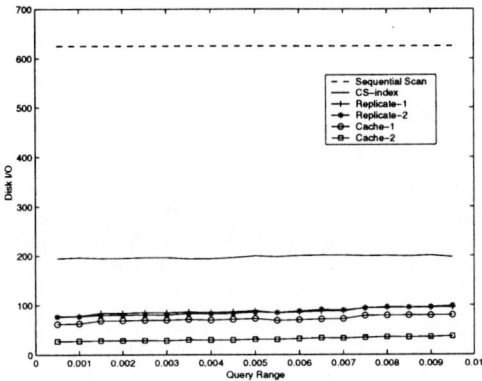

Figure 10. I/O overhead of CS-index versus sequential scan for the 2-attribute stock market dataset.

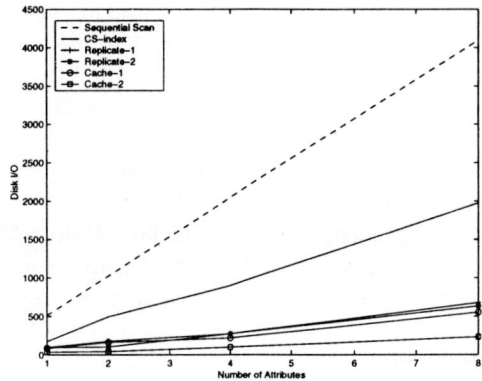

Figure 12. I/O overhead of CS-index versus sequential scan for the sine curve dataset of an increasing number of attributes.

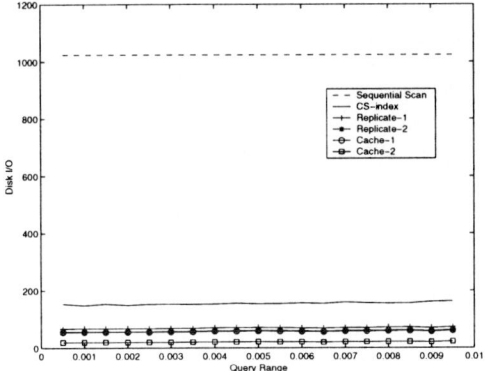

Figure 11. I/O overhead of CS-index versus sequential scan for the 2-attribute motion dataset.

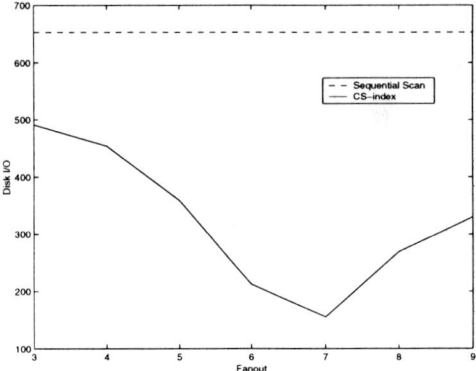

Figure 13. I/O overhead of CS-index versus sequential scan for the 2-attribute sine curve dataset for different fanout values.

structure clusters based on Euclidean distance, the resulting boxes may span a large angular distance. This may lead to a large number of candidate MBRs. Though R-tree index and sequential scan access a similar number of pages, R-tree index accesses pages randomly. Since a random read can cost approximately 10 times a sequential read [18], the performance of the R-tree index is about 10 times worse than sequential scan. Therefore, we conclude that the R-Tree index structure is not appropriate for scale and shift invariant searches. On the other hand, the CS-index structure accesses less than 20% of all the pages. Since, the CS-index structure performs a single sequential disk read, it incurs a seek cost only once (like sequential scan). CS-index is 5 to 10 times faster than sequential scan. Replicating only one one level of the CS-index structure almost doubles the speedup. Replicating two or more levels degrades the performance. If only one level of the CS-index is cached, then the speedup is 15 to 35.

In remaining experiments, we assume that the time sequences in the dataset have independent attributes. We compare the performance of our method to sequential scan. R-Tree results are not presented, because its performance was much worse than sequential scan. Figures 10 and 11 show the experimental results for the stock market dataset and the motion dataset compressed to 4 dimensions. We obtained speedups up to 8 for CS-index with no replication and caching. The speedup doubled when one level is replicated. The speedup increased to 40 when 2 levels of our index structure are cached.

The next experiment reports the scalability of the CS-index structure for an increasing number of attributes. In this experiment, we used the third dataset. Figure 12 plots the number of disk reads for sequential scan and CS-index structure. As the number of attributes increases, the size of the dataset increases linearly. This is reflected in the number of disk reads for sequential scan. The experimental results show that the number of disk reads for the CS-index structure also increases linearly with the number of attributes. Therefore, the speedup of the CS-index structure remains invariant under an increasing number of attributes.

Figure 13 plots the number of disk read for the CS-index structure for different values of fanout for the stock market data set compressed to 4 dimensions. The CS-index performs the best when the fanout is 7 for this data set. We obtained a similar *U-shaped* graph for the other datasets too. The optimal fanout for other datasets varies between 5 and 7. Reason for this is explained in Section 4.4.1.

More comprehensive experimental results for additional datasets and a discussion for subsequence searches are available in the technical report version of this paper [12].

6 Discussion

We considered the problem of similarity search for multi-attribute sequences. First, we considered the simple Euclidean distance metric. Later, we considered more challenging metrics of shift and scale invariance. We formulated a new notion of similarity that is symmetric in allowing transformations on both query and data sequences. Furthermore, we do not impose any restriction on the shifting and scaling constants. We considered both the cases of when the scalings and the shiftings of the attributes are dependent and when they are independent.

We proposed a new index structure called CS-index that clusters time sequences according to their scalings and shiftings. This index structure recursively splits the search space into hierarchical cones and selects a slice of each cone as a disk page. This index structure allows early pruning in the search phase. Later, we considered techniques to improve the performance of the index structure: in-order based placement on disk, choice of right fanout, and caching/replication. Finally, we showed that the method can be extended to multi-attribute time sequences.

According to experimental results with both real and synthetic datasets, our method performs 5 to 10 times faster than sequential scan. We also evaluated the effects of replicating higher levels of the index structure. Replicating only the root level of the CS-index almost doubled the performance of our method. Further replication eventually degraded the performance. We also experimented with caching. According to our experiments, if only the pages at the root level are cached, our method performs 10 to 25 times faster than sequential scan. We obtained speedup up to 45 when we cached one more level.

The techniques presented in this paper can be easily extended to perform shift and scale invariant subsequence searches. According to our experiments, our technique is 3 times faster than sequential scan for subsequence searches.

Multi-attribute time sequences are an important emerging class of applications. They arise ubiquitously and naturally: in medical applications, in control applications, in video and event sequences, and in history-based applications such as the stock market. The ability to query such data under different distance metrics is necessary for understanding and analyzing the characteristics of such datasets.

The index structures presented in this paper are an important first step toward this, and should be widely applicable.

References

[1] R. Agrawal, C. Faloutsos, and A. Swami. Efficient similarity search in sequence databases. In *FODO*, Evanston, Illinois, October 1993.

[2] R. Agrawal, K. Lin, H.S. Sawhney, and K. Shim. Fast similarity search in the presence of noise, scaling, and translation in time-series databases. In *VLDB*, Zürich, Switzerland, September 1995.

[3] N. Beckmann, H.-P. Kriegel, R. Schneider, and B. Seeger. The R*-tree: An efficient and robust access method for points and rectangles. In *SIGMOD*, pages 322–331, Atlantic City, NJ, 1990.

[4] K.-P. Chan and A.W.-C. Fu. Efficient time series matching by wavelets. In *ICDE*, 1999.

[5] K.K.W. Chu and M.H. Wong. Fast time-series searching with scaling and shifting. In *PODS*, Philadelphia, PA, 1999.

[6] G. Das, D. Gunopulos, and H. Mannila. Finding similar time series. In *PKDD*, pages 88–100, 1997.

[7] C. Faloutsos, M. Ranganathan, and Y. Manolopoulos. Fast subsequence matching in time-series databases. In *SIGMOD*, pages 419–429, Minneapolis MN, May 1994.

[8] D. Q. Goldin and P. C. Kanellakis. On similarity queries for time-series data: Constraint specification and implementation. In *CP*, pages 137–153, France, September 1995.

[9] A. Guttman. R-trees: A dynamic index structure for spatial searching. In *SIGMOD*, pages 47–57, 1984.

[10] T. Kahveci and A. Singh. An efficient index structure for string databases. In *VLDB*, pages 351–360, Roma, Italy, September 2001.

[11] T. Kahveci and A. Singh. Variable length queries for time series data. In *ICDE*, Heidelberg, Germany, 2001.

[12] T. Kahveci, A.K. Singh, and A. Gürel. Shift and scale invariant search of multi-attribute time sequences. Technical report, UCSB, 2001.

[13] S.-L. Lee, S.-J. Chun, D.-H. Kim, J.-H. Lee, and C.-W. Chung. Similarity search for multidimensional data sequences. In *ICDE*, San Diego, CA, 2000.

[14] C.-S. Perng, H. Wang, S.R. Zhang, and D.S. Parker. Landmarks: a new model for similarity-based pattern querying in time series databases. In *ICDE*, San Diego, USA, 2000.

[15] D. Rafiei. On similarity-based queries for time series data. In *ICDE*, Sydney, Australia, March 1999.

[16] D. Rafiei and A.O. Mendelzon. Similarity-based queries for time series data. In *SIGMOD*, pages 13–25, Tucson, AZ, 1997.

[17] D. Rafiei and A.O. Mendelzon. Efficient retrieval of similar time sequences using DFT. In *FODO*, Kobe, Japan, 1998.

[18] B. Seeger, P.-A. Larson, and R. McFayden. Reading a set of disk pages. In *VLDB*, pages 592–603, 1993.

[19] C. Shahabi, X. Tian, and W. Zhao. TSA-tree: A wavelet-based approach to improve the efficieny of multi-level surprise and trend queries. In *SSDBM*, 2000.

[20] M. Vlachos, G. Kollios, and D. Gunopulos. Discovering similar multidimensional trajectories. In *ICDE*, San Jose, CA, 2002.

[21] D. White and R. Jain. Similarity indexing: Algorithms and performance. In *SPIE Storage and Retrieval for Image and Video Databases*, 1996.

Panel

Developing Scientific Database Applications in a Grid Environment

— *Panel Discussion* —

Abstract

Whether for astrophysics, bioinformatics or any other discipline, developing a scientific database application in a distributed computing environment remains a challenging task. Part of this challenge lies in having to make critical choices with respect to the underlying computing technologies and the roles they play in such projects. Emerging Grid technologies are adding further to the choices available to developers. Although the Grid promises significant advances in support of e-Science applications, much of the envisaged infrastructure, including data management, is still under development. The panel will discuss the role of data management in the e-Science process and the integration of database services into a Grid necessary to enable the realisation of this role.

1 Introduction

Grid computing is aimed at supporting dynamic integration of services in a distributed heterogeneous computing environment, allowing the formation of a "virtual organisation" which may include one or more organisations collaborating to achieve a particular objective. The Grid plays a key role in enabling the vision of e-Science, where scientists around the world can join large scale collaborative scientific enterprises through the Internet.

The Grid is generally described in terms of three layers. The *data/computation layer* "deals with the way that computational resources are allocated", the *information layer* "deals with the way that information is represented, stored, accessed, shared and maintained", the *knowledge layer* "is concerned with the way that knowledge is acquired, used, retrieved, published and maintained", where *data* refers to uninterpreted bits and bytes, *information* is data and its meaning, and *knowledge* describes information applied to achieve a goal or solve a problem [1]. Of course, this is only a conceptual model and implementations of the Grid will vary in the degree to which they adhere to this model.

Foster et al. [2] propose the *Open Grid Services Architecture* (OGSA) for the realisation of the Grid. All required functionalities are abstracted in terms of *Grid ser-* vices which are described using the Web Services Description Language (WSDL). Hierarchies of such services are envisaged to support various levels of abstractions in the design of distributed systems. Each service specification may be supported by one or more service implementation in a Grid.

Grid applications initially considered were mostly using files for data storage. More recent work ([3][5]), however, has also been looking at the integration of databases and the Grid. Paton et al. [3] have proposed an initial draft of a collection of database services that articulate with the service-based approach of OGSA. A comprehensive evaluation of data requirements for the Grid can be found in [4].

2 Discussion

The panel will discuss the role and current state of data management in different e-Science applications, the limitations and problems of software that is in use today to integrate databases into a Grid environment, and possible future developments, such as alternative software architectures, to address the existing problems.

NOTE: Below references [1,3,4,5] can be found on-line at the UK National e-Science Centre (www.nesc.ac.uk). Reference [2] can be found at www.globus.org/research/papers/ogsa.pdf.

References

[1] D. De Roure, N. Jenings, and N. Shadbolt. Research Agenda for the Semantic Grid: A Future e-Science Infrastructure. *Report Commissioned for EPSRC/DTI Core e-Science Programme*, 2001.

[2] I. Foster, C. Kesselman, J. Nick, and S. Tuecke. The Physiology of the Grid, An Open Grid Services Architecture for Distributed Systems Integration. *Argonne National Laboratory*, 2002.

[3] N. Paton, M. Atkinson, V. Dialani, D. Pearson, T. Storey, and P. Watson. Databases Access and Integration Services on the Grid. *Manchester University*, 2002.

[4] D. Pearson. Data Requirements for the Grid. *Technical Report*, 2002.

[5] P. Watson. Databases and the Grid. *Technical Report CS-TR, University of Newcastle*, 2001.

Invited Speaker

Metadata for Traveling Statistics -

the World of Statistics Meets the Semantic Web

Jostein Ryssevik
Director of Technology and Development
Nesstar Ltd.
Jostein.Ryssevik@nsd.uib.no
Tlf.: +47 55582654

Abstract

The paper is focusing on the requirements to metadata in a world where the Internet and the Web has become the dominant communication and dissemination medium for statistical information. Four important aspects of metadata, (finding, understanding, assessing and sharing) will be discussed in the light of this development. Special focus will be kept on the ideas and concepts of the semantic web – the latest step in the development of the global information system – and how these ideas and concepts are challenging our traditional views on metadata and standards development. Experiences from the parallel development of the NESSTAR system and the DDI (Data Documentation Initiative) will be used as illustrations.

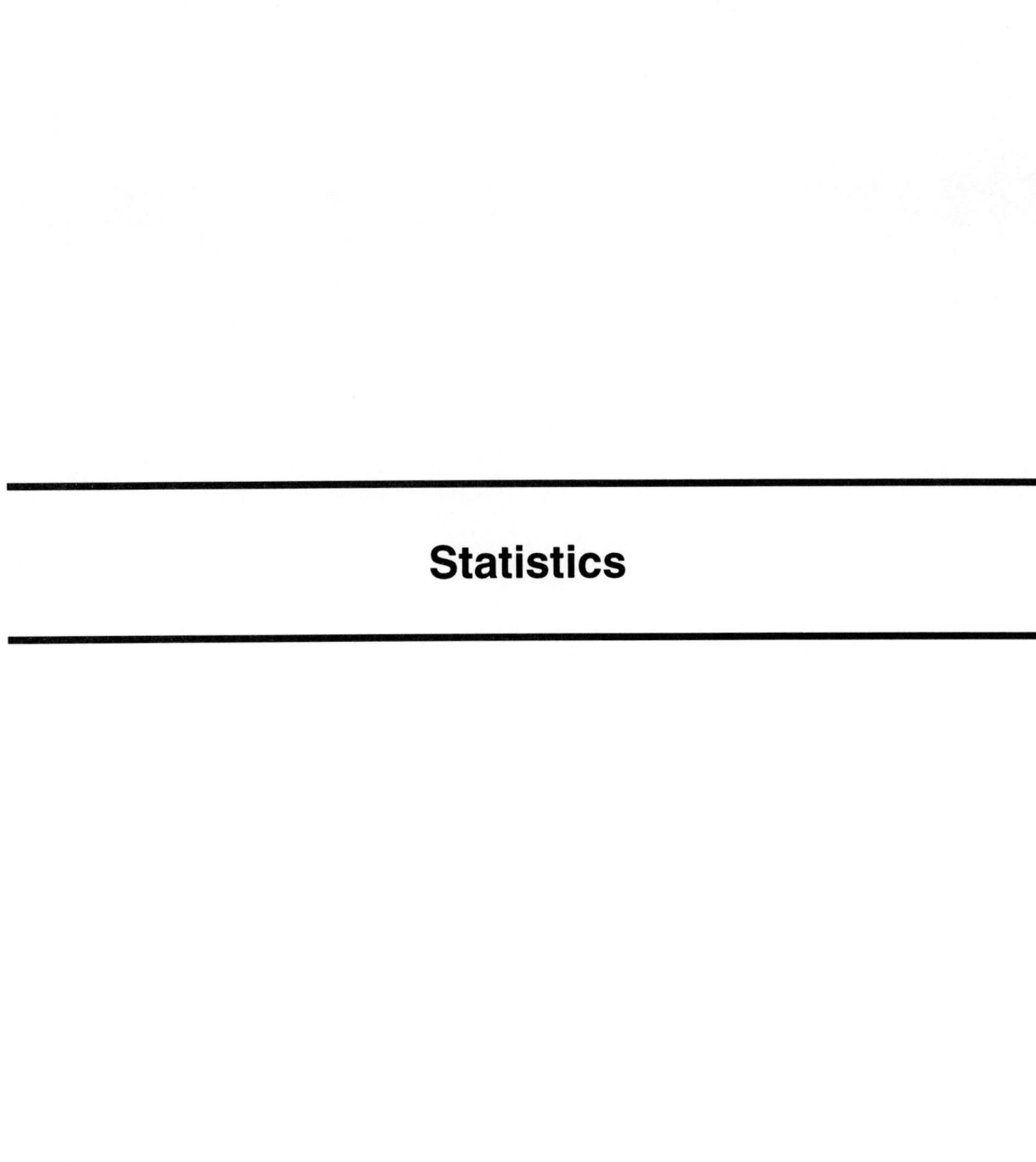

Statistics

XML-Extended OLAP Querying

Dennis Pedersen   Karsten Riis   Torben Bach Pedersen

Department of Computer Science, Aalborg University
Fredrik Bajers Vej 7E, 9220 Aalborg Ø, Denmark
{dennisp,riis,tbp}@cs.auc.dk

Abstract

The rapidly changing data requirements of today's dynamic business environments are not handled well by current On-Line Analytical Processing (OLAP) systems. Physically integrating data from new sources into OLAP systems is a long and time-consuming process, making logical integration the better choice in many situations. The increasing use of Extended Markup Language (XML), e.g. in business-to-business (B2B) applications, suggests that the required external data will most often be available in XML format.

In this paper we present a theoretically well-founded approach to the logical federation of OLAP and XML data sources. The approach allows external XML data to be presented along with dimensional data in OLAP query results and enables the use of external XML data for selection and grouping. Special care is taken to ensure that semantic problems do not occur in the integration process. This opens up many new application areas for OLAP, e.g., in the B2B and scientific domains. A number of effective optimization techniques for OLAP-XML federations are presented. Performance results from the prototype implementation show that the approach is an attractive alternative to physical integration. The approach is exemplified using a real-world case study from the B2B domain.

1 Introduction

On-line Analytical Processing (OLAP) and Extensible Markup Language (XML) are currently two of the most significant database technologies. However, the connection between them has so far received little attention.

OLAP systems enable powerful analysis of large amounts of summary data commonly drawn from a number of different transactional databases. OLAP data are often organized in multidimensional *cubes* containing *measured values* that are characterized by a number of hierarchical *dimensions*. The multidimensional approach offers a number of advantages over traditional types of DBMSs, including automatic aggregation, visual querying, and good query performance due to the use of pre-aggregation [18]. However, it is difficult for OLAP systems to handle changing and unanticipated data requirements as physically integrating data can be a complex and time-consuming process requiring the cube to be rebuilt [18]. In some situations, the required data cannot be integrated into the cube at all, e.g. because interface or copyright restrictions do not allow data to be retrieved and stored locally, but only to be queried in an ad hoc manner. Thus, *logical*, rather than physical, integration of data is desirable, i.e., a *federated* database system [17] is called for. The increasing use of Extended Markup Language (XML), e.g. in B2B applications, suggests that the required external data will mostly be available in XML format. Also, most major DBMSs are now able to publish data as XML. Thus, it is desirable to be able to access XML data from an OLAP system. The hierarchical and sometimes irregular structure of XML data means that problems related to correct aggregation of data can occur.

In this paper we present a theoretically well-founded approach to the logical federation of OLAP and XML data sources. The approach allows external XML data to be used as "virtual" dimensions, enabling three specific uses of XML data. First, OLAP query results may be "decorated" with XML data. Second, external XML data may be used for selection. Third, OLAP data may be grouped by external XML data when aggregation is performed. Special care is taken to ensure that the possibly irregular structure of the XML data does not cause problems w.r.t. correct aggregation of data. A flexible linking mechanism is devised to associate cube data with parts of XML documents. We make no assumptions about the existence of Document Type Definitions (DTDs) or XML Schemas [20]. To demonstrate the capabilities of the approach, we present a data model and a multi-schema query language, *XML-Extended Multidimensional SQL* (SQL_{XM}), based on SQL and XPath [19]. SQL and XPath are chosen for their simplicity, wide-spread use, and compact syntax. We also present a number of effective rule-based and cost-based optimization techniques for the approach, including algebraic query rewriting, inlining, optimizations over limited query interfaces, and caching and prefetching. A prototype implementation and experiments that show the performance of the approach and the effectiveness of the optimizations are also presented. The experiments suggest that the approach is in many cases an attractive alternative to physical integration.

As almost all data sources can be efficiently wrapped in XML format [1], the approach also allows external data from sources such as relational, object-relational, and object databases to be used in a powerful and flexible way, opening up new application areas for OLAP as data need no longer be integrated physically in the OLAP DB.

There has been a great deal of previous work on data integration, e.g., on integrating relational data [10], object-oriented data [16], semi-structured data [4], and a combination of relational and semi-structured data [6]. However, none of these handle the advanced issues related to OLAP systems, e.g., dimensions with hierarchies and the problems related to correct aggregation. This is also true for the combined relational and XML query language xQuery [21], and for nD-SQL [5], which considers the federation of relational sources providing basic OLAP functionality. One previous paper [15] has considered the federation of OLAP and object data. In comparison, our approach is not restricted to object DBs, and their rigid schemas, but can be used on any imaginable data source as long as it allows XML wrapping. Also, we allow irregularities in the external data and offer a more general use of external data when performing decoration, selection, and grouping. Query processing and optimization has been considered for data warehousing/OLAP systems [18], federated, distributed, and multi-databases [17], heterogeneous databases [2, 9], and XML and semistructured data [4]. However, previous work does not address the special case of optimizing OLAP queries in a federated environment.

We believe this paper to be the first to consider the integration of OLAP and XML data, including advanced issues such as dimension hierarchies and correct aggregation of data. Also, we believe to be the first to consider query processing and optimization for this setting.

The rest of the paper is organized as follows. Section 2 motivates our approach and presents the case study used throughout the paper. Section 3 defines the data models and query languages used in the federation components. Section 4 defines the linking mechanism and its use in OLAP-XML federations. Section 5 defines the semantics of the approach. Sections 6 and 7 describes the optimizations, and implementation and experiments, respectively. Section 8 concludes the paper and points to future work.

2 Motivation

Federating OLAP and XML As described in the introduction, this work is aimed at, but not limited to, the use of XML data from autonomous sources, such as the Internet, in conjunction with existing OLAP systems. Our solution is to make a federation which allows users to quickly define their own *logical cube view* by creating *links* between existing dimensions and XML data. This immediately permits queries that use these new "virtual" dimensions in much the same way ordinary dimensions can be used. For example, in a cube containing data about sales, a Store-City-Country dimension may be linked to a public XML document with information about cities, such as state and population. Instead of being restricted to queries that use only the existing dimensions, like "Show sales by month and city", it is now possible to pose queries such as "Show sales by month and state" or "Show sales by month and city population". Thus, in effect the cube data can be *grouped by* XML data residing e.g. on a web page or in a database with an XML interface. In addition, such data can be used to perform *selection* (also known as filtering) on the cube data, e.g. "Show only sales for cities with a population of more than 100.000" or to *decorate* dimensions, e.g. "Show sales by month and city and for each city, show also the state in which it is located".

Many types of OLAP systems may benefit from being able to logically integrate external XML data. In a business setting, consider e.g. an OLAP database containing data about products and their production prices. To aid in determining future sales prices, these products could be decorated with a competing company's prices for the same or similar products. Such prices would typically be available from the competing company's website. Although our examples mostly come from the business world, scientific applications can also benefit heavily from this type of system. Indeed, in many scientific domains, there are already a number of data sources, e.g., the SWISSPROT protein databank, which are primarily accessed over the Internet. We believe that such data sources will publish their data in XML-based formats in the future. Also, statistical database users such as census agencies also have a long tradition of using information published on the internet, e.g., demographic information, in their analyses.

This federated approach where *users* are responsible for defining the federation, has been referred to as a *loosely coupled federation* [17]. There are many reasons why this approach is a good choice for this setting. It provides the ability to do *ad hoc integration*, which may be needed for a number of reasons. First, it is rarely possible to anticipate all future data requirements when designing a database schema. OLAP databases may contain large amounts of data and thus, physically integrating the data can be a time consuming process requiring a partial or total rebuild of the cube. However, being able to quickly obtain the necessary data can sometimes be vital in making the right strategic decision. Second, not all types of data are feasible to copy and store locally even though it is available for browsing e.g. on the Internet. Copying may be disallowed because of copyright rules, or it may not be practical, e.g. because data changes too frequently. Third, attempting to anticipate a broad range of future data needs and physically integrating the data increases the complexity of the system, thereby reducing maintainability. Also, this may degrade the general performance of the system. Finally, ad hoc integration allows *rapid prototyping* of OLAP systems, which can significantly ease the task of deciding which data to physically integrate.

The federated approach also allows components to maintain the *high degree of autonomy* which is essential when data is accessed from sources outside the organisation controlling the federation, e.g., when a component is accessed on the Internet, the federation will typically have no control over the component's structure, naming conventions, access methods, availability, etc. Also, data is always *up-to-date* when using a federated system as opposed to physically integrating the data. This may be crucial for certain types of dynamic data such as price lists, stock quotes, contact information, scheduled dates etc.

Case study The case study concerns the trading of electronic components. It is inspired by the Electronic Component Information Exchange (ECIX)[3], which is a widely adopted initiative to use XML as a means of communicating information about electronic components. The setting, simplified to fit this paper, consists of companies producing electronic components (ECs), and of companies buying these components and integrating them to larger appliances. In the following we refer to them as suppliers and customers, respectively.

Customers use an OLAP database to analyze the purchases they have made over time. Purchases are characterized by an EC dimension, a supplier dimension, and a time dimension, and for each purchase the total cost and the purchased amount are measured. ECs are categorized by their manufacturers and their classes, e.g. flip-flops or latches. For suppliers, we capture the country in which they are located. Purchase dates are categorized according to the regular calendar. This database allows customers to view purchases at different levels of granularity e.g. to calculate the total amount spent on ECs by class and month. Suppliers present their products on the Web at a B2B marketplace. This allows customers and others to access detailed specifications of their ECs. This information is encoded in an industry-wide markup language defined in XML, which makes it easy to limit a search to the relevant parts of specifications. A simplified example of a document contain-

```
<?xml version="1.0" encoding="utf-8"?>
<Components>
  <Supplier SCode="SU13"><SName>John's ECs</SName>
    <Class ClassCode="C24"><ClassName>Flip-flop</ClassName>
      <Component CompCode="EC1234">
        <Manufacturer MCode="M31">
  <MName>Smith Components Inc.</MName>
</Manufacturer>
        <UnitPrice Currency="euro" NoOfUnits="1000">3.00</UnitPrice>
        <UnitPrice Currency="euro" NoOfUnits="10000">2.60</UnitPrice>
        <Description>16-bit flip-flop</Description>
      </Component>
      <Component CompCode="EC1235">
        <Manufacturer MCode="M32"><MName>John's ECs</MName></Manufacturer>
        <UnitPrice Currency="euro" NoOfUnits="1000">4.25</UnitPrice>
        <Description>16-bit flip-flop</Description>
      </Component>
    </Class>
  </Supplier>
  <Supplier SCode="SU15"><SName>Jane's ECs</SName>
    <Class ClassCode="C27"><ClassName>Latch</ClassName>
      <Component CompCode="EC2346">
        <Manufacturer MCode="M31">
  <MName>Smith Components</MName>
</Manufacturer>
        <UnitPrice Currency="euro" NoOfUnits="1000">3.31</UnitPrice>
        <Description>16-bit latch</Description>
      </Component>
    </Class>
    <Class ClassCode="C24"><ClassName>Flip-Flop</ClassName>
      <Component CompCode="EC1234">
        <Manufacturer MCode="M33">
  <MName>Johnson Components</MName>
</Manufacturer>
        <UnitPrice Currency="euro" NoOfUnits="1000">2.95</UnitPrice>
        <Description>D-type flip-flop</Description>
      </Component>
    </Class>
  </Supplier>
</Components>
```

Figure 1. The Components Document

ing information from different suppliers is shown in Figure 1. The fundamental part of an XML document is the *element*. Elements are identified by a *start tag* and an *end tag*, and can contain other elements, text data, and attributes. In the example document the `Component` element has an attribute `CompCode` and contains the elements `Manufacturer`, `UnitPrice` and `Description`. All ECs sold by a particular supplier belong to a component class. ECs are referred to by their code. In addition to this, a document captures the manufacturer, which need not be the same as the supplier, the price per unit, and a textual description.

Several aspects of ECs like textual descriptions and current prices are not included in the Purchases database because their use was not anticipated or because they change too frequently. Despite this, it may sometimes be desirable e.g. to group ECs by their marketplace descriptions, or view only purchases of ECs within a specific price range. By logically integrating the Purchases database and the Components document in a federation this can be handled in an easy and flexible way.

3   Component Models

This section describes the data models and query languages used for the federated components. For the OLAP component, a prototypical model capturing common multidimensional terms such as facts, dimensions, and hierarchies is used, and an OLAP-extended version of SQL is used as the query language. The OLAP data model captures complex multidimensional data, e.g., irregular dimen-

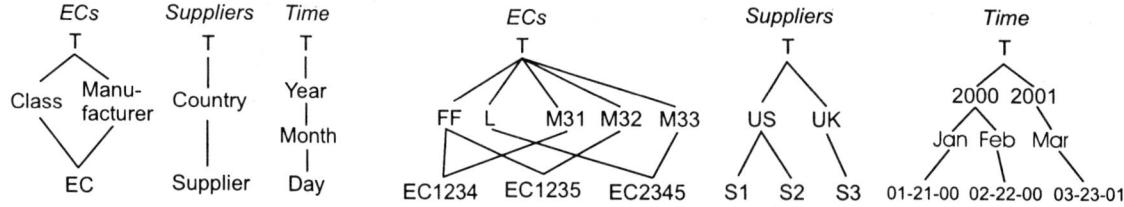

Figure 2. Schema (left) and instance (right) of the Purchases database.

a tree. Each node in the tree has one of the types: root, element, namespace, text, processing instruction, attribute, or comment. For more details about the XML data model and its use in our approach we refer to the XPath specification [19] and the full paper [14].

The basic syntax of an XPath expression resembles a Unix file path where a full path expression is given as a number of locations separated by a "/", e.g. location-step$_1$/. . ./location-step$_n$. The returned set of nodes can also be restricted by applying one or more *predicates* which supports the usual boolean, mathematical, and string operators.

Example 3.3 Select all ECs which are of the flip-flop class and are manufactured by either Johnson Components or by the manufacturer with code M33: /Components/Supplier/Class/Component[Manufacturer/MName = 'Johnson Components' OR Manufacturer/@MCode = 'M33'][../ClassName='Flip-flop']. The "../ClassName" notation finds an element named "ClassName" at any level in the document. □

For our purpose, we can abstract an XPath expression to be a function over a set of nodes:

Definition 3.1 (XPath Expression) Let S be a set of nodes in an XML document. An XPath expression is a function $XP : S \mapsto \mathcal{P}(S)$ The set of all valid XPath expressions over an XML document x is called XP_x, while the subset of XP_x that are absolute XPath expressions is called $AbsXP_x$. That is, $AbsXP_x = \{xp \in XP_x | Dom(xp) = \text{Root}(x)\}$. $RelXP_x$ is the set of expressions in XP_x that are not in $AbsXP_x$. □

4 Federating OLAP and XML

We now describe how *links* between an OLAP DB and external XML data can be used to make it easy for users to reference XML data in OLAP queries. The mechanism provides location transparency, since links can be changed without affecting existing queries. The fundamental linking mechanism is a relation between one dimension value in a cube and one node in an XML document.

Definition 4.1 (Link) A link is a relation $link_L \subseteq \{(e,s)|e \in L \wedge s \in S\}$, where L is a level and S is a set of nodes. □

The basic way of specifying a link is by *enumerated linking*, which explicitly defines the relation by providing a set of three-tuples consisting of a dimension value, the XML document in which a node is to be found, and an XPath expression identifying one or more nodes in the document. Thus, one such tuple can define a number of link tuples for a single dimension value.

Definition 4.2 (Enumerated link) An enumerated link is a function EnLink : $\mathcal{P}(L \times X \times AbsXP_x) \mapsto Links$ where L is a level, X is a set of XML documents, $AbsXP_x$ is a set of absolute XPath expressions over document $x \in X$, and $Links$ is a set of links. The resulting link relation is given by: EnLink($\{(e_1, x_1, locator_{x1}), \ldots, (e_k, x_k, locator_{xk})\}$) = $\{(e_i, s)|e_i \in \{e_1, \ldots, e_k\} \wedge s \in locator_{xi}(root(x_i))\}$, where e_i is a dimension value, x_i is an XML document, $locator_{xi}$ is an absolute XPath expression over x_i called the *locator path*. □

Example 4.1 We want to refer to the suppliers' names in the Components document when querying the Purchases database. Since the codes used for suppliers in the document are different from the ones used in the database, we have no way of identifying the links automatically. Hence we must use an enumerated link: {("S1", "www.comp-org.org/components.xml", "/Components/Supplier[@SCode='SU13']"), ("S3", "www.comp-org.org/components.xml", "/Components/ Supplier[@SCode='SU15']")}

Note that, S2 is not present in the XML document. In this case each of the tuples identify only one node in the document and the resulting link is: $Sup\_link = \{(S1, s_1), (S3, s_3)\}$, where s_1 is the single element pointed to by: "/Components/Supplier[@SCode='SU13']" in the document "www.comp-org.org/components.xml", and similarly for s_3. □

Often, names of dimension values, or a simple transformation of the names, can be found somewhere in the nodes they should be linked to. For example, when decorating countries with their populations, it is likely that the country names can be used to identify the populations. However, there may not be an exact match between the name of a dimension value and a node in the XML document. For example, dimension values may be full country names, while only abbreviated country codes, e.g., UK, are found in the XML document.

Enumerated linking is only necessary in the rather special case when names of dimension values cannot easily be mapped to nodes in the linked XML document, or the nodes occur in different documents. The former situation may e.g. be necessary if also historical population figures are present in the document and the link should only point to the most resent figure. More often *natural links* can be used as a shorthand. Here, the idea is to specify a level and a set of nodes in an XML document, and use the dimension values to identify one or more of these nodes. Optionally, an *alias function* may be supplied, mapping each dimension value to an alias which is used to identify the XML nodes. The set of nodes is defined for each level by a URI identifying the XML document and two XPath expressions. The first one identifies the nodes to which the link will point, and the second one is used to select the subset of these nodes that are linked to the given dimension value. The reason for using two XPath expressions is to facilitate the common case that a link must point to a subtree, but the subtree is identified by some lower node in the subtree. It is not necessary to use two expressions since XPath expressions allow you to move up the tree as well as down, but it makes it easier to use the links.

Definition 4.3 (Natural link) Assume a domain $Aliases$ of string values for XML nodes and an injective function $Alias : L \to Aliases$, mapping dimension values from L to strings in $Aliases$. A natural link is a function: NatLink : $LS \times X \times AbsXP_x \times RelXP_x \times Alias \mapsto Links$. The resulting link relation is given by NatLink($L, x, base, locator, alias$) = $\{(e,s)|e \in L \land s \in base(\text{Root}(x)) \land \exists s' \in locator(s)(\text{StrVal}(s') = alias(e))\}$, where L is a level, x is an XML document, $base \in AbsXP_x$ identifies the nodes, and $locator \in RelXP_x$ identifies the nodes being compared to dimension values in L. □

If no *alias* function is necessary, it may be omitted, i.e. the *identity* function is assumed.

Example 4.2 If we want to create a link between the ECs in the Purchases database and those in the Components document we can make a natural link, since the same codes are used in both places.

From the natural link: ("EC", "www.comp-org.org/components.xml", "/Components/Supplier/Class/Component", "@CompCode", i_{EC}), where i_{EC} is the identity function,

we create the link EC_Link = $\{(EC1234, s_1), (EC1234, s_2), (EC1235, s_3)\}$. s_1 is the first element in the Components document with CompCode="EC1234", s_2 is the second element with CompCode="EC1234", and s_3 is the single element with CompCode="EC1235". □

A flexible linking mechanism must allow both dimension values and nodes to occur more than once in the same link. The *cardinality* of a link $link$ between a level L and an XML document x can be either [1-1], [n-1], [1-n], or [n-n]. A link is [1-1] if $\|link\| = \|\pi_L(link)\| = \|\pi_x(link)\|$, where π denotes relational projection and $\|R\|$ denotes the cardinality of relation R. Similarly, the cardinality of $link$ is [n-1] if $\|link\| = \|\pi_L(link)\| > \|\pi_x(link)\|$, [1-n] if $\|link\| = \|\pi_x(link)\| > \|\pi_L(link)\|$, and [n-n] if $\|link\| > \|\pi_x(link)\|$ and $\|link\| > \|\pi_L(link)\|$. We use the abbreviations [-1] to denote [1-1] or [n-1] and [-n] to denote [1-n] or [n-n]. Note that these cardinalities are not specified in any way, but are merely properties of the links.

Example 4.3 Sup_Link is [1-1] and EC_Link is [1-n]. □

To allow references to XML data in OLAP queries, links are used to define *level expressions*. A level expression consists of a starting level L, a link $link$ from L to nodes in one or more XML documents, and a relative XPath expression xp which is applied to these nodes to identify new nodes.

Definition 4.4 (Level expression) A level expression of the form $L/link/xp$, where L is a level, xp is an XPath expression, and $link$ is a link from L, defines a link $E = \{(e,s)|e \in L \land \exists s'((e,s') \in link \land s \in xp(s'))\}$. The cardinality of a level expression is the link cardinality of E. Also, we say that a level expression *covers* its starting level if $L = \pi_L(E)$. If the starting level is not covered some facts may not be linked to any nodes. We will refer to such tuples as *unconnected* facts. To simplify link usage we assume a function DefaultLink : $L \mapsto Links$, where L is a set of levels and $Links$ is a set of links. The function returns the default link for a given level. □

Example 4.4 The level expression "EC/EC_Link/CompCode" is [1-n] and does not cover its starting level, since "EC2345" is not mentioned in the Components document. □

Assuming that DefaultLink(EC) returns "EC_Link" the above level expression can be written "EC/CompCode". In the following we assume that EC_Link and Sup_Link are default links for the EC and $Supplier$ levels, respectively.

With the linking mechanism in place, we can now define the federated data model consisting of a cube, a set of XML documents, and a set of links between them. We only consider one cube since multiple cubes can be handled by creating a view over the cubes.

sion hierarchies. We present just an overview of the model, mainly through the use of examples, the formal definition can be found in another paper [12]. For the XML component, the XPath data model and query language [19] is used, mainly because of its simplicity and wide-spread use.

The OLAP Data Model and Query Language The model is defined in terms of a multidimensional *cube* consisting of a *cube name*, *dimensions*, and a *fact table*. Each dimension comprises two partially ordered sets (posets). The first poset represent hierarchies of the *levels* which specify the possible levels of detail of the data. Each level is associated with a set of *dimension values*. The second poset represent the ordering of the dimension values, i.e., which values roll up to one another. Dimensions are used to capture the possible ways of grouping data. A *fact table* F is a relation containing one attribute for each dimension, and one attribute for each measure. Thus, An *n-dimensional cube* is a three-tuple consisting of a cube name, a non-empty set of dimensions, and a fact table. The *cube name* describes the type of facts contained in the cube.

Example 3.1 In the case study, we have a Time dimension, an ECs dimension and a Suppliers dimension. The Suppliers dimension consists of the levels Supplier, Country, and $\top_{Sup}$, which denotes *all* of the Supplier dimension. The ordering of the levels and the dimension values can be seen in Figure 2. We have the two measures *Cost* and *Number of Units*. A part of the fact table is represented in Table 1. To save space, only tuples with non-NULL measure values are shown although all combinations are logically present in the relation. This is done throughout the paper. From these parts, we can construct a three-dimensional cube with the cube name *Purchases*, the dimensions, levels, and ordering of dimension values as depicted in Figure 2, and the fact table from Table 1. "FF" and "L" are names of classes denoting "Flip-flops" and "Latches", respectively. □

Cost	No. Of Units	Day	Supplier	EC
2940	1000	01.21.2000	S1	EC1234
6900	2000	01.21.2000	S3	EC1234
9480	3000	02.22.2000	S3	EC2345
14400	4000	02.22.2000	S2	EC1235
17650	5000	03.23.2001	S2	EC1235

Table 1. Fact Table For Purchases Database

Next, we discuss the notion of *summarizability* and discuss how it is used to ensure *correct aggregation* when "rolling up" data from lower to higher levels of granularity. Summarizability is an important cube property as it states when lower-level aggregates, which are often pre-computed, can be used to calculate higher-level aggregates, and when they must be computed from base data. Also, it is

possible to get wrong results from aggregate queries if summarizability is not ensured. Checking for summarizability is even more important in this paper's setting than in normal OLAP systems, as the irregular structure of XML data often will violate the summarizability property. It has been shown that summarizability is equivalent to requiring the aggregate function to be distributive, and the ordering of dimension values to be *strict*, *onto*, and *covering* [11]. A hierarchy is *strict* if no dimension value has more than one parent value from the *same* level, *onto* if all paths from top value to leaf value is of equal length, and *covering* if no path skips one or more levels. Intuitively, this often means that dimension hierarchies must be balanced trees. If this is not the case some lower-level values will be either double-counted or not counted at all. For example, the Purchases cube in Figure 2 is strict, onto, and covering (note that strictness is a property of the data, not the schema and that the two parents of, e.g., EC1234, are from different levels, meaning that the EC dimension is strict). We keep track of what data can be aggregated by using so-called *aggregation types* [12].

A formal algebra has been defined over the OLAP data model presented above. Two operators are defined: a selection operator $\sigma_{Cube}[p]$ for selecting only the desired facts based on the predicate p, and a generalized projection operator, Π_{Cube} for aggregating fact data to the desired level of detail, possibly projecting out un-desired dimensions. The formal definitions can be found in [12].

As the formal algebra is not suitable for end users, we have also defined a SQL-like query language in terms of the algebra. We use a slight extension of a subset of SQL, called "Multidimensional SQL" (abbreviated SQL_M), to query multidimensional cubes. SQL is chosen as the base language for its simplicity and wide-spread use. We illustrate the considered syntax with examples. The complete specification is given in [12].

Example 3.2 Calculate costs by class and supplier for suppliers located in UK where total cost exceeds 10000:

SELECT	SUM(Cost), Supplier, Class(EC)
FROM	Purchases
WHERE	Country(Supplier) = 'UK'
GROUP BY	Supplier, Class(EC)
HAVING	SUM(Cost) > 10000

□

The XML Data Model and Query Language The XPath language is used to refer to parts of XML documents. Although not a full blown query language, this language is sufficiently powerful for our purpose. XPath is also chosen because it has a compact syntax making it suitable for integration into another language. The XML data model underlying the XPath language views an XML document as

Definition 4.5 (Federation) A federation $\mathcal{F}$ of a cube C and a set of XML documents X is a three-tuple: $\mathcal{F} = (C, Links, X)$ where $Links$ is a set of links between levels in C and documents in X. □

When it is clear from the context, we will refer to $\mathcal{F}$ as a cube, meaning the cube part of the federation $\mathcal{F}$.

Example 4.5 The collection of the cube in Example 3.1, the Components document, and the two links Sup_Link and EC_Link is a federation. We will refer to this as the Purchases federation in the following. □

5 Querying Federations

We now present an algebra over federations. The semantics are given by extending the cube algebra to federations, providing a decoration operator, a generalized projection operator, and a selection operator. The algebra is closed, as all operators work on a federation and also return a federation.

Decoration It is often useful to provide supplementary information for one or more levels in the result of an OLAP query. This is commonly referred to as *decorating* the result [7]. For example, products could be decorated with a competitor's prices for the same products, employees with their addresses, or suppliers with their contact person. Such information will often be available to the relevant people as Web pages on the Internet, an intranet, or an extranet. Also, this kind of information will most likely not be stored in an OLAP database because it either changes too frequently, was not expected to be used, is owned by someone else, or for some other reason. The solution suggested in this paper is to allow OLAP queries to reference external XML data using level expressions in the SELECT clause. In Section 5 we consider how to use level expressions in the GROUP BY clause.

Example 5.1 Let "AllTimePurchases" be the aggregation of the Purchases cube to the EC and Supplier levels. The fact table of this cube is shown in Table 3(a). Given the federation consisting of the "AllTimePurchases" cube, the Components document, and the links defined above, the following query decorates all ECs with their descriptions from the Components document:

```
SELECT    SUM(Cost), Supplier, EC, EC/Description
FROM      AllTimePurchases
```
□

There are two important problems with the use of level expressions for decoration which are related to the problems with non-strict and non-covering hierarchies as discussed earlier. First, a dimension value may be associated with more than one node, i.e. when the level expression has cardinality [-n] resulting in non-strictness in the new "decoration" dimension. Second, some dimension values may not be associated with any nodes at all, which is the case if the level expression does not cover its starting level. The first problem allows for a number of different decoration semantics. Consider the following example:

Example 5.2 From the query in Example 5.1 we could get the result shown in Table 3(b), where a fact is created for each different description node resulting from the level expression. Another possibility is the result shown in Table 3(c), where an arbitrary node is picked and at most one fact is created for each EC. A third possibility is shown in Table 3(d), where all description nodes are concatenated. In all cases we use a special "N/A" (Not Available) value to indicate that no description is found for an EC. Note that EC1234 gets several description decoration values as the decoration expression EC/Description is only dependent on the EC dimension. □

Cost	Supplier	EC
2940	S1	EC1234
6900	S3	EC1234
32050	S2	EC1235
9480	S3	EC2345

(a)

Cost	Supplier	EC	Description
2940	S1	EC1234	D-type flip-flop
2940	S1	EC1234	16-bit flip-flop
6900	S3	EC1234	D-type flip-flop
6900	S3	EC1234	16-bit flip-flop
32050	S2	EC1235	16-bit flip-flop
9480	S3	EC2345	N/A

(b)

Cost	Supplier	EC	Description
2940	S1	EC1234	D-type flip-flop
6900	S3	EC1234	D-type flip-flop
32050	S2	EC1235	16-bit flip-flop
9480	S3	EC2345	N/A

(c)

Cost	Supplier	EC	Description
2940	S1	EC1234	D-type flip-flop, 16-bit flip-flop
6900	S3	EC1234	D-type flip-flop, 16-bit flip-flop
32050	S2	EC1235	16-bit flip-flop
9480	S3	EC2345	N/A

(d)

Figure 3. Fact table And Decorations

The problem is which of the nodes to use for decoration when a level expression returns more than one node. Several solutions are possible including picking one arbitrarily,

201

using the first one, concatenating all different nodes, or using all the nodes thereby creating duplicated facts. Note that, concatenating the nodes from an XML document is always possible since all nodes have a string value, though the concatenated string may not make sense to a user. Duplicating facts means that further aggregation may give an incorrect result. This is the case when grouping over decoration values, whereas grouping over other values produces a correct result.

The second problem with the use of level expressions for decoration is how to handle expressions that do not cover its starting level. The solution used in Example 5.2 is to add a special N/A value, indicating that no nodes are available. Alternatives are to remove the facts that are not linked to any nodes or to require the level expression to cover its starting level. Removing the unconnected facts would lead to a non-summarizable result, whereas requiring all values in the starting level to be covered would reduce the practical usefulness of decoration. Thus, we propose to add a special value for all unconnected facts.

Since different semantics are needed in different situations, we allow the user to choose between different types of semantics when decorating a cube with XML data. We have chosen the following because we believe they can all be useful under different circumstances:

ANY: Use an arbitrary node. This is useful when summarizability should be preserved and no node is more important than another, as might be the case e.g. when decorating suppliers with a contact person.

CONCAT: Use the concatenation of string values for all different nodes. Useful when summarizability should be preserved and all nodes are needed, e.g. when decorating products with text descriptions.

ALL: Use all different nodes, possibly duplicating facts. Useful when the decoration is used for grouping or selection.

The user specifies the semantics of a decoration by giving a ANY, CONCAT, or ALL semantic modifier in the level expression, e.g., EC[ANY]/EC_Link/Description. If no semantic modifier is specified ANY semantics are assumed as the default. Notice that, if the cardinality of the level expression is [-1] then decoration with the three semantics will produce the same result. The query in Example 5.1 actually results in Table 3(c) since ANY is the default.

We decorate a cube by adding a new dimension containing only the top level and a level containing all the decoration values. Different approaches could be to attach the decoration data as special attributes of the decorated values, create a new level in the same dimension as the starting level, or to keep the decorated data in an external component [15]. Our approach has the advantage that the external data

can easily be used for aggregation and selection because the decoration data is incorporated into the cube. Also, aggregation is still possible in the original dimensions, as these are not changed by decoration.

Intuitively, only two things are changed when decorating a cube: A new dimension is added and the fact table is updated to reflect this. The new dimension contains only the decoration level and the top level. The new dimension values in the decoration level are created from an arbitrarily chosen node found by following the link from the starting level and then applying the XPath expression. If one or more values in the starting level does not produce any decoration values the special N/A value is used instead. The new fact table is created from the Cartesian product of the dimension values from the old fact table and the new decoration values. Measure values are replaced with NULL values such that no facts are duplicated.

Extending Grouping to Federations Allowing level expressions in the GROUP BY clause makes it possible to group by data from XML documents, without having to physically store this data in the OLAP database. For example, product prices will often be available from a supplier's Web page or an e-marketplace. These up-to-date prices can then be used to group products in an OLAP product database without having to store the prices.

Example 5.3 The following query groups ECs after their text descriptions from the Components document.

SELECT	SUM(Cost), EC[ALL]/Description
FROM	Purchases
GROUP BY	EC[ALL]/Description

GROUP BY queries with level expressions are (logically) evaluated in two steps. First, the cube is decorated as described in the previous section. Second, aggregation is performed by using the already defined generalized projection Π_{Cube} on the new cube.

Example 5.4 The above query is evaluated by first decorating the Purchases cube resulting in the fact table shown in Table 3(b), and then grouping by "Description" using Π_{Cube}. □

When decorating the cube, the new decoration dimension may be non-strict if ALL semantics are used and a bottom value is decorated by more than one decoration value. This is the reason for allowing non-strictness in a cube and for handling it in the generalized projection operator. Consequently, if non-strictness occurs because of the decoration and if aggregation results in duplicate facts, this is handled by setting the aggregation type to c, preventing further aggregation. Formally, the generalized projection operator over federations is defined as follows:

Definition 5.1 (Generalized projection over federations)
Let $\mathcal{F} = (C, Links, X)$ be a federation and $M_{j_1}, \ldots, M_{j_m}$ be measures in C. Also let $L_1, \ldots, L_k$ be levels in C such that at most one level from each dimension occurs. The generalized projection operator Π_{Fed} over federation $\mathcal{F}$ is then defined as: $\Pi_{Fed[L_1, \ldots, L_k] < f_{j_1}(M_{j_1}), \ldots, f_{j_m}(M_{j_m})>}(\mathcal{F}) = (C', Links', X')$ where the new cube is $C' = \Pi_{Cube[L_1, \ldots, L_k] < f_{j_1}(M_{j_1}), \ldots, f_{j_m}(M_{j_m})>}(C)$.

Links for which the starting level no longer exists are removed from the resulting federation. That is: $Links' = \{link \in Links | \exists L \in C'(\exists (e, s) \in link(e \in L))\}$. XML documents to which no links refer are also removed: $X' = \{x \in X | \exists link \in Links'(\exists (e, s) \in link(s \in \mathrm{Nodes}(x)))\}$. □

Extending Selection to Federations XML data can also be used to perform selection over cubes. This makes it possible e.g. to view only products where a certain supplier is cheaper than another supplier by referring to their Web pages. The idea adopted here is to allow level expressions in WHERE and HAVING predicates in places where levels can already be used. For example, level expressions can be compared to constants, levels, measures, or other level expressions.

Example 5.5 Show component costs by supplier and EC but only those available for less than 3.00 euro.

SELECT	SUM(Cost), Supplier, EC
FROM	Purchases
WHERE	EC/UnitPrice[@Currency='euro'] < 3.00
GROUP BY	Supplier, EC

□

As discussed in Section 3 selection semantics are also affected by the cardinality and covering properties of level expressions. As for selection over cubes, we handle this by using *any* semantics.

Selection over federations is (logically) evaluated by first decorating with all the level expressions mentioned in the predicate. The resulting federation is then sliced using the selection operator, and finally, the new decoration dimensions are removed again. The selection operator simply applies the cube selection operator to the cube part of the federation since the link and XML parts should not be affected by selection. ALL semantics are used for the decorations to make sure that all decoration values are available. This is important since *any* selection semantics are used in predicates, and thus, a predicate may be satisfied by any of the decoration values. No facts are duplicated since the ALL decoration is never actually rolled up to the decoration level. The roll-up is handled entirely by the cube selection operator.

Example 5.6 Show only components that are manufactured by the supplier.

SELECT	SUM(Cost), Supplier, EC
FROM	Purchases
WHERE	EC/Manufacturer/MName =
	EC/../../Suppliers/SName
GROUP BY	Supplier, EC

This query is evaluated by first decorating with the two level expressions EC/Manufacturer/MName and EC/../../Suppliers/SName using the ALL semantics. This results in two new columns EC' and EC'' in the fact table both duplicating the EC level. A new predicate is then constructed rolling up to the decoration level: "Manufacturer/MName"(EC') = "../../Suppliers/SName"(EC''), and this is used to select a part of the fact table. Finally, the two new columns are removed again. □

Formally, selection over federations is defined as follows:

Definition 5.2 (Selection over federations) Let $\mathcal{F} = (C, Links, X)$ be a federation, where C has dimensions $D_1, \ldots, D_n$ and measures $M_1, \ldots, M_l$. The selection operator over federations is then defined as: $\sigma_{Fed[p]}(\mathcal{F}) = (C', Links', X')$, where $Links' = Links$, $X' = X$, and the new cube is $C' = \sigma_{Cube[p]}(C)$. □

Hence, selection is performed on a federation by applying cube selection to the cube part using a predicate without the level expressions. The decoration and predicate transformation are not handled by the selection operation but instead by the mapping from SQL_{XM} to the federation algebra as described in the next section.

The formal semantics of a SQL_{XM} query is given in the full paper [14].

An SQL_{XM} query over a federation is (logically) evaluated in four major steps. First, the cube is sliced as specified in the WHERE clause, possibly requiring a decoration with XML data which is then projected away after selection. Second, the resulting cube is decorated with external XML data from the level expressions occurring in the SELECT and GROUP BY clauses. This creates a number of new dimensions in the cube. Third, all dimensions, including the new ones, are rolled up to the levels specified in the GROUP BY clause. Finally, the resulting cube is sliced according to the predicate given in the HAVING clause, which may also require a decoration. Notice that the new decoration dimensions used for selection are not mentioned in the following generalized projection and are therefore removed after use. Since the new dimensions are never aggregated up to the decoration level, no changes are made to the aggregation types.

6 Query Optimization

A naive query processing strategy will process SQL_{XM} queries in three major steps. First, any XML data referenced in the query is fetched and stored in a temporary database as relational tables. Second, a pure OLAP query is constructed from the SQL_{XM} query and evaluated on the OLAP data, resulting in a new table in the temporary database. Finally, these temporary tables are joined, and the XML-specific part of the SQL_{XM} query is evaluated on the resulting table. This strategy will only perform satisfactorily for rather small databases. The primary problems are that decoration operations require large parts of the OLAP and XML data to be transferred to temporary storage before decoration can take place, i.e., the primary bottleneck in the federation will most often be the moving of data from OLAP and XML components. Thus, our optimization efforts have focused on this issue. These efforts include both *rule based* and *cost based* optimization techniques.

The *rule based* optimization uses the heuristic of pushing as much of the query evaluation towards the components as possible. Although not always valid for more general database systems, this heuristic is always valid in our case since the considered operations all reduce the size of the result. The rule based optimization algorithm *partitions* a SQL_{XM} query tree, meaning that the SQL_{XM} operators are grouped into an OLAP part, an XML part, and a relational part. After partitioning the query tree, it has been identified to which levels the OLAP component can aggregate data and which selections can be performed in the OLAP component. Furthermore, the partitioned query tree has a structure that makes it easy to create component queries. The partitioning is based on *transformation rules* for the federation algebra. These include a set of novel rules for moving the decoration operator around the query tree. The rules not involving decoration are close to the rules for Multi-Set Extended Relational Algebra [8].

Three different *cost based* optimization techniques are employed. The use of cost based optimization requires a complete cost model and the estimation of several cost parameters which is described in another paper [13].

The first technique tries to tackle one of the fundamental problems with the idea of evaluating part of the query in a temporary component: If selections refer to data that are not present in the result, much more data than the result needs to be transferred to the temporary component. The proposed solution to this problem is to *inline* literal XML data values into OLAP predicates, i.e., to have the predicates refer to a list of literal constants rather than a XPath expression. However, it is not always a good idea to do so because, in general, a single query cannot be of arbitrary length. Hence, more than one query may have to be used. Whether or not XML data should be inlined into some OLAP query, is decided by comparing the estimated cost of inlining with the estimated cost of not doing so.

The second technique is focused on the special kind of XML queries that are used in the federation. These queries

can easily be expressed in more powerful languages like XQuery, but many XML sources have more limited interfaces, such as XPath or XQL. The special queries needed to retrieve data from XML components cannot be expressed in a single query in these simple languages, and hence, special techniques must be used for this to be practical. The main solution suggested here is to combine these queries, even though more data would have to be retrieved. Again, a cost analysis is used to decide whether or not to employ this technique. In summary, three different strategies are used when evaluating a set of XPath expressions resulting from a level expression: combining none, some, or all of the expressions. For each of these three strategies, the total evaluation cost is estimated and the cheapest one is used.

The third technique is an application of *caching* to this particular domain. The use of caching is important for both OLAP and XML components, as both types of components may cause significant delays for certain kinds of queries. One of the approaches is an efficient way to find a useful cached result for a given OLAP query. *Pre-fetching* is also employed for speeding up queries that have not been posed before. In summary, we perform caching and pre-fetching for component queries only. Intermediate OLAP results stored in temporary tables as well as raw XML data are kept for a certain amount of time, which can specified as a tuning parameter. If adequate storage is available, temporary XML tables are stored to avoid constructing the same tables again. Currently, we do not cache or pre-fetch entire federation queries as the cache space they take up is in most cases too high in comparison with the probability that they can be re-used.

7 Implementation and Experiments

The overall architecture of the prototype is shown in Figure 4. The key component is the *Federation Manager*, which processes SQL_{XM} queries fed to it by the *user interface* by fetching data from the *OLAP* and *XML* components. Intermediate *source components* are inserted between the Federation Manager and the OLAP and XML components to make the Federation Manager independent of the query languages used by these components. The Federation Manager uses three auxiliary components to store meta data, link data, and temporary data used in the evaluation of a SQL_{XM} query. The *meta data component* contains descriptions of the dimensions in the OLAP component, whereas the *link data component* contains link specifications as described in Section 4. The *temporary data component* is used for storing intermediate results during the processing of a query. In the prototype, the OLAP component is based on Microsoft Analysis Services and queried with the MDX language. The XML component is based on Software AG's Tamino XML Database system, which provides an XPath interface. A single Oracle 8*i* system is used for all three auxiliary components.

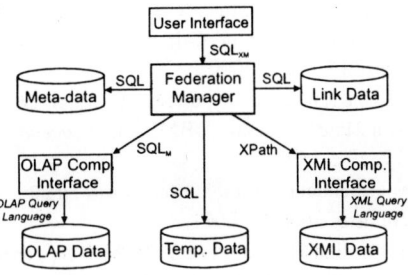

Figure 4. Prototype Architecture

We have performed a set of experiments to evaluate the performance of our federation approach and the effectiveness of the optimization techniques. Here, we only summarize the results, the details can be found in the full paper [14]. The experiments were performed on a 900 Mhz Pentium machine with 512MB of RAM and 20GB of disk. The cube is based on about 50 MB of data generated using the TPC-H benchmark and about 10 MB of pre-aggregated results. The cache size was limited to 10 MB to prevent an unrealistically large part of the data from being cached. About 3 MB of XML data is used for the XML component which is divided into two documents that have been generated from the TPC-H data and public data about nations. Two natural links, NLink and Tlink, are defined to from Nation level in the cube to the Nation elements in the XML data, and from the Type level in the cube to the Type elements in the XML data, respectively.

For pure decoration queries, we found that the overhead of performing decoration was low compared to the time it takes to retrieve the OLAP and XML data. Thus, if it is acceptable to wait for the component data when retrieved independently, it will also be acceptable to wait for the federation query. This low overhead is representative for decoration queries since they do not require additional data to be retrieved from the OLAP component. Hence, the overhead of combining the intermediate results will be low, since typically only small amounts of OLAP data (at most a few thousand facts) will be requested for presentation to a user.

For grouping and selection queries, we found that the overhead caused by the temporary component query is also low for these queries. This is typical for both grouping and certain types of selection because the size of the intermediate OLAP result will often be comparable to the size of the final result. Again, since the final result is mostly presented to a user, it is often relatively small. For grouping, the OLAP and final results are comparable in size unless there is a great overlap in the decoration values which reduces the size of the final result. For selections, the OLAP result is often comparable in size to the final result for queries

where the predicate refers to decorations of levels that must be present in the result. This is true unless the predicate is very selective.

The experiments with the inlining technique compared three alternatives: no inlining, a simple inlining strategy, and cost-based inlining. Where no inlining is used, there is a very large overhead as the OLAP query can only aggregate to lower levels. As a consequence, not only the OLAP query but also the temporary component query take much longer to evaluate. The use of the simple inlining strategy, "Always use inlining if the predicate is simple" is significantly faster because the OLAP query can now aggregate to higher levels. Also, since the WHERE clause has been evaluated entirely in the cube, no work needs to be done after the OLAP result has been returned. However, six OLAP queries are needed to hold the new predicate because the higher level contains a large number of dimension values. Also, the OLAP query cannot be evaluated until all XML data has been retrieved. Consequently, it sometimes faster to inline only some predicate data, i.e., as determined by the cost based inlining strategy. This required only one OLAP query. Thus, by estimating the cost of the federation query for each of the four inlining strategies and picking the fastest one, a better query performance is achieved. In summary, the simple inlining strategy improved performance by a factor of 4.5 over no inlining, while the cost-based strategy was 3 times faster than simple inlining and 14 times faster than no inlining.

The experiments with caching/prefetching showed a performance improvement by factors of 4–25 over no caching or prefetching, depending on the query type, with decoration and grouping queries gaining the most improvement.

We have also performed experiments that compare our approach to the performance-wise ideal situation where the external data is *physically integrated* in the OLAP cube. These experiments were performed with 1GB of TPC-H data in the OLAP cube, plus 100MB used for pre-computed aggregates. In the XML component, we had 10 MB of XML data. The results of these experiments are seen in Figure 5. The "O" bars show the time spent in the OLAP component, while the "T" bars show the time spent in the temporary component. We compared three typical queries that performed decoration (1), selection (2), and grouping (3) using XML data, respectively. The experiments show that for typical queries the overhead of our approach was only 25–50% compared to physical integration. To conclude, the optimizations discussed above suggests that an SQL_{XM} query can in most cases be evaluated with a level of efficiency comparable to that of physical integration, while avoiding the problems related to physical integration in dynamic environments.

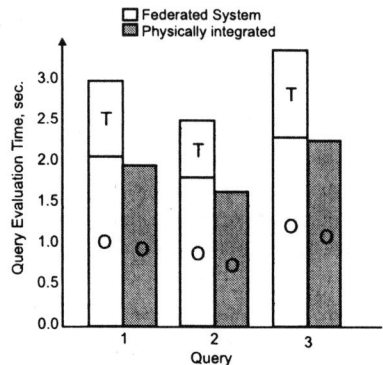

Figure 5. Performance Comparison

8 Conclusion and Future Work

Many OLAP systems operate in highly dynamic environments where changes in data requirements are common and data changes frequently, meaning that logical OLAP federations are far more feasible than physically integrated databases. As external data will most often be available in XML format, it should be possible to integrate OLAP and XML data seamlessly.

In this paper we presented an approach for the federation of OLAP and XML data, given in terms of a formal data model and algebraic query language. To demonstrate our approach we introduced a federated query language, SQL_{XM}, incorporating the XML query language XPath into a subset of SQL adapted to multidimensional querying. SQL_{XM} allows XML data to be used directly in an OLAP query to *decorate* multidimensional cubes with external XML data, and to *group* and *select* cube data based on XML data values. The incorporation of XML data in cubes was made such that semantic problems were avoided, e.g., when aggregation was performed on the resulting cube no double-counting of data could occur. A number of effective optimization techniques for OLAP-XML federations were described. Finally, a prototype implementation and a set of experiments stating the performance of the approach and the effectiveness of the optimization techniques were described, showing the attractiveness of the approach compared to physical integration.

We believe this paper to be the first to consider the integration of OLAP and XML data, including advanced issues such as dimension hierarchies and correct aggregation of data. Also, we believe to be the first to consider query processing and optimization for this setting.

In future work, interesting research issues include how to capture the document order of an XML document in the result of an OLAP query, how to incorporate new XML-based measures into a cube, and how extra structural information about the XML data, such as DTDs and XML Schemas, can be utilized. Also, other query languages than SQL and XPath could be considered for the federation.

References

[1] V. Christophides, S. Cluet, J. Simeon. On Wrapping Query Languages and Efficient XML Integration. In *Proc. of SIGMOD*, pp. 141–152, 2000.

[2] W. Du, R. Krishnamurthy, and M.-C. Shan. Query Optimization in a Heterogeneous DBMS. In *Proceedings of VLDB*, pp. 277–291, 1992.

[3] ECIX Quickdata Architecture. `www.si2.org/ecix/`. Current as of June 15, 2001.

[4] Garcia-Molina H. et al. The TSIMMIS Approach to Mediation: Data Models and Languages. *JIIS* 8(2), pp. 117–132, 1997.

[5] F. Gingras and L. V. S. Lakshmanan. nD-SQL: A Multi-Dimensional Language For Interoperability and OLAP. In *Proc. of VLDB*, pp. 134–145, 1998.

[6] R. Goldman and J. Widom. WSQ/DSQ: A Practical Approach for Combined Querying of Databases and the Web. In *Proc. of SIGMOD*, pp. 285–2.596, 2000.

[7] J. Gray et al. Data Cube: A Relational Aggregation Operator Generalizing Group-By, Cross-Tab, and Sub-Total. In *Proc. of ICDE*, pp. 152–159, 1996.

[8] P. W. P. J. Grefen and R. A. de By. A Multi-Set Extended Relational Algebra - A Formal Approach To A Practical Issue. In *Proceedings of ICDE*, pp. 80–88, 1994.

[9] L. M. Haas, D. Kossmann, E. L. Wimmers, and J. Yang. Optimizing Queries Across Diverse Data Sources. In *Proceedings of VLDB*, pp. 276–285, 1997.

[10] Hellerstein, J. M. et al. Independent, Open Enterprise Data Integration. *Data Engineering Bulletin*, 22(1):43–49, 1999.

[11] H-J. Lenz and A. Shoshani. Summarizability in OLAP and Statistical Databases. In *Proc. of SSDBM*, pp. 39–48, 1997.

[12] D. Pedersen, K. Riis, and T. B. Pedersen. A Powerful and SQL-Compatible Data Model and Query Language for OLAP. In *Proc. of ADC*, pp. 121–130, 2002.

[13] D. Pedersen, K. Riis, and T. B. Pedersen. Cost Modeling and Estimation for OLAP-XML Federations. In *Proc. of DawaK*, 10 pages, to appear, 2002.

[14] D. Pedersen, K. Riis, and T. B. Pedersen. XML-Extended OLAP Querying. *Technical Report TR 02-5001, Department of Computer Science, Aalborg University*, 20 pages, 2002.

[15] T. B. Pedersen et al. Extending OLAP Querying to External Object Databases. In *Proc. of CIKM*, pp. 405–413, 2000.

[16] Roth, M. T. et al. The Garlic Project. In Proceedings of *SIGMOD*, pp. 557, 1996.

[17] A. P. Sheth and J. A. Larson. Federated Database Systems for Managing Distributed, Heterogeneous, and Autonomous Databases. *ACM Computing Surveys*, 22(3):183–236, 1990.

[18] E. Thomsen. *OLAP Solutions: Building Multidimensional Information Systems*. John Wiley & Sons, 1997.

[19] W3C. Xml path language (xpath) version 1.0. `www.w3.org/TR/xpath`. Current as of June 15, 2001.

[20] W3C. Xml schema part 0: Primer. `www.w3.org/TR/xmlschema-0/`. Current as of June 15, 2001.

[21] W3C. XQuery 1.0: An XML Query Language. `www.w3.org/TR/xquery`. Current as of June 15, 2001.

A Negotiation Agent for Distributed Heterogeneous Statistical Databases

Sally McClean, Rónán Páircéir, Bryan Scotney, and Kieran Greer

School of Information and Software Engineering, University of Ulster

{si.mcclean, rpairceir, bw.scotney, krc.greer}@ulst.ac.uk

Abstract

The World-wide Web provides an ever-increasing source of diverse information. Agent Technology provides an intelligent and flexible mechanism for querying and integrating large amounts of statistical data that are distributed among different computing systems on various sites on the Internet.

We focus on query agents, in particular the matching and negotiation agents that are responsible for pre-integration where the matching agent decomposes the query into sub-queries, and then searches metadata to find datasets that match the query fragments. In the case of heterogeneous data, the matching agent utilises a negotiation agent to find datasets that match the query fragments, provides mappings from the data to the query, and constructs the appropriate (sub-)query re-writing rules. Such matching is done by generalising the data and testing if the (sub) query is matchable to the generalised (meta) data: we call this g-matchable; if it is then we can construct an operator stack to transform the data to match the (sub) query.

Such an approach provides a capability of automating the process of executing queries on heterogeneous statistical databases that are distributed over the Internet. The novelty lies in the provision of automated methods for statistical aggregates, where the heterogeneity essentially resides in the classification schemes of categorical data, including both heterogeneity of nomenclature and heterogeneity of granularity. In addition, our solution permits queries to be specified in a goal-driven query-by-example format. Rather than impose an a priori global standard, the user can query through a unified interface where integration is done at run-time.

1. Introduction

The World-wide Web provides an ever-increasing source of diverse information. Software agents have been developed by the Artificial Intelligence community with the aim of providing flexible and pro-active tools for users to access and utilise such information. In particular, Agent Technology provides an intelligent and flexible mechanism for utilising Database tools to manipulate and integrate large amounts of data, including data that are distributed among different computing systems on various sites.

We have previously provided functionality for the management and integration of distributed statistical aggregates. Such functionality supports automated metadata-guided statistical processing; in the conventional approach each operation must be specified explicitly, along with appropriate parameters. Now, by accompanying the aggregate tables with appropriate metadata, we allow the system to hide many of the details of internal data processing from the user.

Our current agent-based solution permits queries to be specified in a goal-driven query-by-example format. Such matching is done by testing if the (sub) query can be answered by generalising the data; we call this g-matching. If such matching is possible then we can construct an operator stack to transform the data to match the (sub) query. Rather than impose an a priori global standard, the user queries through a unified interface and integration is done at run-time. Intelligent Software Agents allow us to adopt intelligent and flexible solutions at various stages of the query process. Such an approach allows us to provide a modular system of software that enables providers of official statistics to publish their data in a unified framework, and allows consumers of statistics to access these data in an informed manner with minimum effort. The vision is of a number of independent organisations publishing their data within a framework that makes comparison and harmonisation possible. The system thus allows suppliers of statistics to subscribe to an integrated network of datastores, while retaining control over access to their own data.

In this paper our focus is on the query agents, in particular the matching and negotiation agents, where the matching agent decomposes the query into sub-queries and then searches metadata to find a dataset that matches the query fragments. In the case of heterogeneous data, the matching agent utilises a negotiation agent to find datasets that match the query fragments, provides mappings from the data to the query, and constructs the appropriate (sub-) query re-writing rules. Such an approach provides a capability of automating the process of executing queries on heterogeneous statistical databases that are distributed over the Internet. The novelty lies in the provision of methods for such problems, which are commonly occurring for statistical data, where the heterogeneity essentially resides in the classification schemes of categorical data, including both heterogeneity of nomenclature and heterogeneity of granularity.

2. Background

Classifications are the foundations on which all statistical systems, national as well as international, are built. However, the existence of a variety of classifications as well as their revisions raises problems of compatibility and comparability of data collected and disseminated in a distributed environment. Having such independently developed classification schemes raises a number of requirements such as:

- The user might pose a query and require the results to use a specific classification and its nomenclature; we call this the *ontology*. This may be a local classification or an internationally recognised classification.

- The system requires to find matches to the sub-queries, by searching through metadata from each local data site stored in *Libraries* (metadata repositories). We here achieve this functionality via a matching agent.

- Once a set of matches is found, the query plan is constructed. This must include mappings generated by the negotiation agent so that the query fragments can be re-written in the local data site's ontology. In addition, the query result must be translated back into the ontology of the user.

Aspects of such problems have been discussed in the literature of other related fields, principally:

- Ontology research in computer science - both in the context of the Internet and also, increasingly, for distributed heterogeneous databases.

- Schema matching, primarily in the database field, where the mappings between heterogeneous schema are learned.

- Information brokers and Information integration over the Internet.

- Strategies that the statistical agencies have employed in dealing with heterogeneous classifications and nomenclatures.

Here we use the definitions of classification provided by the METANET group [16] and the Neuchatel group [21]. A classification is defined as a structured list of mutually exclusive categories, each of which describes a possible value of the classification variable. Such a structured list may be linearly or hierarchically structured.

In [8] an *ontology* is defined to be an explicit specification of a conceptualisation. In [26] an ontology is defined to be a shared understanding of some domain of interest. We will here use the term ontology to refer to a set of variables along with their classifications. Thus, an ontology might be a survey, e.g., UK Labour Force Survey 2001, whereas particular categorical variables within the ontology will have classifications. A set of ontologies, along with some mappings between them, we call a *frame*, e.g., UK Labour Force Survey 2001, UK Labour Force Survey 2000, UK Labour Force Survey 1999, is a frame.

In a broader context, for general information sources, the development of *ontology servers* has been an active research area over recent years and a number of systems have been designed and implemented, e.g., [6, 22]. Typically such systems store ontological information in a knowledge-base that works in tandem with a database system to produce a unified view of heterogeneous distributed data. Current research issues include the management of changing ontologies in a distributed environment [10], support for dynamic and multiple ontologies [11], and resusable ontologies [20]. Issues of schema mapping and integration are also discussed in [9].

Ideally we would like to match the data sources to the query purely on the basis of nomenclatures and classification schemes. Some preliminary work has been done on learning the schema mappings for a restricted domain, e.g. [2]. In this case the application designer must specify a mediated schema and supply the description of the data sources. The schema are then modelled as a tree, the nodes of which are XML tags. The schema matching problem then looks for a 1:1 mapping between the source schema and the mediated schema. In order to achieve this, a number of learners are employed, each using different data, e.g., element names, word frequencies. A meta-learner then combines the different learners to predict the mapping between the source and mediated schema.

However, such systems are necessarily restricted to particular domains, while in our current context we require to process data from different domains. For integration we therefore require that there is a link from a local classification to at least one of the internationally recognised classifications. We therefore need a method of determining if it is possible to map from the local classifications to the query classification possibly via an intermediary (typically a recognised classification) and then performing the mapping via statistical operators.

An approach related to our own uses a web information integration system [7] to answer queries that may require extracting and combining data from multiple web sources. Likewise, [3] employs a query based approach for integrating heterogeneous data sources that is similar in spirit to our own approach. Semantic web data integration systems have to deal with large and evolving numbers of web sources, little metadata about the characteristics of the source, and a large degree of source autonomy. Most web data integration systems take a "virtual approach" where the data remains in the web sources and queries to the data integration system are decomposed at run-time into queries on the sources; such an approach is well suited to statistical data where the base (micro) data may be both high volume and confidential.

Two main features distinguish such a system from a traditional database system:

- The query execution engine communicates with a set of wrappers [27] rather than directly with a local storage manager. A wrapper is a program which is specific to every web site and whose task is to translate the data in the web site to a form that can be further processed by the data integration system.
- The second difference is that the user does not pose queries directly in the schema in which the data is stored. Instead, the user poses queries on a mediated schema. As a consequence, the data integration system must first reformulate a user query into a query that refers directly to the schemas in the sources; the query is therefore in a common format. In order to perform the reformulation step, the data integration system requires a set of source descriptions. Such a description specifies the contents of the source, the attributes that can be found in the source, constraints on the contents of the source, and completeness and reliability measures associated with the source.

Such transparency for the user is provided by the StEM (Statistical database Expert Manager) query processing mechanism developed in [1]. However, in this paper we are engaged with a distributed and open system in which there is no concept of a global manager, and mediation is an essential feature provided by our agent-based approach. A mediator is a system that refines in some way information from one or more sources. When the mediator receives a query, it knows which sources to forward the query to. The mediator may also process answers before forwarding them to the user. Each localised query is therefore passed by the integration system to the appropriate wrapper.

In our case, we use metadata in the Libraries in collaboration with *classification servers* to translate heterogeneous statistical data from the local ontology into the ontology of the query. The Libraries contain metadata relating to specific datasets, while the classification servers contain conceptual metadata, including mappings between ontologies. Mediation is carried out by negotiation agents that use the information in the classification server to learn the appropriate mappings from the local data ontology to the user's. We thus employ a mediator/wrapper approach where the Negotiation Agents mediate while the data are wrapped by Information Agents. Our focus in this paper is therefore on mediation.

Statistical agencies have long been aware that the existence of a variety of nomenclatures, as well as their revisions, raises issues of compatibility and comparability of data collected and disseminated. Recent work has proposed Classification Servers as a way of providing information on classifications, along with appropriate mappings, available to the user on a web site. An example is RAMON Classification Server, developed by EUROSTAT [24]. An important feature of RAMON is its multilingual character. Whenever possible or available, the nomenclature information is presented in all languages in which it exists.

Previous work [23 , 17, 18] has developed a global data model that enables a universal, harmonised analysis of the data and the metadata. Specific mappings, procedures, functions and algorithms have been developed to enable transformations from local to global views of the data; this permits local-as-view processing, along with query re-writing at run-time to transform to global-as-view. However, this approach requires the data providers to map their data to a global ontology which may be quite laborious. Our current approach, on the other hand, has a number of advantages over previous research, namely:

- The data providers can choose to map their data to (different) classifications, available on the Internet. This is a less laborious solution.
- The query may be posed in a local ontology defined by the user – a query-as-view solution. There is therefore no global ontology, as such. Instead, the ontology mappings and query re-writing rules are computed dynamically. This is clearly more flexible than previous approaches.
- By employing a Negotiation Agent, the task of constructing the query plan is substantially automated compared with previous approaches.

3. The Agent Architecture

As we have discussed, the central functionality of an agent-based distributed database system may be described in terms of *mediator agent*s. They are supported by a set of *wrapper agents* that in addition collaborate with *user interface agent*s, *matching agent*s, and *ontology agent*s. The main purpose of such an architecture is to enable intelligent interoperability among heterogeneous sources and avoid centralised control of the system. So-called *wrapper agents* provide access to local information sources, extract content from that source, and perform appropriate data conversion. In addition, a mediator may collaborate with other *information agents* such as *broker* or *matchmaker agents*, *ontology agents*, and *user interface agents* [12, 13]. In general, the mediator:

- translates between local ontologies (variable names, value labels, and classifications),
- decomposes and executes complex queries on distributed relevant sources with the help of a matching agent,
- composes the partial responses obtained from multiple information sources, and
- returns the query result to the user.

Our agent solution has been implemented as part of the EU funded MISSION (Multi-agent Integration of Shared Statistical Information over the (inter)Net) system. In MISSION the mediator agent is called the negotiation agent, which is called in by the matching agent to decide how the sub-query is best covered by the candidate datasets. The covering agent then composes the best combination of sub-queries and the planning agent converts the optimal cover into a plan (defined in terms of statistical operators) and manages the execution of that plan. If only a partial match is made, the matching agent may use a negotiation agent to determine if a full integration is possible. The negotiation agent is utilised to determine if different classification schemes can be mapped onto each other via a common ontology. This task is carried out using classification servers. These tasks have been termed *pre-integration*; they are the focus of the current paper.

The corresponding wrapper agents comprise the brokering agents in liaison with the costing agents and information agents. Here brokering agents have the capability of learning other retrieval strategies if there is a problem with the optimal strategy (as constructed by the covering agent). Costing agents (including authentication) are responsible for determining costs for retrieving various data fragments from possible data sources; possible costs are monetary, Internet transportation time, and processing time. Information agents act on behalf of the data sources. These tasks comprise the *integration* and are carried out once the integration strategy has been formulated by the pre-integration process.

The query is communicated to the system via interface agents. Interface agents keep user profiles at the client site allowing the system to construct queries in an intelligent way tailored to the user's characteristics. Such a system provides a flexible and elegant way of providing access to heterogeneous distributed statistical databases, allowing the common ontology to be constructed on-the-fly once a query has been specified. The Agent Architecture is summarised in Figure 1. Further details are provided in [19].

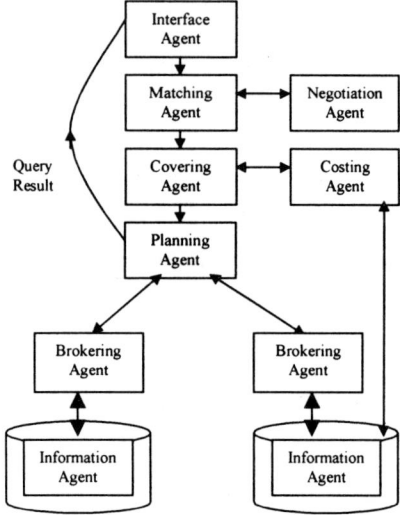

Figure 1. The Agent Architecture

4. The Matching and Negotiation Agents

4.1 Introduction

Within any system or organisation, a number of options exist in terms of having local and/or global ontologies or classifications [25]. The options include:

- A single global reference classification that everyone agrees to use
- Multiple local classifications with no global classification
- Multiple local classifications along with a global reference classification.

From an architectural point of view there are two ways to define mapping rules to indicate how terms in one classification correspond to terms in another.

- *Direct point to point mapping* - a collection of mapping rules are defined for each pair of classifications which directly map from one classification into another.
- *Mapping via a global reference classification* - where mapping rules are defined for each local classification into a neutral global reference classification, and from this global reference classification into the local classifications.

If local classifications are changing over time, then periodically there will need to be some way for the changes in the local classification to be reflected in the global classification. There can be additions, modifications, deletions, changes in the generalisation hierarchy, or mergers.

In our system, there are multiple local classifications with no single global ontology/classification. However, each local classification may be matched by direct point to point matching to one of possibly many internationally recognised classifications that are stored in the classification server. Further direct point to point mappings may exist from one internationally recognised classification to another, so that mappings may be possible from one local classification to another local classification through an intermediary of mappings between the internationally recognised statistical classifications.

The negotiation agents provide (ontology) mappings between heterogeneous data sources, thus allowing us to determine if data integration is possible. This is a general problem for such mediation systems; in our case we intend to provide such mappings via classification servers. A negotiation agent first ascertains if the classification schemes can be mapped onto each other. This task will be carried out by processing the knowledge in the classification servers and manipulating correspondence matrices to ascertain if mappings are possible, and, if so, derive them. In general terms the Matching Agent facilitates access to distributed heterogeneous data by presenting a uniform query interface, thus providing the user with the "illusion" that information is stored in a single global database [4, 5, 15].

The net result of this architecture is that the ontology can be constructed on-the-fly, once the query has been specified. This dynamic shared ontology may be implicitly specified by the user, as part of the query, through the user specifying classification schemes as generalisations of those held in the classification server. However, if the user is not too concerned about the exact ontology, an alternative is that the system may derive the shared ontology, either as the sum of the available classification schemes, or, if estimation is permitted, as their product. This process would utilise the schema mappings between different classification schemes. For the moment we implement a system where the query ontology is specified by the user as part of the query process.

4.2 The Theoretical Framework

We now define a number of relevant terms and derive some results that are needed to develop the algorithms used by the negotiation agent. Our strategy is to start with some simple definitions for the one (categorical) dimensional case and then extend these to cover the general situation involving a Cartesian product of a number of categorical variables.

Definition 4.1: We define a *conceptual variable* (dimension) D to consist of a set of *base values* contained in the set $B = \{v_1,...,v_m\}$. Then a classification C is a partition: $C=\{c_1,...,c_k\}$ where the c_i's are subsets of B s.t.

$$c_i \cap c_j = \varnothing \ \forall i,j \text{ and } \bigcup_{i=1}^{k} c_i = B$$

e.g., $B = \{PT, FT, U\}$, $C = \{\{PT, FT\}, U\}$.

Definition 4.2: A classification $C_1 = \{c_1^{(1)},...,c_{k_1}^{(1)}\}$ of a conceptual variable D is *finer* (less general) than classification $C_2 = \{c_1^{(2)},...,c_{k_2}^{(2)}\}$ of D if $\forall i \ \exists j$ s.t. $c_i^{(1)} \subseteq c_j^{(2)}$ and $\forall j \ \exists i$ s.t. $c_i^{(1)} \subseteq c_j^{(2)}$; NB at least one of the subset inclusions must be strict.

We write $C_1 \leq C_2$ ($C_1 \geq C_2$) if C_1 is finer (coarser) than C_2.

Definition 4.3: A *sum classification* C of two, or more, classifications $C_1,...,C_r$ is the finest partition such that $C_1,...,C_r$ are all finer than C.

Definition 4.4: A *product classification* C of two, or more, classifications $C_1,...,C_r$ is the coarsest partition such that $C_1,...,C_r$ are all coarser than C.

Definition 4.5: A *summary additive function* σ is a set function that maps a set onto the integers or real numbers s.t. for two disjoint sets A_1 and A_2:
$$\sigma(A_1 \cup A_2) = \sigma(A_1) + \sigma(A_2).$$
Cardinality of a set is an example of a summary additive function.

Definition 4.6: A *classification summary* S is a vector defined on a classification $C = \{c_1,...,c_k\}$ for a summary additive function σ s.t.
$$S = (C : \sigma(c_1),...,\sigma(c_k)).$$
For example, for the conceptual variable JOB, $C = \{FT, PT, U\}$, $\sigma(FT) = n_1$, $\sigma(PT) = n_2$, $\sigma(U) = n_3$, $S = (FT:n_1, PT:n_2, U:n_3)$.

Theorem 4.1: If C_1 and C_2 are two classifications such that $C_1 \leq C_2$, with associated additive summary function σ, then:
$$S(C_2) = (\sum_{c_i^{(1)} \subseteq c_1^{(2)}} \sigma(c_i^{(1)}),.........,\sum_{c_i^{(1)} \subseteq c_{k_2}^{(2)}} \sigma(c_i^{(1)}))$$

Proof: $c_j^{(2)} = \bigcup_{i:c_i^{(1)} \subseteq c_j^{(2)}} c_i^{(1)}$ for $j=1,...,k_2$ from the definition of a finer (coarser) classification.

Hence, $\sigma(c_j^{(2)}) = (\sum_{c_i^{(1)} \subseteq c_j^{(2)}} \sigma(c_i^{(1)})$ from the definition of a summary additive function.

For example: $C_1 = \{FT, PT, U\}$, $C_2 = \{\{FT, PT\}, U\}$.

Then $S(C_1) = (FT{:}n_1, PT{:}n_2, U{:}n_3)$ implies that $S(C_2) = (\{FT,PT\}{:}n_1 + n_2, U{:}n_3)$

On the basis of this result we have previously defined a *reclassification operator* as a sequence of SQL commands.

Definition 4.7: For classifications C_1 and C_2, with $C_2 \geq C_1$, defined as generalisations on the same set of base values B (i.e. $C_1 \geq B$ and $C_2 \geq B$), we define a *correspondence matrix (correspondence table)* $M = \{m_{ij}\}$ s.t.

$$m_{ij} = \begin{cases} 1 & \text{if } c_i^{(1)} \subseteq c_j^{(2)} \\ 0 & \text{otherwise} \end{cases}.$$

For example: $C_1 = \{FT, PT, U\}$, $C_2 = \{\{FT, PT\}, U\}$,

$$M = \begin{pmatrix} 1 & 0 \\ 1 & 0 \\ 0 & 1 \end{pmatrix}$$

In the situation where the base id reduced, i.e., subsetting, this is reflected in the correspondence matrix having columns of zeros corresponding to those base values (or sets thereof) not included in the reduced base.

Theorem 4.2: A classification C_1 is finer than classification C_2 if $M.I_2 = I_1$ where I_1 is a column vector of 1's with k_1 elements, I_2 is a column vector of 1's with k_2 elements and M is the correspondence matrix. In other words, for $C_1 \leq C_2$, we require that the row sums of M are equal to 1, and the column sums of M are ≥ 1.

Proof: $C_1 \leq C_2$ if $\forall i \exists j$ s.t. $c_i^{(1)} \subseteq c_j^{(2)}$ and $\forall j \exists i$ s.t. $c_i^{(1)} \subseteq c_j^{(2)}$.

i.e. each subset of C_1 is contained in a unique subset of C_2 and each subset of C_2 has at least one subset of C_1 contained in it.

Hence, if $C_1 \leq C_2 \forall i=1,\ldots,k_1 \exists ! j$ s.t. $m_{ij}=1$, and $\forall j=1,\ldots,k_2 \exists i$ s.t. $m_{ij}=1$. This theorem is the basis of algorithm *Gentest*, which tests if one classification can generalise to another. We note in passing that such mappings for statistical classification servers have been discussed in [14].

Algorithm *Gentest*
Input: Correspondence matrix M
IF M has all row sums equal to 1
and all column sums ≥ 1
    THEN Generalisation = true
       ELSE Generalisation =FALSE
Output: Generalisation

Theorem 4.3: Given a correspondence matrix M_1 between classifications C_1 and C_2 and a correspondence matrix M_2 between classifications C_2 and C_3, the correspondence matrix M_3 between C_1 and C_3 is given by:

$$M_3 = \delta(M_1 . M_2)$$

where we define $\delta(A) = \{\delta(a_{ij})\}$ and δ is the Kronecker delta s.t.

$$\delta(x) = \begin{cases} 1 & x > 0 \\ 0 & \text{otherwise} \end{cases}.$$

Proof: $m_{ij}^{(3)} = \delta(\sum_{r=1}^{c} m_{ir}^{(1)} m_{rj}^{(2)}) > 0$ if $\exists$ at least one r s.t.

$m_{ir}^{(1)} = 1 = m_{rj}^{(2)}$

$\Rightarrow c_i^{(1)} \subseteq c_r^{(2)} \subseteq c_j^{(3)}$

i.e. $c_i^{(1)} \subseteq c_j^{(3)}$.

So, $m_{ij}^{(3)} = \begin{cases} 1 & \text{if } c_i^{(1)} \subseteq c_j^{(3)} \\ 0 & \text{otherwise} \end{cases}$

i.e. M_3 is the correspondence matrix between C_1 and C_3.

We should note that in the case where M_1 and M_2 each define mappings between a classification C_i and a coarser classification C_j, i.e., $C_j \geq C_i$, then the use of δ is redundant by Theorem 4.2. However, Theorem 4.3 allows for the more general situation where a mapping may be obtained from one classification to a coarser one via an intermediate classification that is not necessarily coarser than C_i.

Theorem 4.3 is the basis of algorithm *Correspond*, which constructs a correspondence matrix between classifications C_1 and C_3 from the correspondence matrices between C_1 and C_2, and between C_2 and C_3.

Algorithm *Correspond*
Input: Correspondence matrices $M_1(C_1, C_2)$
and $M_2(C_2, C_3)$
$M_3 = M_1 . M_2$ (matrix multiplication)
$M_3 = delta(M_3)$
Output: Correspondence matrix $M_3(C_1, C_3)$

Definition 4.8: A *datacube* D is defined on the scheme $(C_1, \ldots, C_n, \sigma_1, \ldots, \sigma_m)$, where $C_1, \ldots, C_n$ are classifications of categorical variables $D_1, \ldots, D_n$ respectively, and $\sigma_1, \ldots, \sigma_m$ are additive summary functions. A datacube is then a classification summary defined on a Cartesian product of $D_1, \ldots, D_n$.

For example: $D = ((M, FT/PT){:}n_1), ((M, U){:}n_2), ((F, FT/PT){:}n_3), ((M, U){:}n_4)$.

We note that a datacube is a general structure with the same logical content as a macrodata object.

Definition 4.9: A datacube D_1 is *generalisable* to a datacube D_2 ($D_1 \leq D_2$) if:
$D_1 = (C_1^{(1)}, \ldots, C_n^{(1)}, \sigma_1, \ldots, \sigma_m)$

$D_2 = (C_1^{(2)}, \ldots, C_n^{(2)}, \sigma_1, \ldots, \sigma_m)$
and $C_i^{(1)} \leq C_i^{(2)}$, $i = 1, \ldots, n$, for all variables D_i, with at least one of these inclusions being strict, and the common categorical variables having common summaries.

Theorem 4.4: If $D_1 \leq D_2$ we can compute the summaries of D_2 from those in D_1.

Proof: The proof follows from the definition of an additive summary function and the result of Theorem 4.1.

Definition 4.10: We define D_1 as *g-matchable* to D_2 if we can generalise D_1 to obtain D_2 and/or the set of categorical attributes in D_1 is a subset of the set of categorical attributes in D_2, where the common categorical variables have the same classifications.

N.B. this definition includes 1:1 mappings (e.g., nomenclature mappings/ translations) and projection (removing a categorical variable from the datacube).

If D_1 is g-matchable to D_2 then we can construct an operator stack (using the statistical operators defined in ADDSIA) that constructs the datacube D_2 from D_1. The job of the negotiation agent is then to use algorithms *Gentest* and *Correspond* to test if the datacube requested by a query is g-matchable to a dataset described in the metadata.

Definition 4.11: We define an *ontology* O as a set of categorical variables and their classifications, along with corresponding numerical variables and their summaries. Ontologies are grouped into *frames,* which have a common set of variables. An ontology therefore provides the schema for a set of datacubes within a frame.

We further define an ontology O_1 to be generalisable to ontology O_2 in the same way as we define datacube D_1 to be generalisable to datacube D_2.

4.3 The Matching and Negotiation Agents

During the query process the user specifies the ontology of a query, along with a frame, and categorical and numerical variables. Also included in the query are some keywords such as geo-reference (country) and temporal reference (year). This information on ontology and keywords is passed to the matching agent in the library. Initially, the matching agent looks for a homogeneous match, matching on frame, ontology and keywords. If it is not possible to match the (sub) query in this way then the negotiation agent is called ; its role is to seek a (heterogeneous) match using its knowledge base – the classification server. In this case, the negotiation agent's task is to ascertain if the (sub) query is g-matchable to the (meta) data; if it is then we can construct an operator stack to transform the data to match the (sub) query.

Matching Agent
Input: OQ
Search for Match with OQ
 Is there a cached query result that can be retrieved?
  If so, return to requesting Matching Agent.
 Is there an OD that homogeneously matches OQ (including the ontology)?
  If so (matching on *Frame, Variable Names, Location, Time, Ontology*) then this is returned to the requesting Matching Agent
 else
 Is there an OD that might give a heterogeneous match (matching on *Frame, Location, Time*)?
  If so, then call the Negotiation Agent
 If Negotiation Agent reports that map=TRUE then correspondence table and mapped variable names from OD to OQ are added to the *Rough Query Object* for further processing.

Negotiation Agent
Input: OQ, OD
Search for Mapping
 Is there a set of correspondence tables that map OD to OQ?
 If not, is there a correspondence table that maps OD to O* and another correspondence table that maps O* to OQ?
 If mapping exists then map=TRUE, else map=FALSE
If map= TRUE then Find Mappings:
 Extract the correspondence tables from OD to OQ
 else
 Begin
  extract the correspondence tables from OD to O* and O* to OQ
  Multiply the matrices and adjust the elements to give the OD to OQ correspondence table
 End
 For each variable
 Begin
  Use algorithm *Correspond* to test the OD to OQ correspondence table to see if the mapping is generalisable, using *Gentest*
  Map the variable names from OQ to OD
 End
Output:
 If map = TRUE return correspondence table and mapped variable names from OD to OQ to requesting Matching Agent
 else
 report that negotiation is impossible

For such heterogeneous matching the matching agent initially matches on the frame and keywords in the query; it is not possible to match on the ontological information since we cannot do this until it has been mapped by the negotiation agent.

Once the initial match has been made, the matching agent calls in the negotiation agent, passing it the query and (meta) data ontologies (OQ and OD, respectively). The negotiation agent's task is to determine if OD is g-matchable to OQ and, if so, to return the mapping information to the requesting matching agent.

We now describe the process by which the matching agent interacts with the negotiation agent and the operation of the negotiation agent itself. The negotiation between ontologies O1 and O2 in Table 1 is carried out using the correspondence table from JOB-STATUS to JOB.

Table 1. Negotiation between the Query Ontology (O1) and (meta)Data Ontology (O2)

Query	Metadata
LFS	LFS
JOB	SEX
France	JOB-STATUS
O1	France
	O2

Correspondence Table from JOB-STATUS (O2) to JOB (O1)

	JOB		
		W	NW
JOB-	PT	1	0
TITLE	FT	1	0
	U	0	1

Initially O1 is matched to O2 on the frame: LFS (Labour Force Survey), and Country: France. After negotiation, the variable name mappings (JOB to JOB-STATUS) and correspondence table are returned to the matching agent. Here the metadata, containing information about specific datasets, is held in the Library, while the correspondence table, containing conceptual information, is held in the Classification Server.

Using the results from the negotiation agent, the matching agent translates the query from O1 to O2.

4.4 The Classification Server

Within the MISSION system we have implemented a classification server that contains all the essential features we have described but that is more suitable for processing than currently available classification servers, which are primarily designed for web browsing. Therefore we do not explicitly define hierarchies. Instead, both hierarchies and nomenclatures are described within the same framework, where we specify when ontolgies may be mapped and then give the appropriate variable name mappings and correspondence matrices. This strategy allows us to encompass 1:1 nomenclature mappings, hierarchy generalisations and variable deletion (projection) within the same scheme, which we have termed g-matchability. The eventual aim, when clasification servers are more widespread, would be to have functionality to access third party servers supplied by "trusted" statistical providers. The structure of the current MISSION classification server is described in Figure 2.

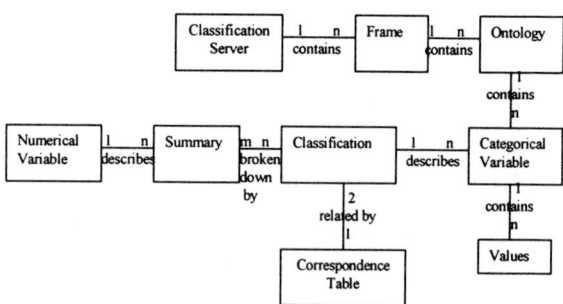

Figure 2. Structure of a classification server

Example of the contents of a classification server

Frame: LFS
Ontologies: O1, O2, O3
Mappings: O1↔O2
Variables:
O1: JOB(W, NW), SEX(M,F);
O2: JOB_STATUS(PT,FT,U),
　　GENDER(male,female);
O3: SEX(M,F), REGION(1, 2, 3)
Correspondence:
O1↔O2:
　　JOB, JOB-STATUS

$$\begin{pmatrix} 1 & 0 \\ 1 & 0 \\ 0 & 1 \end{pmatrix}$$

　　SEX, GENDER

$$\begin{pmatrix} 1 & 0 \\ 0 & 1 \end{pmatrix}$$

5. Summary and Further Work

We have developed query agents for pre-integration of heterogeneous distributed statistical data, in particular the matching and negotiation agents. The matching agent decomposes the query into sub-queries, and then searches metadata to find a dataset that matches the query fragments. In the case of heterogeneous data, the matching agent utilises a negotiation agent to find datasets that match the query fragments, provides mappings from the data to the query, and constructs the appropriate (sub) query re-writing rules. Such matching is done by testing if the (sub) query is g-matchable to the (meta) data; if it is then we can construct an operator stack to transform the data to match the (sub) query.

Such an approach provides a capability of automating the process of executing queries on heterogeneous statistical databases that permits queries to be specified in a goal-driven query-by-example format. Rather than impose an a priori global standard, the user can query through a unified interface, and integration is done at run-time. Further work will extend this aspect of the query process to allow for an inexperienced user to make an imprecise query, without specifying an ontology. The system then automatically constructs a dynamic shared ontology by analysing the correspondence graphs that relates the heterogeneous classification schemes [18].

Acknowledgement

This work was partially funded by MISSION- Multi-agent Integration of Shared Statistical Information over the (inter)Net (IST project number 1999-10655) within EUROSTAT's EPROS initiative.

References

[1] Basili C., and Meo-Evoli L., A "Deductive Query Processor for Statistical Databases", Proc. DEXA 1992 (Tjoa, Ramos, eds.), 1992, pp. 390-395.

[2] Doan A., Domingos P., and Levy A.Y., "Learning Source Description for Data Integration", Proc. WebDB, 2000, pp. 81-86.

[3] Domenig R., and Dittrich K.R., "A Query Based Approach for Integrating Heterogeneous Data Sources", Proc CIKM. 2000.

[4] Duschka O.M., "Query Optimisation using Local Completeness", Proc. AAAI-97, 1997, pp. 249-255.

[5] Duschka O.M., and Levy A.Y., "Recursive Plans for Information Gathering", Proc. IJCAI-97, 1997, pp. 778-784.

[6] Farquhar A., Fikes R.,and Rice J., "The Ontolingua Server: a Tool for Collaborative Ontology Construction", Proc. KAW96, 1996.

[7] Florescu D., Levy A.Y, and Mendelzon A.O., "Database Techniques for the World-Wide Web: A Survey", SIGMOD Record, 27(3), 1998, pp. 59-74.

[8] Gruber T., "A Translation Approach to Portable Ontology Specifications", Knowledge Acquisition, 5, 1993.

[9] Häder T., Sauter G., and Thomas J., "The Intrinsic Problems of Structural Heterogeneity and an Approach to their Solution", The VLDB Journal, 8, 1999, pp. 25-43.

[10] Heflin J., Hendler J., and Luke S., "Coping with Changing Ontologies in a Distributed Environment", Proc. AAAI-99 Workshop on Ontology Management, 1999.

[11] Heflin J., and Hendler J., "Dynamic Ontologies on the Web", Proc. 17th National Conference on Artificial Intelligence (AAAI-2000), 2000, pp. 443-449.

[12] Huhns M.N., and Singh M.P., "Ontologies for Agents", IEEE Internet Computing, Nov/Dec 1997.

[13] Huhns M.N., and Singh M.P., "Managing Heterogeneous Transaction Workflows with Co-operating Agents. In Jennings N.R. and Wooldridge M. (Eds.), Agent Technology, Springer, 1998.

[14] Hulliger B., "Linking of Classifications by Linear Mappings", Journal of Official Statistics, 14(3), 1998, pp. 255-266.

[15] Levy A.Y., Rajaraman A., and Ordille J.J.,), "Querying Heterogeneous Information Sources using Source Descriptions", Proc. VLDB-96, 1996.

[16] METANET, http://www.epros.ed.ac.uk/metanet/ , 2000

[17] McClean S., Páircéir R., Scotney B., and Zhang Y., "Adding Context to the Retrieval of Aggregate Data from Distributed Databases via the Internet", submitted to the International Journal of Information Systems.

[18] McClean S., Scotney B., and Greer K., "A Scalable Approach to Integrating Heterogeneous Aggregate Views of Distributed Databases", IEEE Transactions on Data and Knowledge Engineering, forthcoming, 2002.

[19] McClean S., Kareli I., Scotney B., Greer K., Kapos G.-D., Pairceir R., Hong J., Bell D., and Hatzopolous M., "Agents for Querying Distributed Statistical Databases over the Internet", International Journal on Artificial Intelligence Tools, 11(1), 2002.

[20] Musen, M.A. "Modern Architectures for Intelligent Systems: Reusable Ontologies and Problem-Solving Methods", In Chute C.G. (Ed.), AMIA Annual Symposium, Orlando FL, 1998, pp. 46-52.

[21] Neuchatel paper, Version 2, available from claude.macchi@bfs.admin.ch, 2000

[22] Noy N.F., and Mason M.A., "PROMPT: Algorithms and Tools for Automated Ontology Merger and Alignment", Proc. AAAI'00, 2000.

[23] Papageorgiou H., Pentaris F, Theodorou E., Vardaki M., and Petrakos M., "A Statistical Metadata Model for Simultaneous Manipulation of both Data and Metadata", International Journal of Intelligent Systems, forthcoming.

[24]   Pongas G., and Reiter S., "RAMON: Eurostat's
        Classification Server", Proc. NTTS'2001, 2001.

[25]   Uschold M., "Creating, Integrating and Maintaining
        Local and Global Ontologies", Proc. ECAI, 2000.

[26]   Uschold, M., and Grüninger, M., "Ontologies: Principles,
        Methods and Applications", *Knowledge Engineering
        Review*, 11(2), 1996.

[27]   Wiederhold G., "Intelligent Integration of Information",
        Proc. SIGMOD-93, 1993.

Statistical Composites: A Transformation-bound Representation of Statistical Datasets

MICHAELA DENK, KARL A. FROESCHL, WILFRIED GROSSMANN
Dept. of Statistics and Decision Support Systems
University of Vienna, Austria
A-1010 Wien, Universitaetsstrasse 5
{Michaela.Denk,Karl.Anton.Froeschl,Wilfried.Grossmann}@UniVie.ac.at

Abstract

Statistical data processing makes use of data matrices and tables as primary structures for data representation. Embedding these structures into processing-relevant context information gives rise to enhanced data structures linking data and metadata. The paper describes a framework for statistical data processing utilising metadata computationally.

1. Introduction

Statistical data processing is driven by context knowledge in its aim to disclose (by way of approximation) intensional features of scrutinised statistical populations. Generally speaking, statistical processing in connection with real-world problems can be described as follows: given a population P of well-defined individual units, one is interested in the determination of a statistical parameter, $\theta(P)$. To this end, usually, information about the population P is collected on a sample S related to P (in simple cases P itself being the sampling population from which S is taken). The units, then, are actually observed in terms of a "measuring device" – such as a questionnaire, for example – by defining (random) variates with appropriate value structures. Based on this evidence, unknown $\theta(P)$ is inferred through some suitable estimate $\hat{\theta}$.

Thus far, this is barely new, and there are several approaches seeking to represent the involved information structures in formal terms. Far less attention, however, has been dedicated to extending transformations beyond data and variates, respectively, to encompass also the transformation of populations – an issue of considerable practical relevance thinking, for example, of the increasing role of joining register data [14].

Based on a formal representation of statistical data production, this paper proposes a data representation scheme named *statistical composites* destined to align both statistical data and transformation-relevant metadata in compound operands of a statistical algebra so as to always keep data and metadata congruent. Contrary to the predominating understanding of metadata in a database context, statistical metadata encompasses both structural description and statistically meaningful entities, such as statistical variables, value sets, populations, model parameters, etc., building a statistical ontology.

In section 2, a formalised model of statistical processing is introduced. Various approaches to the representation of statistical data and metadata are reviewed, concluding with the specification of metadata requirements for statistical transformations. As a solution to these needs, section 3 discusses statistical composites and transformation structures. The versatility of this approach is exemplified for weighted estimation in section 4. Finally, section 5 presents an overview of implementation experience and future research.

2. Statistical Process Model

From an intensional point of view, the taking of statistical measurements or observations can be expressed as a *statistical function* mapping *statistical units* (the domain of the statistical function) to *values* (the co-domain of the statistical function) [7]. Customarily, statistical domains are structured, i.e. *population unit types* and *populations* (collectives of statistical units of same type) are discerned. Unit types (and, thus, populations) may or may not be related formally, yet this will not be considered any further here. Likewise, the co-domain (i.e., values) is structured into sets (of *modalities*) arranged in enumerations, groups, partitions (taxonomies), and (classification) hierarchies (cf. [11]). Statistical functions (or "measures"), in turn, are expressed in terms of *characteristics* applying to the statistical units considered: each characteristic actually maps statistical units into a pre-defined space of (coded) values. Practically, a characteristic ('variable') is stated intensionally by linking it to (i) a statistical unit type, (ii) a value/code set (including set structure), and (iii) some interpretation of the value set ('variable definition'). Opposite to the intensional definition of a statistical function, actually observing or taking measurements on a (sub-)set of some population in a census or survey produces one possible extension of this function, i.e. *statistical data* [29]. In what follows, the terms 'intension' and 'extension', respectively, apply to all types of statistical entities, the intensions capturing the entities' descriptions (both structurally and semantically) and the extensions representing physical enumerations (cf. the definitions used in set theory).

The "statistical triad", population (domain)–variate (function)–value (co-domain), induces a general descriptive framework of statistics in the sense of a statistical meta-ontology comprising principal classes of discerned entities (a "meta-nomenclature", if you like) interrelated by pre-defined associative link types in a generic semantic network [12]. Complementary to this ontological dimension, however, also a processing dimension capturing the "statistical dynamics" needs to be considered. Resorting to a rough sequencing of statistical processing (deemed sufficient for present purposes, though), three successive phases come to mind:

- *statistical production* generating the statistical data;
- *statistical transformation* concerning the application of (mainly) statistical algorithms, and
- *statistical dissemination* conveying derived information (both data and inferred estimates) to the sphere of information usage.

In this sequencing, 'production' is deliberately restricted to primary production (for instance, using questionnaires for data capture) as opposed to a looser interpretation subsuming also part of the transformation phase. Combining both ontological and processing dimensions results in a two-way arrangement as depicted in Table 1.

Table 1: Statistical description framework

Dynamics	*Ontology*		
	domain	function	co-domain
Production	S' →	M' →	d
Transformation			$\hat{\theta}_{S',M'}(d)$
Dissemination			

For the sake of clear distinctions, in Table 1 and forthwith the following notation is used. First, to single out several salient levels of notions,

- *capital* symbols, such as X, refer to "integral" notions encompassing both intension and extension(s) of the concept denoted, whereas
- *lower-case* symbols, such as x, always refer to concept extensions (i.e., the code level).

Furthermore, symbols carrying a prime, such as X' or x', respectively, refer to the *semantic adjunct* of the notion denoted by the symbol; if written lower-case, this again represents the adjunct's extension (commonly termed *metadata* [17]). Symbol usage is summarised in Table 2, introducing also the separation of 'reference' and 'description' layers of the statistical meta-ontology proposed.

Generally, lower-case symbols refer to coded context elements, the so-called context foreground, while the "residual" part of statistical context (populated with context elements denoted by capital symbols) remains informal [13]; this context background thus gathers all intensional context elements lacking explication (for whatever reason).

Table 2: Meaning of symbol types

Layer	*Symbol*	capital	lower-case
reference	bare	"notion *per se*"	extension of notion
description	prime	description of notion ("meta-notion")	extension of *meta*-notion

Now, using this symbolism, let P denote a statistical population as before; then, P' denotes its description, i.e. the statistical *unit type* inducing the whole (homogeneous) collective intended. Conversely, p is the extension of P (if any) and, plainly, making up a (statistical) *register*, or listing, of all population units included. Finally, p' can be conceived of as a (partly) formal characterisation of the statistical unit type P' relative to, say, some family of related unit types.

Analogously, S denotes *sample* structure and S' its description in terms of sampling methodology, stratification, etc.; s encodes the sample drawn (selection of population units, $s \subseteq p_s$ where p_s refers to some sampling population related to P in a well-defined way), and s' provides the (meta-)data about the sample (strata, sampling fractions, etc.). It is suggestive to think of S as a function of P, i.e. $S \equiv S(P)$. Likewise, M refers to the *measuring* (observation) structure, with M' providing variable definitions and variate/covariate structure at function level; m denoting the actual observation function and m' the actual observation structure. Again, let $M \equiv M(S)$.

While S maps, formally speaking, to an event (or $\sigma-$) algebra in little need of any elaboration, M maps to value structures of varying complexity and diversity (as, e.g., in hierarchical classification systems), making a formal characterisation highly desirable. Letting D intensionally represent all possible outcomes of M on S, D' corresponds to the description of the observation structure at value level (attribute domains, in database terms). Extensionally, d' is *data* information, and, finally, d the actual realisation of the sample (the "detailed observed structure" of m, if you like), in other words, $m(s_i) = d_i$ the observation for sampled case i (i.e., $s_i \in s$) having $d_i \in d$.

Based on the statistical sample and function an estimate $\hat{\theta}_{S'M'}(d)$ for the parameter of "interest" $\theta(P)$ is computed. This notation indicates that $\hat{\theta}$ in fact depends on the semantic context (i.e., the meaning with respect to P and $\theta(P)$) of the coded observations as well. In other words, $\hat{\theta}_{S'M'}$ is determined, in its proper application, by S' and M', and hence implicitly also by P, i.e. knowledge – the *statistical meta-information* – about the statistical domain and the statistical function (hence, also value structure) generating the data. Hence, any useful information structure must link d, m and s together in order to facilitate formalized determination of $\hat{\theta}_{S'M'}$.

Additionally, in general, $\hat{\theta}_{S'M'}$ is brought about in a *sequence* of steps transforming d, implying that $\hat{\theta}_{S'M'}$ in fact becomes decomposed into sub-transformations each depending not only on the initial S' and M' but the intermediary transformation states as well [9]. Depending on attained processing stage $t \equiv t(d)$, different parameters' estimates $\hat{\theta}_{S'M'T'}^{(1)}$, $\hat{\theta}_{S'M'T'}^{(2)}$, ... (in favour of "interests" $\theta^{(1)}(P)$, $\theta^{(2)}(P)$, ...) may still be feasible. Hence, succinct information on T – i.e. T' – needs to be communicated, too.

2.1 Process-Representation Structures

By theory, strong quality statements about $\hat{\theta}_{S'M'}$ (like *consistency*, *efficiency* or *robustness*) can be ascertained on condition that a number of rather strict prescriptive assumptions on S and M hold. Usually, assumptions are stated with respect to the probability distribution on P.

Despite recognisable progress in developing less "demanding" methods (for example non-parametric models), statistical methodology still often owes definite receipts how to proceed (say, in case of precondition mismatch) because either the translation of substantive matter meta-information into the statistical semantics of a model is unclear, or decisive knowledge about S and M is lacking (cf. [10]). Since, in traditional analysis settings data context rarely ever gets stated explicitly, as a consequence, checking of method preconditions against application contexts takes place informally (that is, in the analyst's mind only). This, however, runs counter specifically to widespread access (e.g. DDI [6]) to "remote" data from linked-up resources with its inherent propensity to separate data from context [13], thus obstructing appropriate method choice/application and result interpretation, respectively, unless sufficient meta-information (i.e., S', M', T') is provided along-side the data so accessed.

Metadata-enhanced data communication offers one conceivable counteraction to this problem. Typically, ensuing encoding schemes derive directly from serial (i.e., file) transfer, or message, formats (such as Sundgren's [30] "e-messages", GESMES [31], PC-AXIS [22], [23]; more recently, the focus changed towards *resource description* [33] emphasising marked-up exchange formats (such as Triple-S [20] and, especially, DDI [6]; one might also have a look at Bisdorff's [3] approach). Little surprise, all these representations deal with one dataset at a time, attaching structural and semantic information to the coded observation data conveyed but doing so without reference to any broader "external" data context. With respect to the above-defined statistical process model, such approaches are useful in facilitating meta-information *exchange*.

Another brand of approach draws on relational database models (for instance, [25]–[28]). Naturally, this promotes algebraic transformations close to "standard" database query processing, including aggregation while, for the very same reason, statistically crucial features are often not taken care of at all. To illustrate, database semantics sticks to "closed worlds" and, particularly, fails to tell statistical (P) from sampling (P_s) populations. Another cardinal difference concerns the distinction between functional and *probabilistic* dependence: while the first is a matter of (schema) design and, hence, not of statistical significance, the latter is one of correlation and statistical inference, respectively.

A by far more elaborated description framework for statistical datasets is provided by "system approaches" seeking to integrate a multitude of datasets either for administrative purposes (such as Gillman's [16] CMR, based on ISO 11179 [21]) or in regard of keeping and documenting datasets related by intervening transformations (such as [5] or, in a sense, "system files" of statistical software packages). Targeted at context sharing, it comes as no surprise that especially NSOs and supra-national statistical institutions favour an integrated representation of multi-source data (for example, [4]); however, support of representation structures for keeping track of statistical transformations typically remains limited to linking primary data production (or import) to output systems.

Statistical software (i.e., "packages") focuses on computing $\hat{\theta}_{S,M}(d)$, and, hence, tends to abstract from data semantics (the meaning of d) while, of course, providing strong support in data transformations. Generally, analysis software is biased towards managing single datasets, though often offering powerful means of expression (such as, e.g., the 'S-plus' system drawing on OO methodology for both data representation and transformation); increasingly, these tools also incorporate essential statistical concepts (like *factor dependent variable*, *covariate* etc.) necessary for preserving transformation semantics.

From an integrated statistical data and process documentation perspective, all of the hitherto proposed representation schemes apparently fall short of supplying the meta-information necessary, in one way or another, in statistical transformations and dissemination phases.

2.2 Metadata-modelling Requirements for Statistical Transformations

Whether man or machine, on the data receiver's side, to proceed reasonably with statistical data, some representation of data context D – wrapped up in terms of P, S, M, and T – is called for. This holds even more so if the data receiver is to carry out further statistical analyses. Clearly, the benefit of context representation rises by its degree of formalisation since, then, data and metadata are prepared to undergo all transformations side by side. Furthermore, in doing so, metadata – as but higher-order data – can assume a *machine-active* data processing role [2] in both checking and controlling transformation method application.

Apparently, in order to fully encompass statistical data contexts with all its possibly relevant interrelations and interactions between modelled entities, a sort of comprehensive representation framework becomes indispensable (cf. [32] with respect to practical implications). Yet, to date, there seems to be little agreement upon both, the scope of the representation structures, and the adequate concepts and terminology to be used. At any rate, however, as highlighted by Table 1, it is *not* only the data undergoing transformation (rightmost column). Rather, also variates (in the "data step" of processing) and populations (e.g., in fixing under-coverage) get transformed, yet often

in subtly differing ways, and by affecting and modifying the interrelations of represented entities to varying degree. Correspondingly, the type of transformation changes with the type of operands it is applied to. To illustrate, in addition to "plain" aggregation turning case data into summary statistics, by change of statistical unit type (say, from 'person' to 'nation' defined as national population), previous aggregates may be viewed again as *case* data at the 'nation'-level.

The representation of statistical transformations has, in fact, to be considered from two angles [9], [18], viz.

- *plan formation*: the drawing-up of a transformation plan actually deriving $\theta(P)$ given D (with d possibly not yet available, or produced) by composing elementary operations on P, S, M, and D into a compound expressions T called ;
- *plan execution*: the application of T at the code level, i.e. on p, s, m, and d, respectively.

It must be noted that plan formation amounts, prevalently, to metadata computing in that transformations affect semantic adjuncts p', s', m', and d', in fact calling for the explicit definition of (algebraic) transformation operators at the data description level (cf. Table 2).

In concluding, metadata modelling in support of statistical transformations has to take account of:

- *subject matter* aspects important for building the bridge to the 'real' world (ontological semantics);
- *methodological* aspects setting up the statistical model and the statistical computing logic (operative semantics);
- *algorithmic* and *representation* (i.e., syntax) aspects geared towards machine-supported manipulation, storage and retrieval;
- *administrative* aspects addressing management and book-keeping of all the structures.

Of course, relative importance and impact of these aspects depends specifically on the type of transformation and its position in the transformation phase, respectively. Take, for example, the case of (multi-) dimensional data: despite syntactical identity, the data-generating process (the statistical model, if you prefer) determines whether we deal with statistical summary data or just a dimensional number arrangement.

To the best of our knowledge, there have only been few (and partial) attempts to devise encompassing metadata models meeting all the specifications outlined. A model concentrating mainly on variables (the 'statistical function'-part of statistics is discussed in [7]; another point of reference is IDARESA [15] highlighting the case of documentation and retrieval of distributed statistical resources (cf. Papageorgiou et al. [24] reporting on a recent application of this model).

3. Statistical Composites

In fact, transformation plan T is a compound expression $T_1 \circ T_2 \circ \cdots \circ T_K$ made up of elementary transformations T_k, suggesting a self-contained operand structure used as input and created as output of all transformations. All T_k operate on different extensional data- and metadata-components and, at the same time, admissibility of T_k is in general subject to these components. Thus, it seems quite obvious to pool extensional components usually transformed by T_k into integrated operands – termed *statistical composites* [9] or SC, for short – in order to keep data and metadata representations always congruent during statistical processing.

Since elementary transformations T_k do not just transform data d but also parts of their semantic adjuncts d' in consideration of M and S, the entire D is transformed to $T(D)$ successively, traversing different intermediate processing levels $T_1 \circ T_2 \circ \cdots \circ T_l(D)$, for $l = 1,\ldots,L$ such that each T_l is, in turn, composed of elementary transformations. I.e., a "life-cycle" of statistical composites is established [9], where composites of each processing level must be properly embedded into the transformation context, or, more precisely, into transformation invariant parts of the context foreground.

3.1 Operand Structure

As generic syntactic building block of SCs, a relational *container* structure is used, receiving different interpretations depending on content, viz.

- *buckets* holding all data extensions (in particular, data matrices and multidimensional tables);
- *bucket schemas* describing formal bucket structure, and
- *directories* listing SC components.

Each SC comprises an *attributes directory*, a *container directory*, and, subject to its processing level, a different number of buckets and corresponding schemas as shown in Figure 1.

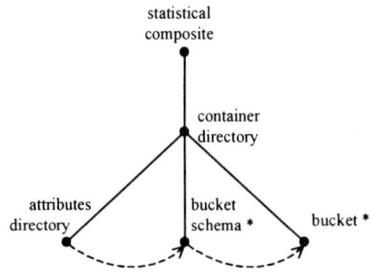

Figure 1: Structure of statistical composites

The basic structure of a bucket varies depending on the interpretation of its content. On the one hand, with regard to format, *case level* and *summary level* buckets are discerned, the former containing relational ("case-by-variate") data at statistical unit-level, the latter holding dimensional ("tabular") data.

On the other hand, the following bucket classes are distinguished:

- *data* ... observation data or aggregates thereof,
- *sampling* ... description of the sampling plan,
- *weighting* ... weighting information for different purposes, such as multipliers for grossing up sample data to population figures, or for adjustment of non-response,
- *method* ... estimates $\hat{\theta}_{S,M}(d)$ and further information on estimates arising from statistical procedures, such as estimated parameters or a coefficient of determination in a regression model,
- *annotation* ... textual descriptions attached to any selected subset of bucket cells.

Essentially, these bucket classes are discerned according to the attribute classes that may be included in the bucket, which obviously is reflected in the composition of the respective schema.

With regard to conventional *data* attributes, corresponding to statistical formats 'case' and 'summary', we distinguish attribute classes *data variables* (D) and *summary attributes* (S). The former are further divided into *categorical* (C), *quantitative* (Q), *textual* (T) and *key* (K) variables, where, of course, textual and key variables belong to the class of C-variables. As a subclass of Q-attributes, U attributes are introduced, representing variables normed to [0,1]. Weighting information (lump factors, or bound to particular data variables) is encoded in *weight* attributes (W). A special case, *method* attributes (M) bear sets of statistical variables rather than sets of values.

Obviously, K or Q attributes are only allowed in case buckets, while S attributes, resulting from statistical aggregation, may solely occur in summary buckets. During computation, W attributes should always be included in the weighting bucket; yet, a lump weight might also be attached to a data bucket. M attributes can only be included in method buckets.

Ensuing bucket schema structures are summarised by regular expressions in Table 3 overleaf (as usual, '*' indicates an arbitrary number, '+' at least a single occurrence, and '?' optional occurrence of the preceding symbol); in any case, empty schema structures are not admissible.

SC attributes come in two ways, viz. as

- *bucket attributes* listed in bucket schemas describing bucket structure, and
- *composite attributes* establishing equivalence relations over all bucket attributes used in a SC.

Composite attributes are gathered in the attributes directory of the SC. Clearly, attributes shared among several SCs – "context attributes" – need to be equated analogously at (transformation) context foreground level via references to explicit measurement information M' not described here for shortage of space.

Table 3: Bucket schema structures

b_format	b_class	Schema structure (a_class)
case	*data*	$K^?;C*T*Q*W^?$
	sampling	$K;C*U$
	weighting	$K;W^+$
	method	$K;C*Q*$
	annotation	$K;T^+$
summary	*data*	$C*;S^+W^?$
	sampling	$C*;S*U^?$
	weighting	$C*;W^+$
	method	$M*C*;S^+$
	annotation	$C*;T^+$

Using *semi-structured* data modelling [1], the formal set-up of SCs is briefly sketched below. Prefixing with '&' denotes references to sharable (typed) instances; ellipses indicate desirable extensions of SC definition. Primary production delivers 'source' SCs (cf. Table 1); otherwise SCs are 'derived' by applying some intervening transformation (cf. Section 3.2) "lifting" the SC to a specific processing level.

---

type STATISTICALCOMPOSITE =
    {
        (Label : string) ?,
        (Description : string) ?,
        Origin : *source* | *derived*,
        Context : *input* | *transformation* | *output*,
        Format : b_format,
        ProcessingLevel : *raw* | *micro* | *macro* | *adjusted* | ...,
        Components : CONTAINERDIRECTORY,
        StatisticalPopulation : &POPULATION |
                            &POPULATIONCOMPOSITE,
        SamplingPopulation : &POPULATIONCOMPOSITE,
        GeneratedBy : &SOURCE | &TRANSFORMATIONSTEP
    }

Figure 2: Type – statistical composite

In this definition, 'CONTAINERDIRECTORY' refers to a container component listing all containers the SC consists of by container role. In the notation used, 'R' denotes a relational schema definition; parenthesised names are the relation's attribute with key attributes (left) being separated from non-key attributes (right) by a semicolon.

---

type CONTAINERDIRECTORY =
    {
        Attributes : &ATTRIBUTESDIRECTORY,
        Contains : R(Role : *data* | *sampling* | *weighting* |
                            structure | *corrmat* | *footnote* | ...;
                Schema : &SCHEMA,
                Bucket : &BUCKET)
    }

type ATTRIBUTESDIRECTORY =
    {
        Contains : R(ID : ca_id;
                Class : a_class,
                Role : *count* | *sum* | *mean* | *min* | *var* | ... |
                        corr_p | *corr_k* | *corr_s* | ... |
                        sel_prob | *strat_count* | *strat* | ... |
                        weight_base | *weight_cal* | ...,
                CorrespondsTo : &CONTEXTATTRIBUTE,
                ...),
        (UsedBy : &STATISTICALCOMPOSITE) +
    }

Figure 3: Types – container and attributes directory

The 'ATTRIBUTESDIRECTORY' referenced in the container directory consists of a container component listing all composite attributes and a reference to the SCs it is used by.

Similarly, a bucket schema consists of a container listing bucket attributes and references to the SCs it is used by. In addition, the class and format of the schema, or more specifically, of the bucket(s) it describes, are stated.

In the bucket type definition '*BUCKETDATA*' is to be substituted for the actual data relation as determined by bucket schema '&SCHEMA' according to Table 3.

---

type SCHEMA =
    {
        Format : b_format,
        Class : b_class,
        Contains : R(ID : ba_id;
                CorrespondsTo : ca_id,
                ...),
        (UsedBy : &STATISTICALCOMPOSITE) +
    }

type BUCKET =
    {
        Schema : &SCHEMA,
        (UsedBy : &STATISTICALCOMPOSITE) +,
        Contains : *BUCKETDATA*
    }

Figure 4: Types – bucket and schema

3.2 Transformation Structures

SCs provide the operand structure of statistical transformations. Now, metadata structures for both operator and operator application representation are required. Essentially, for each operator, (i) a structure holding *parameters* of the operator, (ii) *preconditions* concerning these parameters as well as (iii) *post-conditions* on the structure of the output SC, must be defined. For the description of parameters, again, semi-structured data typing [1] is used, and conditions are represented as path constraints.

---

type TRANSFORMATIONSTEP =
   {
      (Label : string) ?,
      (Description : string) ?,
      Applies : &OPERATOR,
      (Uses : *PARAMETERS*) ?,
      (Input : &STATISTICALCOMPOSITE) +,
      Output : &STATISTICALCOMPOSITE,
      (Transforms : &CONTEXTATTRIBUTE) *,
      (Generates : &CONTEXTATTRIBUTE) *
   }

Figure 5: Type – transformation step

Each individual transformation step, i.e. each call to an operator, creates a log-entry (Figure 5 above) containing references to the applied operator, to input and output composite(s) as well as to further parameters and, incidentally, transformed / generated variables of the shared transformation context. This way, a transparent record of transformation sequences is established.

As an example of operator representation, consider the case of correlation. Figure 6 defines a structure – actually a subtype of the (union) type definition '*PARAMETERS*' of Figure 5 – for specifying the variables to be correlated:

---

type CORRPARS =
   {
      (Correlate : &CONTEXTATTRIBUTE) 2..*,
      (ClassifiedBy : &CONTEXTATTRIBUTE) *,
      CorrelationMethod : *pearson | spearman | kendall | …*
   }

Figure 6: Type – correlation parameters

To be applicable, several preconditions have to be met; for instance, each specified variable must be contained in the data bucket (i.e., listed in the *data* schema) of the "input" SC. In the path constraints below, *ts* ranges over type TRANSFORMATIONSTEP, *cp* over CORRPARS and *v* over CONTEXTATTRIBUTE.

$\forall$ *ts, cp, v* (Uses(*ts, cp*) $\wedge$
      (Correlate(*cp, v*) $\vee$ ClassifiedBy(*cp, v*)) $\rightarrow$
  $\exists$ *x, y* (Input.Components.Attributes.
      Contains$_{CorrespondsTo(x)}$(*ts, v*) $\wedge$
      Input.Components.Contains$_{Schema(data)}$.
      Contains$_{CorrespondsTo(y)}$(*ts, x*)))

Furthermore, Pearson-type correlation can only be determined for **Q** attributes:

$\forall$ *ts, cp, v* (Uses(*ts, cp*) $\wedge$ Correlate(*cp, v*) $\wedge$
      CorrelationMethod(*cp, pearson*) $\rightarrow$
  $\exists$ *x* (Input.Components.Attributes.
      Contains$_{CorrespondsTo(x)}$(*ts, v*) $\wedge$
      Input.Components.Attributes.
      Contains$_{Class(x)}$(*ts,* Q))

As an example of post-conditions, correlation output must comprise a method bucket (and schema) containing two method attributes referring to the *same* composite attribute (the set of variables correlated) as well as the attribute holding the correlation coefficients computed:

$\forall$ *sc* (GeneratedBy.Applies.OpClass(*sc, correlation*) $\rightarrow$
  $\exists$ *x, c, s, b, y, u, t, v* (
      Components.Attributes.Contains$_{Role(x)}$(*sc, c*) $\wedge$
      c $\in$ {*corr-p, corr-s, corr-k*} $\wedge$
      Components.Contains$_{Schema(corrmat)}$(*sc, s*) $\wedge$
      Components.Contains$_{Bucket(corrmat)}$(*sc, b*) $\wedge$
      Contains$_{CorrespondsTo(y)}$(*s, x*) $\wedge$
      Contains$_{CorrespondsTo(u)}$(*s, t*) $\wedge$
      Contains$_{CorrespondsTo(v)}$(*s, t*) $\wedge$ *u* $\neq$ *v* $\wedge$
      Components.Attributes.Contains$_{Class(t)}$(*sc,* M)))

Now, as an example, consider the application of the correlation operator to a composite "SC3" in order to compute the Pearson correlation matrix of three **Q** variables represented as context attributes CA4, CA5 and CA6, grouped by **C** variable CA1, which amounts to the invocation of

$$T^{correlation}_{\{\&CA4,\&CA5,\&CA6\},\{\&CA1\},pearson}(\&SC3)$$

Essentially, the output composite is a copy of the input composite, with – according to the post-condition stated above – an additional method bucket and appropriate schema attached. Table 4 below shows the container of the method bucket. BA1 corresponds to the **C** attribute, **M** attributes BA2 and BA3 indicate correlated **Q** variables (thus, having as value domain the corresponding set of references as stated in the correlation parameters), and BA4 holds correlation coefficients.

223

Table 4: Container of result method bucket

BA1	BA2	BA3	BA4
1	&CA4	&CA5	-0.68
1	&CA4	&CA6	0.74
...	...	...	...
2	&CA5	&CA6	0.52

Apparently, the created bucket attributes must also be taken into account in the attributes directory of the composite; i.e., an M and an S attribute are added to a copy of the input attributes directory. In analogy, a new container directory must be created, additionally listing the new method bucket and schema.

4. Application Case: Weighting

Weighting is a statistical transformation technique to account for unequal selection probabilities for sampling units (due to sampling design, nonresponse or shortcomings with respect to coverage) or for stratification mismatch. In general, the weighting process consists of three transformation steps: First, base weights compensating for unequal sampling probabilities are calculated. In the second step, base weights are adjusted for unit non-response, and in a final weighting step sample estimates are adjusted according to the population structure. Base weights and weights for unit non-response usually apply to all sample data whereas adjustment weights may be specific for the population parameter of interest.

In what follows, for the sake of clarity, *case_id* dubs the SC's key (K) attribute while *sel_prob* and *strat_count* are bucket attributes providing selection probability per stratum and stratum size, respectively; *weight_base* denotes the resulting W bucket attribute. Moreover, assume that for a SC with case data bucket containing relation R_c and summary sampling bucket containing relation R_s

$$R_c (case\_id; y_1, ..., y_k, ...)$$

$$R_s (y_1, ..., y_k; sel\_prob, strat\_count, ...)$$

base weights have to be calculated. In place of actual bucket attributes, the y_i denote corresponding C composite attributes linking cases to sampling information. Basically, this produces a new SC with additional weighting schema and bucket containing relation

$$R_w (case\_id; weight\_base)$$

Numerically, the classical Horvitz–Thompson estimator calculates

$$weight\_base = 1/ (sel\_prob*strat\_count)$$

That is to say, there are three preconditions to be satisfied in case of application of the *HT_weight* operator. First, the data bucket must comprise a K attribute. Secondly, the (summary) sampling bucket must contain an attribute of role *sel_prob* and another one of role *strat_count*:

$$\forall\ \textit{ts, sc}\ (\ Input(\textit{ts, sc}) \wedge Applies.OpClass(\textit{ts, HT\_weight}) \rightarrow$$
$$\exists\ x, y, u, v\ (\ Components.Attributes.$$
$$Contains_{Role(x)}(\textit{sc, sel\_prob}) \wedge$$
$$Components.Contains_{Schema(sampling)}.$$
$$Contains_{CorrespondsTo(y)}(\textit{sc, x}) \wedge$$
$$Components.Attributes.$$
$$Contains_{Role(u)}(\textit{sc, strat\_count}) \wedge$$
$$Components.Contains_{Schema(sampling)}.$$
$$Contains_{CorrespondsTo(v)}(\textit{sc, u}\ )\ )$$

Moreover, all attributes $y_1, ..., y_k$ defining the strata in the sampling bucket must also be contained in the data bucket (the y_i are "collected" in the $\forall\ x$ clause):

$$\forall\ \textit{ts, sc}\ (\ Input(\textit{ts, sc}) \wedge Applies.OpClass(\textit{ts, HT\_weight}) \rightarrow$$
$$(\forall\ x \exists\ y\ (\ Components.Contains_{Schema(sampling)}.$$
$$Contains_{CorrespondsTo(x)}(\textit{sc, y}) \wedge$$
$$Components.Attributes.$$
$$Contains_{Role(y)}(\textit{sc, strat}) \rightarrow$$
$$\exists\ u\ (\ Components.Contains_{Schema(data)}.$$
$$Contains_{CorrespondsTo(u)}(\textit{sc, y}\ )\ )\ )\ )$$

If any of these conditions is violated execution of the transformation has to be deferred until other transformations (for example, determination of strata sizes or re-design of the data bucket) are applied in order to obtain a suitably shaped input SC.

Note that, here, in order to keep the description short, only summary level sampling data has been considered (details of a complete analysis including case level sampling buckets may be found in [19]).

If preconditions are satisfied, the operator

$$T^{HTWeight} (\&\ SCk)$$

can be invoked for statistical composite *SCk* without any additional parameters. The output SC is obtained from the input SC by the following augmentations: to the attributes directory a new composite attribute with role *weight_base* is added; weighting bucket and schema are adapted to contain R_w (optionally, *weight_base* could also be included in the data bucket); all this has to be reflected in the container directory. Furthermore, a reference to the transformation step is given in the 'GeneratedBy' statement of the output SC.

Calibration weights are used mainly for the adjustment of estimates to known population structures as, for example, marginal counts of the population. To be more precise, the algorithmic solution of calibration weighting is based on a regression type computation using, for some target attribute CA*t*, attributes {CA1, CA2, ..., CA*k*} defining a cross-classification of constraints on weights, i.e. cell values are weighted sums of cell counts in the sample to conform to the corresponding population marginals. Consequently, in analogy to SCs, required information on the underlying population is represented in terms of a POPULATIONCOMPOSITE type (cf. Figure 1). Population composites comprise *data* and *method* buckets. Case data buckets only exist for registers, *p*; method buckets result from population transformations, such as set operations, and equivalence relations for population units. In calibration weighting, a summary data bucket (role *structure*) provides all the weight constraints. Hence, 'PARAMETERS' for calibration weighting may be specified in the following way:

---

type CALWPARS =
  {
    Target : &CONTEXTATTRIBUTE,
    (ConstraintBy : &CONTEXTATTRIBUTE) +,
    Marginal : &POPULATIONCOMPOSITE,
    DistanceFunction : *linear* | *rakingratio* | *logit* | *trunclin*
  }

Figure 7: Type – calibration parameters

Note that the population referenced in the 'Marginal' clause, &P*i*, could differ from the 'StatisticalPopulation' (cf. Figure 2) referenced in the input SC. For example, the statistical population of the SC may be a register, yet, for calibration, a population composite with a summary data bucket holding the population marginals with respect to the cross-classification attributes constraining the weights is necessary.

According to the parameter definition above, the calibration operator can be invoked by

$$T^{CalWeight}_{\&CAt,\{\&CA1,...,CAk\},\&Pi,G}\left(\& SCj\right)$$

where *G* refers to the distance function between original and new calibration weights.

Besides the existence of specified attributes as stated in the correlation example in Section 3.2, the following preconditions have to be verified:

- the input SC already contains base weights,
- all attributes defined in the 'ConstraintBy' clause of the parameters occur in the data buckets of both, statistical and population composite,
- the population composite is of summary format and derived from the statistical population of the input SC.

Post-conditions have to assure that, apart from an additional method bucket, the output SC is created as in the case of base weighting. This method bucket contains a relational table with attributes {CA1, CA2, ..., CA*k*} (serving as primary key), and an attribute for the deviation of the estimate, allowing an evaluation of the calibration procedure.

5. Conclusion

Generally speaking, the present paper introduces strictly formalized, ready-to-implement representations of statistical data and metadata, especially emphasising the aspect of statistical processing. By design, the devised architecture is open to insertion of new transformation procedures as corresponding structures and operators may be developed.

Previous approaches rather attempted to represent a statistical ontology (at least partially); the inclusion of transformations has not yet got beyond simple metadata "throughput" mechanisms just perpetuating input documentation or adding explanatory texts to results of statistical processing, for instance in terms of table footnotes. Hence it should be quite obvious, that the proposed SC-methodology may serve as a foundation for future integrated statistical (meta-) information systems, supporting clear and transparent documentation of even complex statistical data transformations and thus enabling the correct interpretation of transformed data.

A first prototype of the proposed structures has been implemented in SAS®, providing – apart from simple transformations such as selection, projection or univariate statistics – correlation and weighting operators as discussed above [19], as well as several data combination operators applying, for instance, record linkage methods [8]. Actually, any programming environment supporting the management of relational tables as well as the switching between data and schema layer could have been used for the implementation. Yet, SAS® has been chosen because of its versatile statistical functionality and its wide spread among high-level users of statistical systems. Hence, compatibility with the prototype is given for systems holding data in relational format and using SAS® for statistical analysis.

References

[1] S. Abiteboul, P. Buneman, and D. Suciu, *Data on the Web / From Relations to Semistructured Data and XML*, Morgan Kaufmann Publishers, San Francisco, 2000.

[2] J. Bethlehem, J.-P. Kent, A. Willeboordse, and Ypma W., "On the use of metadata in statistical data processing",

Working Paper No. 23, *Work Session on Statistical Metadata (METIS), Conference of European Statisticians*, UN/ECE, Geneva, 1999.

[3] R. Bisdorff, "The Conceptual Model of a Documented Statistical Database", *Proc. New Techniques and Technologies for Statistics*, Bonn, 1992, pp. 310 – 319.

[4] M.J. Colledge, "Statistical Integration Through Metadata Management", *International Statistical Review* 67 (1), ISI, 1999, pp. 79–98.

[5] P.L. Darius, et al., "Modeling Metadata", *Statistical Journal of the United Nations Economic Commission for Europe* 10 (2), 1993, pp. 171–179.

[6] Data Documentation Initiative (DDI), *Codebook Document Type Definition (DTD)*, http://www.icpsr.umich.edu/DDI/CODEBOOK/, 2001.

[7] V. DelVecchio, *Tematiche haziendali (Statistical data and concepts representation)*, Internal Report, Banca d'Italia, 1997.

[8] M. Denk, *Statistical Data Combination: A Metadata Framework for Record Linkage Procedures*, Doctoral thesis, Dept. of Statistics, University of Vienna, 2002.

[9] M. Denk, and K.A. Froeschl, "The IDARESA data mediation architecture for statistical aggregates", *Research in Official Statistics* 3 (1), Luxembourg, 2000, pp. 7–38.

[10] K.A. Froeschl, "A Formal Model Evaluation Approach to the Analysis of Treatment Effects in Paired Sample Data", *Computational Statistics & Data Analysis* 19, 1995, pp. 493–517.

[11] K.A. Froeschl, *Metadata management in statistical information processing*, Springer, Wien–New York, 1997.

[12] K.A. Froeschl, "Metadata Management in Official Statistics – An IT-Based Methodology Approach", *Austrian Journal of Statistics* 28 (2), 1999, pp. 49–79.

[13] K.A. Froeschl, "On Standards of Formal Communication in Statistics", Working Paper No. 16, *Work Session on Statistical Metadata (METIS), Conference of European Statisticians*, UN/ECE, Geneva, 1999.

[14] K.A. Froeschl, and W. Grossmann, "The Role of Metadata in Using Administrative Sources", *Research in Official Statistics* 3 (1), Luxembourg, 2000, pp. 65–82.

[15] K.A. Froeschl, W. Grossmann, J. Lamb, S. McClean, and H. Rutjes, "IDARESA: An Integrated Documentation and Retrieval System for Statistical Aggregates", Project DOSIS No.20478 (EUROSTAT) Summary Deliverable 0.1/5, Annex I, 1998, pp. 50–67.

[16] D.W. Gillman, M.V. Appel, and S.N. Highsmith, Jr., "Building statistical metadata repository at the U.S. Bureau of the Census", Working Paper No. 11, *Work Session on Statistical Metadata (METIS), Conference of European Statisticians*, UN/ECE, Geneva, 1998.

[17] W. Grossmann, "Metadata", in: *Encyclopedia of Statistical Sciences*, Update Vol. 3 (Kotz S. ed.), Wiley, New York, 1999, pp. 811–815.

[18] W. Grossmann, *SUPCOM 98 / Lot 14 Final Report: Metadata Tutorial*, Dept. of Statistics, Vienna, 2001.

[19] W. Grossmann, and P. Ofner, "A self-documenting programming environment for weighting", to appear in COMPSTAT 2002.

[20] K. Hughes, S. Jenkins, and G. Wright, "triple-s XML", *Social Science Computer Review* 18 (4), 2000, pp. 421–433.

[21] ISO/IEC 11179, Specification and Standardization of Data Elements (1994-1996).

[22] E. Malmborg, "Matrix-based interchange of aggregated statistical data", *Proc. Scientific and Statistical Database Management*, Zürich, 1992, pp. 259–273.

[23] L. Nordbäck, "The PC-AXIS Vision, the Liberation of Official Statistics", *Proc. New Techniques and Technologies for Statistics*, Bonn, 1992, pp. 218–225.

[24] H. Papageorgiou, F. Pentaris, E. Theodorou, M. Vardaki, and M. Petrakos, "A statistical metadata model for simultaneous manipulation of both data and metadata", *Journal of Intelligent Information Systems* 17, 2001, pp. 169–192.

[25] M. Rafanelli, "Data Models", in: Michalewicz, Z. (ed.) *Statistical and Scientific Databases*, Ellis Horwood, Chichester, 1991, pp. 109–166.

[26] M. Rafanelli, and F.L. Ricci, "A Functional Model for Macro-Databases", *ACM SIGMOD Records* 20 (1), 1991, pp. 3–8.

[27] M. Rafanelli, and F.L. Ricci, "Mefisto: A Functional Model for Statistical Entities", *IEEE Transactions on Knowledge and Data Engineering* 5 (4), 1993, pp. 670–681.

[28] M.H. Sadreddini, D.A. Bell, and S. McClean, "Framework for query optimisation in distributed statistical databases", *Information and Software Technology* 34 (6), 1992, pp. 363–377.

[29] H. Sato, "Statistical Data Models: From a Statistical Table to a Conceptual Approach", in: Michalewicz Z. (ed.) *Statistical and Scientific Databases*, Ellis Horwood, Chichester, 1991, pp. 167–200.

[30] Bo Sundgren, *Guidelines for the modeling of statistical data and metadata*, Report UN/ECE (Conference of European Statisticians – Methodological Material), New York/Geneva, 1995, 24pp.

[31] UN/EDIFACT (EEG6), *GESMES Version 2.1 Reference Guide 1.0*, 1997. http://forum.europa.eu.int/Public/irc/dsis/eeg6/library?l=/reference_implementation/gesmes_statistical/e6w1d009_pdf/_EN_1.0_&a=d

[32] Ad Willebordse, C. van Duin, and J.W. Altena, "Theme Building by the Art of Cubism", paper presented at the *Int. Seminar on Statistical Output Data Bases and Marketing* in Ottawa, Statistics Netherlands, Voorburg, 2001.

[33] World Wide Web Consortium (W3C), Resource Description Framework, http://www.w3.org/RDF/, 2001.

Information-Theoretic Disclosure Risk Measures in Statistical Disclosure Control of Tabular Data*

Josep Domingo-Ferrer
Universitat Rovira i Virgili
Dept. of Comp. Eng. and Maths
Av. Països Catalans 26
E-43007 Tarragona, Spain
e-mail jdomingo@etse.urv.es

Anna Oganian
Universitat Rovira i Virgili
Dept. of Comp. Eng. and Maths
Av. Països Catalans 26
E-43007 Tarragona, Spain
e-mail aoganian@etse.urv.es

Vicenç Torra
IIIA - CSIC
Campus UAB
E-08193 Bellaterra, Spain
e-mail vtorra@iiia.csic.es

Abstract

Statistical database protection is a part of information security which tries to prevent published statistical information (tables, individual records) from disclosing the contribution of specific respondents. This paper shows how to use information-theoretic concepts to measure disclosure risk for tabular data. The proposed disclosure risk measure is compatible with a broad class of disclosure protection methods and can be extended for computing disclosure risk for a set of linked tables.

Keywords: *Statistical database protection, Tabular data protection, Information theory, Disclosure risk.*

1 Introduction

The most typical output offered by national statistical agencies is tabular data. Tables are central in official statistics: many survey and census data are categorical in nature, so that their representation as cross-classifications or tables is a natural reporting strategy. Tabular data being thus aggregate data, one is tempted to think they are not supposed to contain information that can reveal the contribution of particular respondents. However, as noted in [6], in many cases table cells do contain information on a single or very few respondents, which implies a disclosure risk for the data of those respondents. In these cases, disclosure control methods must be applied to the tables prior to their release.

A number of disclosure control methods to protect tabular data have been proposed (see [10, 3] for a survey). We next list the main principles underlying those methods:

Cell suppression If a table cell is deemed sensitive, then it is suppressed from the released table (primary suppression). If marginal totals or other linked tables are also to be published, then it may be necessary to remove additional table cells (secondary suppressions) to prevent primary suppressions from being computable. Secondary suppressions should be chosen in a way such that the utility of the resulting table is maximized.

Rounding A positive integer b (rounding base) is selected and all table cells are rounded to an integer multiple of b. Controlled rounding is a variant of rounding in which table additivity is preserved (*i.e.* rounded rows and columns still sum to their rounded marginals).

Table redesign Categories used to tabulate data are recoded into different (often more general) categories so that the resulting tabulation does not contain sensitive cells any more. A simple redesign could be to combine two rows containing sensitive cells to obtain a new row without sensitive cells.

Sampling A table is released which is based on a sample of the units on which the original table was built.

Swapping and simulation In data swapping, units are swapped so that the table resulting from the swapped data set still preserves all k-dimensional margins of the original table. A more elaborate version of swapping was proposed in [5], whereby the original table is replaced by a random draw from the exact distribution under the log-linear model whose minimal sufficient statistics correspond to the margins of the original table. Further extensions of this idea would lead to drawing a synthetic table from the full distribution of all possible tables with the same margins of the original table.

*Work partly funded by the European Union under project "CASC" IST-2000-25069

227

As noted by [3], any attempt to compare methods for tabular data protection should focus on two basic attributes:

1. *Disclosure risk*: a measure of the risk to respondent confidentiality that the data releaser (typically a statistical agency) would experience as a consequence of releasing the table.

2. *Data utility*: a measure of the value of the released table to a legitimate data user.

A first approach to measuring data utility is to take generic measures such as the reciprocal of the mean squared error between the original and the released tables [3]. While this may be useful as a crude approach, a more accurate utility assessment must necessarily take into account the specific data uses the user is interested in. Thus, strictly speaking, there is no universal data utility measure.

The situation for disclosure risk is quite different. There is a number of sensitivity rules which are used to decide whether a particular table cell can be safely released. However, these rules operate on an *a priori* basis: the original data are examined *before* they are protected and the rules are used to determine whether the data can be released as they stand or should rather be protected. Note that the disclosure risk incurred if a particular protected table is released is not actually measured by sensitivity rules.

1.1 Our contribution

We will show in this paper that it is possible to use information-theoretic concepts to define a general disclosure risk measure which takes protected information into account. This measure can be termed *a posteriori*, as it measures disclosure risk *after* table protection has been used: the protected table is taken as input to compute disclosure risk and will only be released if disclosure risk is deemed low enough by the data protector. The proposed measure applies to a broad class of disclosure protection methods and is computable in practice.

Section 2 describes some sensitivity rules currently used. In Section 3, a new measure of disclosure risk based on the reciprocal of conditional entropy is proposed as an *a posteriori* alternative to sensitivity rules. Section 4 describes an application of the proposed disclosure risk measure to different table protection methods, both for simple tables and for linked tables. Section 5 is a conclusion.

2 Background on sensitivity rules for tables

For magnitude tables (normally related to economic data), there are two widely accepted sensitivity rules:

$n - k$**-dominance** In this rule, n and k are two parameters with values to be specified. A cell is called sensitive if

the sum of the contributions of n or fewer respondents represents more than a fraction k of the total cell value.

pq**-rule** The prior-posterior rule is another rule gaining increasing acceptance. It also has two parameters p and q. It is assumed that, prior to table publication, each respondent can estimate the contribution of each other respondent to within less than q percent. A cell is considered sensitive if, posterior to the publication of the table, someone can estimate the contribution of an individual respondent to within less than p percent. A special case is the $p\%$-rule: in this case, no knowledge prior to table publication is assumed, *i.e.* the pq-rule is used with $q = 100$.

For tables of counts or frequencies (normally related to demographic data), a so-called **threshold rule** is used. A cell is defined to be sensitive if the number of respondents is less than a threshold k.

According to [7, 8, 4], the $n - k$ dominance rule is the most popular one for magnitude tables, followed by the $p\%$-rule and the pq-rule. Yet, it is significant to note that the U.S. Census Bureau switched in 1992 from the $n - k$ rule to the $p\%$-rule, and the German Statistisches Bundesamt did the same in 2001.

Due to their nature, the above-mentioned rules are strongly oriented toward certain methods, in particular cell suppression (a sensitivity rule is used to decide which cells should be primarily suppressed) or table redesign. Their usefulness for other metods like rounding is less clear. This is confirmed by the survey [4]: sensitivity rules are nearly always used in conjunction with cell suppression. Furthermore, being *a priori*, sensivity rules do not always correctly reflect the disclosure risk caused by the release of a particular table. The following example reported in [9] is an illustration.

Example 1 (Robertson and Ethier, 2002) *In the dominance rule, let $n = 1$ and $k = 60\%$. Then a cell with value 100 and contributions 59, 40, 1 is declared* not *sensitive, while a cell with value 100 and contributions 61, 20, 19 would be declared sensitive. Assume now that the second largest respondent of both cells knows the total 100 and is interested in estimating the contribution of the largest respondent. Then, for the $(59, 40, 1)$ cell, she removes her contribution and gets an upper bound $100 - 40 = 60$ for the largest contribution. For $(61, 20, 19)$ the upper bound she gets is $100 - 20 = 80$, much farther from the real largest contribution. So the cell declared non-sensitive by the rule allows better inferences than the cell declared sensitive!*

3 Conditional entropy as a general measure of disclosure risk

The discussion in Section 2 points out that, as useful as sensitivity rules can be in combination with specific table protection methods, they may fail to capture the notion of disclosure risk in a correct way. Our proposal here is to use the reciprocal of Shannon's conditional entropy to express disclosure risk in a unified manner which takes protected data into account.

Entropy-based measures were already discussed in [10] for computing information loss at the table level, but not for computing disclosure risk. However, the authors of [10] do not believe entropy is a practical information loss measure. We support their opinion with the following example.

Example 2 *Assume we use rounding with integer base b to protect a table. The entropy-based information loss measure defined in [10] is the reciprocal of the number of original tables whose rounded version matches the published rounded table (i.e. the number of original tables "compatible" with the published one). The number of compatible tables depends on the rounding base b, but is independent on how close the published rounded values are to the original values. Thus, the entropy-based information loss measure is the same when the original table exactly corresponds to the rounded table (which happens when all cell values in the original table are multiples of b) and when all differences between corresponding cell values in the original and rounded tables are close to $b/2$. This does not seem to adequately reflect the utility of the published data.*

In [3], the reciprocal of Shannon's entropy (not conditional entropy) is suggested as measure of disclosure risk at the cell level. If p_ω is the probability (as seen by the intruder) that the value of a specific cell is ω, then the disclosure risk for that cell is measured as

$$DR(X) = 1/(-\sum (p_\omega \log_2 p_\omega)) \qquad (1)$$

The summation in Expression (1) extends over all possible values of cell X. What is not clear here is how to compute p_ω, that is, what distribution should be chosen. In fact, the particular distribution for an intruder depends on the knowledge held by that intruder: if the intruder is an outsider, then the only information she has is the released table; if the intruder is an insider (one of the respondents who contribute to the particular cell), then she knows her own contribution and it may be easier for her to estimate the contribution of other respondents.

The information held by an intruder does not only depend on her being outsider or insider; it clearly depends also on what information has previously been published and on how that information has been protected. The following example illustrates this.

Example 3 *Assume we have an n-dimensional table whose cells are deemed sensitive, and therefore cannot be released. Only some 2-dimensional (or $(n - i)$-dimensional) tables are released, which have been obtained as projections of the n-dimensional table. Due to their origin, the released tables are linked tables, so the uncertainty about a cell value in the n-dimensional table is conditional to the particular tables released so far.*

The above discussion suggests that a natural measure for disclosure risk is the reciprocal of conditional entropy

$$DR(X) = 1/H(X|Y = y) = 1/(-\sum_x p(x/y) \log_2 p(x/y))$$
$$(2)$$

where X is a variable representing an original cell and Y is a variable representing the intruder's knowledge (which is suposed to be equal to some specific value y). The intuitive notion behind Expression (2) is that, the more uncertainty about the value of the original cell X (which depends on the constraints $Y = y$), the less disclosure risk (and conversely). There are two practical problems to computing Expression (2):

1. Finding the set $S_y(X)$ of possible values of X given the constraints y.

2. Estimating the probabilities $p(x|y)$, *i.e.* the probability of the cell X being x given that Y is y.

As noted by [10] when discussing entropy-based information loss measures, taking the uniform probability distribution over the set $S_y(X)$ can make sense for some disclosure control methods. Using the uniform distribution, Expression (2) is simplified to

$$DR_{unif}(X) = 1/\log_2 m(S_y(X)) \qquad (3)$$

where $m(S_y(X))$ is the number of possible values of the cell in $S_y(X)$.

Note 1 (On $m(S_y(X))$) *We assume in what follows that table cells take values in a discrete domain: either integer values or real values with a fixed number of decimal positions. This is the usual case in published statistical tables: count tables consist of integer values and magnitude tables consist of either integer values or real values with limited precision. Thus the set $S_y(X)$ of possible values is enumerable and it makes sense to speak of $m(S_y(X))$ as the number of cell values in $S_y(X)$.*

4 Application to several table protection scenarios

We show in this Section how to compute Expression (3) for several disclosure control methods for tables; the case of linked tables will also be discussed.

Table 1. A table with suppressed cells

Economic activity	Size class					Total
	4	5	6	7	8	
2,3	80	253	54	0	0	387
4	641	3694	2062	746	0	7143
5	592	x_1	329	x_2	1440	3898
6	57	x_3	946	x_4	2027	4281
7	78	0	890	1719	1743	4430
Total	1148	4353	4281	4847	5210	20139

Table 2. A table with two rows combined

Economic activity	Size class					Total
	4	5	6	7	8	
2,3	80	253	54	0	0	387
4	641	3694	2062	746	0	7143
5,6	649	406	1275	2382	3467	8179
7	78	0	890	1719	1743	4430
Total	1148	4353	4281	4847	5210	20139

4.1 Cell suppression

The disclosure risk computation for cell suppression is illustrated by extending an example provided in [10]. Let Table 1 be a table from which four cells x_1, x_2, x_3 and x_4 have been suppressed. Assume that the suppressed values are integer.

According to the definition given in Section 3, the disclosure risk for each suppressed cell is the reciprocal of one of the following conditional entropies:

$$H(x_1|x_1 + x_2 = 1537, x_1 + x_3 = 406, x_i \geq 0)$$
$$H(x_2|x_1 + x_2 = 1537, x_2 + x_4 = 2382, x_i \geq 0)$$
$$H(x_3|x_1 + x_3 = 406, x_3 + x_4 = 1251, x_i \geq 0)$$
$$H(x_4|x_2 + x_4 = 2382, x_3 + x_4 = 1251, x_i \geq 0)$$

Expressions (4) contain constraints y_i for each suppressed cell x_i which allow $m(S_{y_i}(x_i))$ to be computed by solving two linear programming (LP) problems (one maximization and one minimization) and subtracting the solutions. In the case of Table 1, minimizations and maximizations bound every cell as follows: $0 \leq x_1 \leq 406$, $1131 \leq x_2 \leq 1537$, $0 \leq x_3 \leq 406$ and $845 \leq x_4 \leq 1251$. By substracting the bounds we obtain $m(S_{y_i}(x_i)) = 407$ for $i = 1, 2, 3, 4$. Using Expression (3), we can compute $DR_{unif}(x_i) = 1/\log_2 407 = 0.115$ for every cell.

4.2 Rounding

When the table is protected by rounding, the cell entropy conditional to the rounded table depends on the rounding base b. In a rounded table without marginals, if the value of a cell x_i' is n_ib (i.e. n times the rounding base), then we know that the original cell x_i must lie in the interval $I_i = [(n_i - 1/2)b, (n_i + 1/2)b)$. Thus, $DR_{unif}(x_i) = 1/\log_2 m(I_i)$, where $m(I_i)$ is the number of possible cell values in I_i (keep in mind that cell values are either integer or with a fixed number of decimal positions).

4.3 Table redesign

This case is very similar to cell suppression. Imagine that the sensitive cells in Table 1 are protected by combining rows with $Economic\_activity = 5$ or 6. This yields Table 2.

Let us label the six cells in the original row with $Economic\_activity = 5$ as x_1 through x_6 and the six cells in $Economic\_activity = 6$ as x_7 through x_{12} (x_6 is the marginal of the first row and x_{12} is the marginal of the second row). Then the following equalities hold:

$$
\begin{aligned}
x_1 + x_2 + x_3 + x_4 + x_5 - x_6 &= 0 \\
x_7 + x_8 + x_9 + x_{10} + x_{11} - x_{12} &= 0 \\
x_1 + x_7 &= 649 \\
x_2 + x_8 &= 406 \\
x_3 + x_9 &= 1275 \quad (4) \\
x_4 + x_{10} &= 2382 \\
x_5 + x_{11} &= 3467 \\
x_6 + x_{12} &= 8179 \\
x_i &\geq 0 \text{ for } i = 1, \cdots, 12
\end{aligned}
$$

From the above, $m(S_{y_i}(x_i))$ and $DR_{unif}(x_i)$ are computed in a way analogous to the case of cell suppression.

4.4 Linked tables

Let us consider the three-dimensional table ASR formed by cells $z_{a_i s_j r_k}$, where each cell denotes the total turnover of businesses with activity a_i and size s_j in region r_k. Assume that table ASR is not released because every cell in it is considered sensitive. Instead of ASR, some of the following tables obtained by bidimensional projection are released: $AS = \{z_{a_i s_j}\}$, which breaks down turnover by activity and business size, $AR = \{z_{a_i r_k}\}$, which breaks down turnover by activity and region, and $SR = \{z_{s_j r_k}\}$, which breaks down turnover by size and region. Assume three scenarios: 1) only AS is released; 2) AS and AR are released;

3) AS, AR and SR are released. The disclosure risk of cell $z_{a_i s_j r_k}$ in each scenario can be expressed as:

$$DR_{unif}(z_{a_i s_j r_k}|AS) = 1/H(z_{a_i s_j r_k}|z_{a_i s_j} = \sum_k z_{a_i s_j r_k})$$
$$(5)$$

$$DR_{unif}(z_{a_i s_j r_k}|AS, AR)$$
$$= 1/H(z_{a_i s_j r_k}|z_{a_i s_j} = \sum_k z_{a_i s_j r_k}, z_{a_i r_k} = \sum_j z_{a_i s_j r_k})$$
$$(6)$$

$$DR_{unif}(z_{a_i s_j r_k}|AS, AR, SR) = 1/H(z_{a_i s_j r_k}|$$
$$z_{a_i s_j} = \sum_k z_{a_i s_j r_k}, z_{a_i r_k} = \sum_j z_{a_i s_j r_k}, z_{s_j r_k} = \sum_i z_{a_i s_j r_k})$$
$$(7)$$

The released tables impose constraints on the possible cell values of the table ASR. Such constraints actually determine the simplexes $S_{AS}(z_{a_i s_j r_k})$, $S_{AS,AR}(z_{a_i s_j r_k})$ or $S_{AS,AR,SR}(z_{a_i s_j r_k})$ where $z_{a_i s_j r_k}$ should lie. By solving one LP maximization and one LP minimization for each $z_{a_i s_j r_k}$, an interval where the cell lies can be determined. Then, the cell disclosure risk is computed using Expression (3). If a cell is too closely bounded, then its disclosure risk is too high and disclosure control methods must be used.

When the disclosure control method chosen is cell suppression, it is important to take into account the number of independent constraints imposed by the released linked tables.

Another point we have to take into account is that disclosure risk is different for different users. Let us imagine that, when solving one LP maximization and one LP minimization for $z_{a_i s_j r_k}$, we find that $995 \leq z_{a_j s_j r_k} \leq 1004$. If company A is the second largest contributor to this cell with a turnover of, say, 400, then company A knows that the largest contributor (company B) has a turnover between 595 and 604. Thus, company A is able to estimate the turnover of company B within 1% of its value. However, the uncertainty of an outsider about the turnover of company B is almost 170% of its value: the outsider only knows that the turnover of the largest contributor is between $\epsilon > 0$ and 1004. Therefore, for an insider (respondent contributing to the cell), the measure of disclosure risk $1/H(z_{a_i s_j r_k}|$released tables$)$ should be replaced by $1/H(z_{a_i s_j r_k}|$released tables, insider's contribution$)$.

5   Conclusion

We have given arguments in favor of using the reciprocal of Shannon's conditional entropy as an *a posteriori* disclosure risk measure for tabular data. While Shannon's entropy may not be suitable to evaluate the impact of disclosure control on table utility, it turns out to be extremely useful to quantify disclosure risk. As shown in Section 4,

computing disclosure risk in this way can easily be done for different disclosure control methods, both with simple tables and linked tables. *A priori* risk assessment through sensitivy rules is indeed useful to locate table cells to be protected, but we claim that the final decision to release a table should be based on an *a posteriori* measure of disclosure risk like the one discussed in this paper, which takes the actual protected table into account.

Acknowledgments

Thanks go to Sarah Gießing for useful comments on earlier versions of this paper. Comments by three anonymous referees are also gratefully acknowledged.

References

[1] L. H. Cox, "Disclosure risk for tabular economic data", in *Confidentiality, Disclosure and Data Access*, eds. P. Doyle, J. Lane, J. Theeuwes and L. Zayatz. Amsterdam: North-Holland, pp. 167-183, 2001.

[2] D. E. Denning, *Cryptography and Data Security*. Reading, MA: Addison-Wesley, 1982.

[3] G. T. Duncan, S. E. Fienberg, R. Krishnan, R. Padman and S. F. Roehrig, "Disclosure limitation methods and information loss for tabular data", in *Confidentiality, Disclosure and Data Access*, eds. P. Doyle, J. Lane, J. Theeuwes and L. Zayatz. Amsterdam: North-Holland, pp. 135-166, 2001.

[4] F. Felsö, J. Theeuwes and G. G. Wagner, "Disclosure limitation methods in use: Results of a survey", in *Confidentiality, Disclosure and Data Access*, eds. P. Doyle, J. Lane, J. Theeuwes and L. Zayatz. Amsterdam: North-Holland, pp. 17-42, 2001.

[5] S. E. Fienberg, U. E. Makov and R. J. Steele, "Disclosure limitation using perturbation and related methods for categorical data", *Journal of Official Statistics*, 14: 485-512, 1998.

[6] S. Gießing, "Nonperturbative disclosure control methods for tabular data", in *Confidentiality, Disclosure and Data Access*, eds. P. Doyle, J. Lane, J. Theeuwes and L. Zayatz. Amsterdam: North-Holland, pp. 185-213, 2001.

[7] J. Holvast, "Statistical dissemination, confidentiality and disclosure", in *Proceedings of the Joint Eurostat/UNECE Work Session on Statistical Data Confidentiality*. Luxembourg: Eurostat, pp. 191-207, 1999.

[8] T. Luige and J. Meliskova, "Confidentiality practices in the transition countries", in *Proceedings of the Joint Eurostat/UNECE Work Session on Statistical Data Confidentiality*. Luxembourg: Eurostat, pp. 287-319, 1999.

[9] D. Robertson and R. Ethier, "Cell suppression: Theory and experience", in *Inference Control in Statistical Databases*, LNCS 2316, ed. J. Domingo-Ferrer. Berlin: Springer-Verlag, pp. 9-21, 2002.

[10] L. Willenborg and T. de Waal, *Statistical Disclosure Control in Practice*. New York: Springer-Verlag, 2001.

Use of Metadata Registries for Searching for Statistical Data

Chris Nelson
Dimension EDI Ltd.
chris@dimension-edi.com

Abstract

The Internet has spawned an insatiable appetite for information. This appetite cannot be satisfied without useful search tools. These tools rely on metadata in order to give targeted and meaningful results to searches, and the Internet world is realising that metadata, its management, and the means to search these metadata, are an integral part of the Internet world.

The EC 5th Framework project COSMOS [1] aims to demonstrate the use of metadata in this way by building a common model for describing a statistical data set in terms of metadata, populating a metadata registry with descriptions consistent with the model, and searching the registry to find the data sets, and finally to access the data sets in the relevant repository.

The Problem

How can a user of statistics find the data she requires when she is not sure exactly what data is available, which organisation has the data, how she can search for these data, and what tools to use to manipulate these data when they are retrieved?

For the publisher of statistics, how can he ensure he is reaching the maximum target audience for the statistics, and how can he ensure that he can disseminate the data in a way that is most useful to the user.

The answer for the user is to access, not a textual search engine, but a search facility based on a metadata registry.

The answer for the publisher is to populate the metadata registry so that the data can be found, to process a standardised request message for the data, and to publish a standardised data dissemination message.

This scenario requires, of course, standards. In fact, it requires three different types of standards.

1. Internationally agreed (web) services definition language

2. Internationally agreed metadata registry access, update, and retrieval
3. Domain specific data structures for data and metadata

The first two are already in place. The third needs to be developed by the statistical community.

The New Internet Paradigms

There are two new technologies that could influence dramatically the way statistical organisations publish information on the web.

The first new paradigm is called web services. For some applications web services will replace traditional web access functionality because the web services paradigm offers the functionality of interactive services using standard web infrastructure.

The second paradigm is the separation of metadata from data and placing the metadata in a registry. The registry is a metadata database. There are international standards for registry services and registry access (for update, query and retrieve functions).

Both of these new technologies will be supported by a range of software from major vendors, some of it free or open source. The statistical world can take advantage of these technologies, and supporting software, to change the way they disseminate data.

Web Services

Web services are a result of the confluence of three streams of influence - the emergence of the infrastructure of the Internet, the emergence of XML as a standard, and the adoption of common standards by infrastructure providers in place of proprietary standards.

At the technical level, web services are built using:

*   Universal connectivity by a ubiquitous standard (Internet, HTTP).

*   Declarative specifications of protocols accepted and interfaces exposed
    *   SOAP –simple object access protocol [2]

o WSDL – web services definition language [3]

- Registration of the web service in a web based registry (such as UDDI – Universal Description, Discovery, and Integration [4])

Both SOAP and WSDL standards making are W3C activities.

The web services paradigm is set to replace existing inter-operable technologies such as CORBA, JMI, and COM *for appropriate applications*. Web services are delivered by co-operating loosely coupled components. The tight bindings between objects in a COM or CORBA environment, are replaced by loose bindings in an XML environment. SOAP message are encoded using XML. In other words, the API interface is realized as an XML schema. The client applications communicate with the server application by means of XML documents.

Registry

There is now an international standard for metadata registries [6] [7] comprising a conceptual model, logical APIs, and physical XML DTDs for implementing the APIs. These standards are integrated with the Internet standards for integrating loosely coupled components - Simple Object Access Protocol (SOAP) for method invocation, and HTTP for transmission. Together, the Registry XML APIs, SOAP, and HTTP can deliver the new Web Services paradigm for searching statistical metadata and retrieving statistical data.

The OASIS Registry model

Whilst this registry model was built with the specific intention of facilitating business to business partnerships and transactions (by allowing the storage and retrieval of XML schema documents, process descriptions, core components and other objects used to facilitate e-business), and describing these objects using metadata, the registry concept is no different if the target objects are statistical data sets and tables, which are described using metadata.

Having said this, the focus of software development on commercial registries has to date concentrated on their use as registries of metadata to support the discovery of potential buyers and sellers, products and services, and XML schemas to support business processes. This leaves a challenge for the statistical world to see if these same registries, or other registries that are built to the same OASIS architecture, can be used to support the discovery of statistical data sets.

COSMOS

COSMOS is an EC 5th framework project, which is, in fact, a Cluster of a number of existing 5th Framework projects, each of which is building a metadata repository.

The main objectives of COSMOS are to:
- build better metadata repositories by exchanging ideas and experiences in using statistical metadata systems for the individual projects;
- identify a common set of metadata objects, with agreed definitions, attributes and methods;
- implement a demonstration subset of these objects to show interoperability of the developed systems;
- define a methodology for further developing this interoperability.

Within the project a variety of sub projects are underway. One such sub project is the linking of a number of the metadata repositories using the web services paradigm and investigating whether the OASIS/ebXML registry services specification can be used to discover statistical data and to retrieve these data from the appropriate repository. In simple terms, we intend to implement some of the XML interfaces defined in the OASIS/ebXML registry specification and to use these interfaces to offer a single point of entry for a search application, regardless of which repository hosts the target data.

The schematic of the system is shown below.

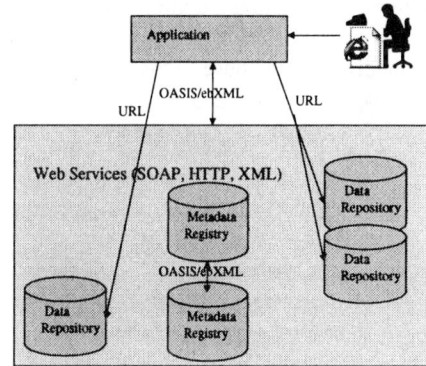

Schematic of Registry Interoperability in the COSMOS project

The first stage of the demonstration is to use a single metadata registry to discover relevant data sets, and to retrieve these from the correct repository, and to view, manipulate, or store the data retrieved.

The second stage will be for the first registry to spawn a query to one or more additional registries and to pass the results back to the user.

The software demonstration shows this first stage scenario working, and also shows what is happening "under the covers" such as the XML passing between the application and the registry in order to achieve this.

The Role of the Metadata Model

A key ingredient in this sub project is the development of a model, or adoption of an existing model, for the metadata that will be held in the registry. Some of the 5th Framework R&D projects related to COSMOS are developing such models, and a separate 5th Framework project, Metanet [8], intends to develop a statistical metamodel which embraces all stages of the statistical lifecycle. The COSMOS project also has a sub project which aims to define a common model for data set and cube.

The role of this metadata model is to allow registry access applications to populate the registry, and to search for and retrieve the metadata. It is probably necessary at this stage to distinguish between the two models – the registry model and the "metadata" model, which I will call the "search" model.

The OASIS registry model is a conceptual model that is used to define the interfaces to the registry, and can be a useful aid for people designing the storage for a registry. The registry can be implemented in any technology – database, XML storage, flat files. The key to the registry standard is the interface specification: these interfaces are exposed to the applications accessing it for update or retrieval. These must comply to the ebXML/OASIS Registry Services Specification [7].

So, for example, the registry model has a RegistryObject class, an Association class, a Slot class and a Classification class. There are many other classes, but we will concentrate on these. This part of the model is reproduced below (note that class attributes are not shown).

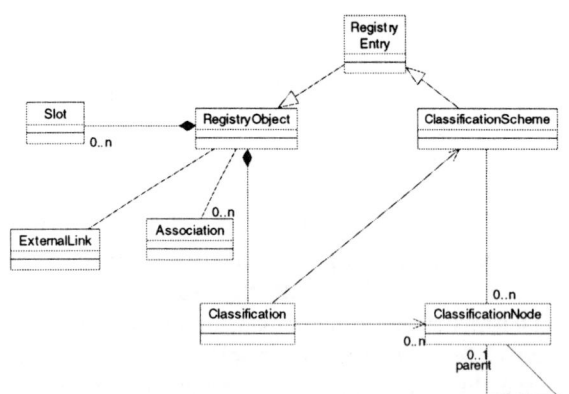

Part of Registry Information Model [6]

Note that a RegistryObject can be associated to any other RegistryObject via the Association class. This makes the registry model highly extensible and it is possible to build virtually any type of domain model as an extension of this model. The RegistryObject has an objectType attribute which identifies the underlying (extended) class

name. The Slot is used to add additional attributes to a RegistryObject. The ExternalLink is used to give a URI of an object associated to the RegistryObject, but which is not in the registry itself. Any RegistryObject can be classified and this Classification can be used as a part of a search function supported by the registry.

On its own, the registry model is not sufficient to give end user functionality, and this is not its purpose. Its purpose is to enable registries to be built that will give sufficient support for an application to use such a registry for metadata storage and retrieval. For instance, this model has no concept of a business process, a core component, an XML schema, and how these are related to each other (e.g. which business process uses which schema). Yet this is how the registry will be used to support the metadata concerned with e-business – the processes, the XML specifications and who uses them.

Similarly, the registry will have no semantic knowledge of the objects that are registered to support the search for statistical data, where they are located and how they can be accessed and retrieved. This semantic knowledge is in the application which will use the registry and which will be built according to the specification of the Search model

A Simple Search Model

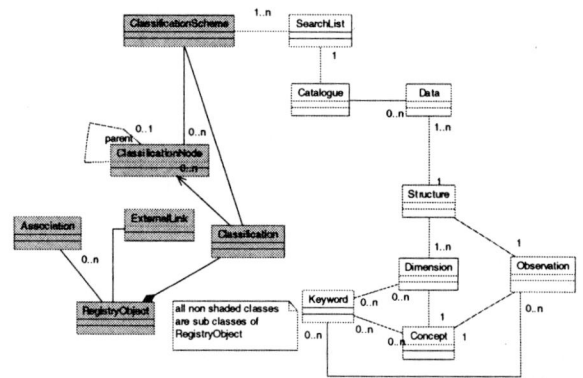

A Simple Model to Support the Search for Statistical Data

The model can be used to search for statistical datasets. It is a very simple model, too simple for a real system, but sufficiently defined to give a flavour of the role of the metadata registry.

The classes that are shaded are the classes in the OASIS/ebXML Registry Information Model. Therefore, each of the "search" model classes (not shaded) is a RegistryObject and is associated to other classes by the fact that a RegistryObject can have an association with another RegistryObject. The idea of this model is to support a search for a dataset that contains the data required. This can be done in two ways in this model. The first way is to classify a dataset in terms of one or more ClassificationNodes in one or more ClassificationSchemes. The second is to register metadata that describes the data (e.g. Concept, Keyword).

The search engine can load the relevant ClassificationSchemes (i.e. the schemes that support the search). These schemes are identified in the model as the SearchList. The search engine allows the user to choose a ClassificationNode from each of the lists, or to choose a wildcard. The Data that are classified according to the criteria chosen are identified, and the ExternalLink to them is returned to the search engine. Note that this is a metadata registry and the actual data resides in a repository, probably on another web site.

Note that the ebXML model and registry services specification [7&8] specifies how objects can be classified, and how objects can be searched based on the way they are classified. The ebXML specification also allows searching of textual or numerical metadata attached to the object (e.g. the value in the Slot object).

A registry built according to this search model could be used to find data that is in multi-dimensional data sets, time series data, and even non aggregated data. The client search application would read the ClassificationSchemes linked to the SearchList and populate drop down lists, which themselves can be hierarchic. These can be used to give a "drill down" approach to finding data. In addition the client application can allow the user to make a contextual search based upon values she enters for one or more of Keyword, Dimension, and Concept.

An Example

The registry contains the following classifications as part of its search list:

- Time,
- Country
- Topic

```
<QueryObject>
    <Object objectType = "Dataset"/>
</QueryObject>
<SelectionCriteria>
    <ClassificationSelection
informationBankName = "Classifications"
classificationSchemeId = "TopicList"
classificationNodeId = "Retail Price Index"/>
    <ClassificationSelection
informationBankName = "Classifications"
classificationSchemeId = "CountryList"
classificationNodeId = "USA"/>
    <ClassificationSelection
informationBankName = "Classifications"
classificationSchemeId = "TimeList"
classificationNodeId = "Jan-2000:Dec-2001"/>
</SelectionCriteria>
```

Fragment of XML for Querying a Registry

The XML query that could be generated for such a model could look something like the one shown above.

This XML fragment would be generated by a client application, possibly hosted in the browser, and sent using HTTP to a web server. The web server would host a registry service and would initiate a registry search and return the results to the client application.

Advantage of Registries

However, these mechanisms are not unlike the facilities offered by statistical organisations today on their web sites. So why do we need a registry?

Well, let's go back to the original problem statement – "How can a user of statistics find the data she requires when she is not sure exactly what data is available, which organisation has the data, how she can search for these data"? The answer is that a search portal service can provide a global search capability. Such a service can be based on a metadata registry. A central registry, or a set of linked registries, can solve this problem. The registry contains only the metadata – the data still resides in the repository (this could be a prepared data set or it could be created dynamically by the publisher from a database).

The key to the success of such a registry is for the statistical community to develop and agree the underlying Search model, and the standards that will be used to access the registry and the data repositories. The use of such a facility and the recognition of the need to adopt standards in this area is already recognised by major international statistical organisations such as the IMF, as part of the SDMX initiative [8].

References

[1] Cluster of Systems of Metadata System for Official Statistics (www.epros.ed.ac.uk/cosmos/)
[2] SOAP: www.w3.org/2000/XP/Group
[3] WSDL: http://www.w3/TR/wsdl
[4] UDDI: http://www.uddi.org
[5] Metadata Network http://www.epros.ed.ac.uk/metanet/
[6] OASIS/ebXML Registry Information Model Version 2.0 18 December 2001 http://www.oasis-open.org/committees/regrep/documents/2.0/specs/ebRIM.pdf
[7] OASIS/ebXML Registry Services Specification v2.0 6 December 2001 http://www.oasis-open.org/committees/regrep/documents/2.0/specs/ebRS.pdf
[8] Report of the International Monetary Fund on Common Open standards for the Exchange and Sharing of Socio-economic Data and Metadata: the SDMX Initiative: 1 March 2002 - http://www.sdmx.org

Posters

The Edinburgh Mouse Atlas and Gene-Expression Database: A Spatio-Temporal Database for Biological Research

Albert Burger*†, Richard Baldock*, Yiya Yang*,
Andrew Waterhouse*, Derek Houghton*, Nick Burton*, Duncan Davidson*
Human Genetics Unit, Medical Research Council, Edinburgh

Abstract

The Edinburgh Mouse Atlas Project (EMAP) has developed a digital atlas of mouse development which provides a bioinformatics framework to spatially reference biological data. The EMAP core database contains 3D grey-level reconstructions of the mouse embryo at various stages of development [3], a systematic nomenclature of the embryo anatomy, and defined 3D regions (domains) of the embryo.

The reconstructions define a spatial framework for mapping data. Software has been developed to re-align serial sections and create a 3D voxel model of a mouse embryo which can be resectioned at any arbitrary 3D orientation and position so that experimental data can be mapped onto it [2]. The anatomical nomenclature is used as a controlled vocabulary for annotating and describing gene-expression patterns and is currently organised as a "part-of" hierarchy. The notion of groups of anatomical terms is used to support alternative structuring of the hierarchy. Anatomical domains are defined 3D regions and provide the mapping between the voxel model and the anatomical terms in the nomenclature.

Data from an in-situ gene-expression database is spatially mapped onto the atlas allowing the users to query gene-expression patterns using the 3D embryo model as a reference. The mouse atlas and gene-expression databases are publicly accessible through a set of Web-based tools, which also provide some level of integration with other bioinformatics resources on the Internet, such as the Mouse Genome Informatics (MGI) database at the Jackson Laboratory, USA.

The system consists of a set of tools and databases, some of which reside locally on the Mouse Atlas hosts, others are remote. The development of tools has been separated from the database side through a layered software architecture approach. The middleware layer is primarily CORBA-based, but also makes use of Java servlets. Data is primarily stored in an object-oriented database system (ObjectStore), though the anatomical nomenclature is also exported in XML and GO formats. The voxel models and the underlying anatomy data are fairly static, therefore users can obtain a version of it on CD, thus reducing much of the network overhead required when accessing the voxel models and the anatomical data. The Proxy design pattern [1] has been used in the design of remote client software access to local and centrally stored data.

As a public bioinformatics resource, the Mouse Atlas system must be easily accessible to researchers all over the world, both for submission of data, e.g. gene-expression patterns, and for querying data. Hence, interoperability is a key issue. At the tools level this is achieved through the use of Web-based technology, e.g. Web pages with embedded Java applets and Java applications that can be deployed through the Web using Java Web Start technology. At the data level, the project makes increasingly use of XML. The project team are also closely monitoring activities relating to the Grid, Semantic Web and formal ontologies (using DAML+OIL), as these technologies may be incorporated into the Mouse Atlas to further develop its interoperability with other resources.

The Mouse Atlas is an on-going research and development project at the Medical Research Council, Human Genetics Unit, in Edinburgh. Access to its databases and further information is available through its Web site (see genex.hgu.mrc.ac.uk).

References

[1] E. Gamma, R. Helm, R. Johnson, and J. Vlissides. *Design Patterns*. Addison Wesley, 1994.

[2] M. Kaufman, R. Brune, R. Baldock, J. Bard, and D. Davidson. Computer-aided 3D reconstruction of serially sectioned mouse embryos: its use in integrating anatomical organization. *Int.J. Dev. Bio.*, 41, 1997.

[3] K. Theiler. *The House Mouse: Atlas of Embryonic Development*. Springer Verlag, 1989.

*Human Genetics Unit, Medical Research Council, Edinburgh

†Dept. of Computing and Elec. Eng., Heriot-Watt University, Edinburgh

Abstract and Discrete Models for Uncertain Spatiotemporal Data

Erlend Tøssebro and Mads Nygård

Department of Computer Science, Norwegian University of Science and Technology,
NO-7491 Trondheim, Norway
tossebro, mads @idi.ntnu.no

The theme of this presentation is uncertainty in spatial and spatiotemporal databases. Due to lack of accurate measurements, or rapid changes in time, spatial and spatiotemporal data are often uncertain. Our work presents new abstract and discrete models for uncertain spatial and spatiotemporal information. The models are based on the principle that one knows that the uncertain object, regardless of type, must be within a certain area.

The first part of this presentation concerns an abstract model. This is to the authors' knowledge the first attempt to create a general type system that is capable of modelling positional uncertainty with spatial data. Individual uncertain types have been modelled before, but no paper has studied points, lines and regions and used the same principles to model all three. It also seems to be the first model to handle temporal as well as spatial uncertainty. Our work contains mathematical definitions of uncertain points, lines, regions and temporal versions of these. Our work also contains definitions of relevant operations on these types. These operations are also evaluated for their usefulness with regard to uncertain data.

The second part of this presentation concerns three discrete models which are all based on the abstract model mentioned earlier. One of these is an advanced model that manages to model almost all of the aspects of the abstract model, but at the cost of increased need for storage space. It is also difficult to compute probabilities in a consistent manner for this model.

The second model is of medium complexity, and balances storage use and modelling power. It also has the advantage that computing probabilities in a consistent manner is much easier than for the advanced model. The third model is an attempt to bring the storage space needed as low as possible. It therefore has somewhat limited modelling power. Unlike the two other discrete models, it cannot be extended to handle spatiotemporal data.

The handling of spatiotemporal data is based on how it is handled for crisp data in [1] and [2]. Our work makes two important additions to these models so that they can handle uncertain data. First, it presents ways of generating

a sliced representation when the times the snapshots were taken are uncertain. Second, it details how operations change as a result of uncertainty. The *Initial* and *Final* operations, which in the crisp case return the initial and final shapes of an object, cannot be defined in the uncertain case. Our work discusses how these operations can be replaced in the uncertain case.

The medium complexity model has been partially implemented. This shows that it is possible to create an implementable model for uncertainty in spatial and spatiotemporal data.

However, analysis of storage requirements in the discrete models shows that storing uncertain data requires significant amounts of storage space. Even the simple model increases storage space by 1.75 times for curves and regions, and by 2 times for base types compared to similar models for crisp data. In the advanced model, regions require 2 times as much space, curves 3.75 times as much space and points N times as much space as the corresponding crisp types. This shows that there is a trade-off between modelling capability and storage space needed. The more complete models require more storage space.

Another conclusion from our work with the discrete models is that there is a trade-off between modelling capability and how easy it is to compute probability functions consistently. For the advanced model, special additions must be made if one wants consistent results from the probability functions. Some of these additions require extra storage space and impose limitations on the modelling capabilities. The less complex models do not require such special considerations.

1. R. H. Güting, M. F. Böhlen, M. Erwig, C. S. Jensen, N. A. Lorentzos, M. Schneider, M. Vazirgiannis: A Foundation for Representing and Querying Moving Objects. In *ACM Transactions on Database Systems* 25(1), 2000, pp. 1-42

2. L. Forlizzi, R. H. Güting, E. Nardelli, M. Schneider: A Data Model and Data Structures for Moving Objects Databases. *Proc. ACM SIGMOD Int. Conf. on Management og Data* (Dallas, Texas), 2000, pp. 319-330.

Joint Queries Estimation from Multiple OLAP Databases

Elaheh Pourabbas*, Arie Shoshani+

* Istituto di Analisi dei Sistemi ed Informatiac "Antonio Ruberti"a-CNR, Viale Manzoni, 30 00185 Roma, Italy, e-mail: pourabbas@iasi.rm.cnr.it

+ Lawrence Berkeley National Laboratory, Mailstop 50B-3238 1 Cyclotron Road, Berkeley, CA 94720, USA, e-mail: shoshani@lbl.gov

Abstract

Given an OLAP query expressed over a collection of source OLAP databases, we study the problem of computing the result OLAP target database when the underlying detail data of the source databases are not available. The problem arises when the source databases do not have the same category attributes, and it is not possible to derive the result from a single database. The method we use is the linear indirect estimator. We consider two obvious methods for computing such a target database, called the Full-cross-product (F) and the Pre-aggregation (P) methods. As for the accuracy and computational performance of these methods, while the method F provides the most accurate estimate possible, it is more expensive computationally than P. Our contribution is in proposing a third new method, called the Shortcut method (S), which is less expensive than F, but is as accurate.

1. Joint queries estimation

In this paper, we address the problem of responding to queries based only on the summary-level data of OLAP databases. That is, given that the base data is not available and that a query cannot be derived from a single summary database, we examine the process of estimating the desired result from multiple summary databases. This is typically done by a method of interpolation, called *linear indirect estimator* [1]. In this model, the population is supposed to be partitioned into large domains d formed by cross classification of demographic variables such as age, sex, race. Let i denotes a small area. For each of the domain d, the variable of interest Y denoted by $Y(d) = \sum_i Y(i,d)$ is calculated from the survey data. It is assumed that the variable of interest and the auxiliary information are respectively available in the form of $Y(d)$ and $X(i,d)$. A synthetic estimator of Y for i is defined by

$$\hat{Y}^s(i) = \sum_d (X(i,d)/X(d))Y(d) = \sum_d \hat{Y}(i,d).$$

Formally, a *joint query* formulated on summary databases $M_i(C_{i1},\ldots,C_{in})$, $M_j(C_{j1},\ldots,C_{jm})$ will be indicated by $M^T(C^T,C^C,C^N)$ where M^T is a selected *target summary database* that can be one of M_i, or M_j;

C^T, C^C, and C^N represent a set of respectively *target*, *common*, and *non-common* category attributes.

Definition: Let $M_i(C_{iu}^C, C_{iv}^N)$ and $M_j(C_{jr}^C, C_{js}^N, C_{jt}^T)$ be summary databases, and let $\hat{M}_i^T(C_{jt}^T)$ be the target database. It can be calculated by the following methods:

(i) *Full cross-product (F)* it is obtained by summarizing all common and non-common category attributes in the full cross product summary database as follows:

$$\hat{M}_i^T(C_{jt}^T) = \sum_{C_{iu}^C, C_{iv}^N, C_{js}^N} M_i(C_{iu}^C, C_{iv}^N) \frac{M_j(C_{jr}^C, C_{js}^N, C_{jt}^T)}{\sum_{C_{js}^N, C_{jt}^T} M_j(C_{jr}^C, C_{js}^N, C_{jt}^T)}$$

(ii) *Pre-aggregation (P)* it is estimated by pre-summarizing all common and non-common category attributes in the summary databases then applying the linear estimator: $\hat{M}_i(C_{jt}^T) = M_i(\bullet)\left(M_j C_{jt}^T \middle/ \sum_{C_{jt}^T} M_j(C_{jt}^T)\right)$

(iii) *Shortcut (S)* it is obtained by pre-summarizing all the non-common category attributes in the summary databases and then estimating the cross product and summarizing over the common attributes as follows:

$$\hat{M}_i^T(C_{jt}^T) = \sum_{C_{js}^N}\left(\left(M_i(C_{iu}^C)\right)\left(\frac{M_j(C_{jr}^C, C_{jt}^T)}{\sum_{C_{jr}^C, C_{jt_t}^T} M_j(C_{jr}^C, C_{jt}^T)}\right)\right)$$

In [2], we have shown that methods F and S yield the same results while F and P yield different results. The different results stem from the presence of common category attributes in the summary databases. In order to evaluate the accuracy of these methods, their average relative errors are calculated. It is stated that F (o S) provides the most accurate estimate w.r.t. P, but S is less expensive than F.

References

[1] M. Ghosh, J. N. K. Rao. Small Area Estimation: An Appraisal. *Journal of Statistical Science*, 9(1): 55-93, 1994.

[2] E. Pourabbas, A. Shoshani. Joint Queries Estimation from Aggregate OLAP Databases. Lawrence Berkeley National Laboratory-University of California, Berkeley, USA, Technical Report, LBNL-48750, August 7, 2001.

Author Index

Press Operating Committee

IEEE Computer Society Publications

The world-renowned IEEE Computer Society publishes, promotes, and distributes a wide variety of authoritative computer science and engineering texts. These books are available from most retail outlets. Visit the CS Store at *http://computer.org* for a list of products.

IEEE Computer Society Proceedings

The IEEE Computer Society also produces and actively promotes the proceedings of more than 160 acclaimed international conferences each year in multimedia formats that include hard and softcover books, CD-ROMs, videos, and on-line publications.

For information on the IEEE Computer Society proceedings, please e-mail to csbooks@computer.org or write to Proceedings, IEEE Computer Society, P.O. Box 3014, 10662 Los Vaqueros Circle, Los Alamitos, CA 90720-1314. Telephone +1-714-821-8380. Fax +1-714-761-1784.

Additional information regarding the Computer Society, conferences and proceedings, CD-ROMs, videos, and books can also be accessed from our web site at *http://computer.org/cspress*

Revised November 7, 2001